PANDORA'S LOCKS

Michigan State University Press • East Lansing

JEFF ALEXANDER

PANDORA'S LOCKS

THE OPENING OF THE GREAT LAKES–ST. LAWRENCE SEAWAY

Publication of *Pandora's Locks* was made possible by the generous support of Peter M. Wege.

⊜ The paper used in this publication meets the minimum requirements of ANSI/NISO Z39.48-1992 (R 1997) (Permanence of Paper).

Michigan State University Press
East Lansing, Michigan 48823-5245

Printed and bound in the United States of America.

18 17 16 15 14 13 12 11 1 2 3 4 5 6 7 8 9 10

Paperback ISBN 978-0-87013-872-0

Cloth edition was originally catalogued by the Library of Congress as:
LIBRARY OF CONGRESS CATALOGING-IN-PUBLICATION DATA
Alexander, Jeff, 1962–
Pandora's locks : the opening of the Great Lakes–St. Lawrence Seaway / Jeff Alexander.
p. cm.
Includes bibliographical references and index.
ISBN 978-0-87013-857-7 (cloth : alk. paper) 1. Biological invasions—Great Lakes (North America) 2. Nonindigenous aquatic pests—Great Lakes (North America) 3. Discharge of ballast water—Environmental aspects—Saint Lawrence Seaway. 4. Discharge of ballast water—Government policy—North America. 5. Nonindigenous aquatic pests—Control—Great Lakes (North America) I. Title.
QH353.A44 2009
577.63'2720977—dc22
2008048261

Cover design by David Drummond
Book design by Sharp Des!gns, Inc., Lansing, Michigan
Cover image: A ship passes through the St. Lawrence Seaway's Eisenhower Lock in northern New York. Courtesy St. Lawrence Seaway Development Corp.

green press INITIATIVE Michigan State University Press is a member of the Green Press Initiative and is committed to developing and encouraging ecologically responsible publishing practices. For more information about the Green Press Initiative and the use of recycled paper in book publishing, please visit *www.greenpressinitiative.org.*

Please visit the Michigan State University Press on the World Wide Web at *www.msupress.msu.edu.*

To Martha, who introduced me to the Great Lakes and their myriad wonders.

To Eric, Kelsey, and Jessica, for all the sandcastles, dune climbs, and bone-chilling swims that enriched my relationship to these freshwater seas.

CONTENTS

ACKNOWLEDGMENTS

This book is the culmination of a personal dream nearly a decade in the making. I had long considered writing a book about the Great Lakes. The conquest of the lakes by invasive species provided the motivation I needed. My dream would have never come to fruition without the help and support of many friends, relatives, colleagues, and scientists. I am particularly indebted to the following people: Dave Dempsey, for persuading me to write this book; Julie Loehr, my editor at Michigan State University Press, for maintaining a sense of humor while shepherding this project from concept to reality; Peter Wege and his eponymous Wege Foundation, for their generous financial support; Paul Keep, my editor at the *Muskegon Chronicle,* who bent company rules so I could take an unpaid sabbatical to conduct research; my colleagues at the *Chronicle,* who graciously covered for me while I spent three months gallivanting around the Great Lakes; and the staff at the Duluth Seaway Port Authority, who allowed me to shadow an international trade delegation as it toured the massive shipping facilities in Duluth-Superior Harbor. I am deeply appreciative of the many scientists, civil servants, shipping-industry officials, anglers, and lake lovers who patiently educated me about Great Lakes ecosystems, the logistics of shipping, the complexities of invasion ecology, and the challenges of dealing with shipborne invasive species. On a personal level, I am forever indebted to my late mother, Judy Alexander, a child of these lakes who instilled in me a love of books; and my father, Ken Alexander, for showing me the value of hard work and calluses. I am especially grateful to my wife, Martha, for her friendship and for believing in me when I stopped believing in myself. Thank you, one and all, for helping me realize my dream of writing about these incomparable lakes. I hope this book does the Great Lakes justice.

The ultimate invasive species is, of course, man himself. Wherever he has gone, he has either deliberately or accidentally taken other species with him.

—Brian Morton, ecology professor, University of Hong Kong

PREFACE

The genesis of this narrative dates to 1989, the year the *Exxon Valdez* ran aground and poisoned Alaska's Prince William Sound with 11 million gallons of crude oil. The *Valdez* oil spill was the worst environmental disaster in American history, a catastrophe that killed a half million water birds as well as countless seals and otters, poisoned salmon streams, and stained 1,500 miles of scenic shoreline. It also became the standard against which future environmental disasters were measured. A few months after the *Valdez* calamity, I attended a conference where a distinguished government scientist drew a stunning parallel between the notorious oil spill and an emerging environmental crisis in the Great Lakes. Jon Stanley, then director of the U.S. Geological Survey's National Fisheries Research Center, stood before a group of journalists and predicted that the damage wrought by zebra

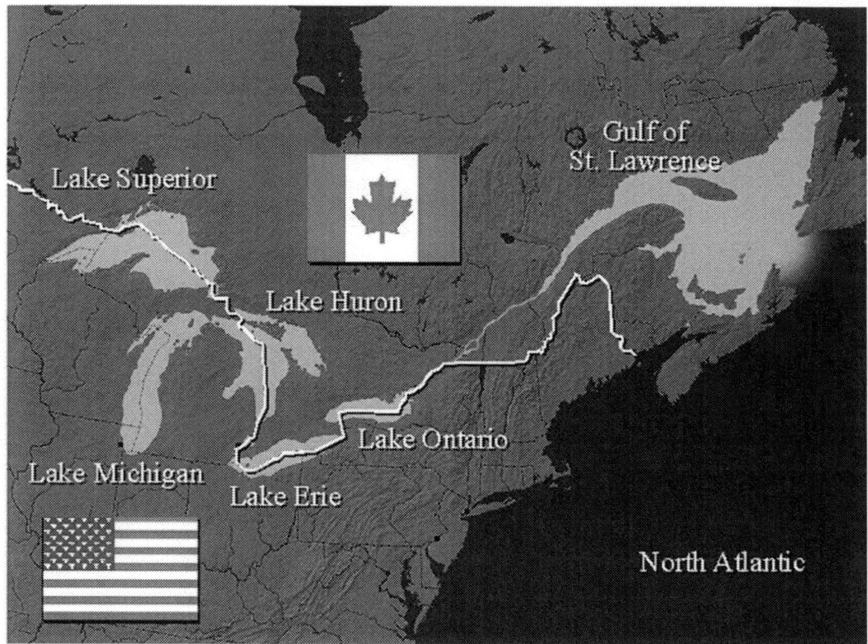

U.S. Army Corps of Engineers

The Great Lakes–St. Lawrence Seaway allows ocean freighters to access the Great Lakes via the St. Lawrence River.

Great Lakes System Profile

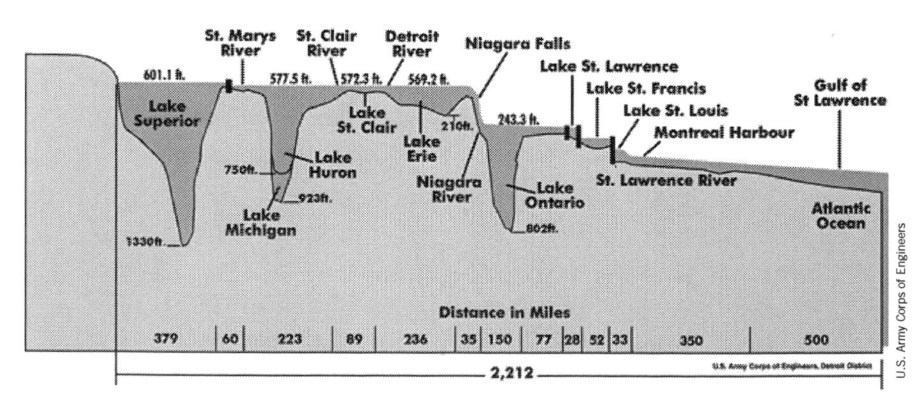

U.S. Army Corps of Engineers

A series of locks in the Seaway allows ocean freighters to reach the highest of the Great Lakes, Lake Superior, which is 601 feet above sea level.

mussels—the European mollusks that snuck into the lakes in the bowels of ocean freighters—would someday eclipse the harm caused by the *Exxon Valdez*. "The zebra mussel is an environmental disaster for the Great Lakes that will be worse than the Exxon Valdez oil spill. Unlike the Exxon oil spill, which will be cleaned up in a few years, the zebra mussel is not going to go away," Stanley said. "They are going to be with us for at least 100 years."[1] Many of the journalists who covered Stanley's speech, myself included, figured his comments were wild hyperbole. It turned out he wasn't crying wolf.

Exxon spent $3.5 billion cleaning up its horrific mess, and another $1.1 billion to settle civil lawsuits filed by commercial anglers whose livelihoods were ruined by the oil spill. The $4.6 billion price tag made the *Valdez* incident the single most expensive environmental disaster in U.S. history. A new type of pollution in the Great Lakes would challenge for that dubious honor: biological pollution from zebra mussels and dozens of other invasive species that ocean freighters accidentally imported into the freshwater seas.

Scientists discovered zebra mussels in the Great Lakes in 1988, the year before the *Valdez* disaster. The mussels initially generated little media buzz. No wonder. Unlike an oil slick that sent waves of dead birds and otters washing onto beaches, the mussels carried out their destructive activities beneath the water's surface. The magnitude of the mussel invasion finally came into focus in December 1989, when the mollusks clogged a municipal water intake and brought the city of Monroe, Michigan, to a standstill. Over the next decade, zebra mussels caused more than $1.5 billion in damages to municipal and industrial water intakes in the United States and Canada. That was far less than the damage caused by the *Valdez*. But there was a catch: Oil spills could be cleaned up, to a certain extent. Biological pollutants like zebra mussels reproduced and extended their range, causing ever-greater ecological harm and economic losses.

The *Valdez* cleanup was finished, more or less, by the late 1990s. Prince William Sound's ecosystem recovered, though it may never be as healthy as before the oil spill. Problems caused by zebra mussels, on the other hand, continued to escalate as the invaders spread across the United States and southern Canada. Worse, an even more pernicious relative of the zebra mussel—the quagga mussel—invaded all five Great Lakes and spread from New York to California. Power plants and water utilities in the Great Lakes region spent millions of dollars annually to cope with zebra and quagga

Photo by the Associated Press, courtesy of the *Muskegon Chronicle*

The *Exxon Valdez* caused more than $3 billion in ecological damage after it ran aground in 1989 and dumped 11 million gallons of crude oil into Alaska's Prince William Sound. Invasive species have caused more damage to the Great Lakes.

U.S. Environmental Protection Agency/*Bay City Times*

Zebra mussels were discovered in the Great Lakes in 1988. The prolific mussels have caused billions of dollars in damage by fouling beaches (like this one on Lake Huron), clogging water intakes, eliminating native mussels, and disrupting ecosystems.

mussels. Those invaders, and others, caused about $5 billion damage annually to the Great Lakes ecosystem, according to Cornell University ecology professor David Pimentel.[2] It was just a matter of time before foreign species in the Great Lakes eclipsed the *Valdez* oil spill as the nation's worst environmental disaster.

Let me be clear: This should not be interpreted as an attempt to downplay the effects of Exxon's oil spill. It was a catastrophic event, one borne of negligence, that caused tremendous harm and suffering for fish, wildlife, and humans. My point is simply that invasive species have already caused more damage in the Great Lakes than the *Valdez* spill inflicted on Alaska. As shipborne invasive species spread beyond the Great Lakes and infested lakes and rivers across North America, the ecological and economic toll would continue to escalate.

In 1999, a decade after Stanley's bold prediction, retired U.S. Coast Guard commander Eric Reeves reached a similar conclusion. "Without minimizing the scope of that [*Exxon Valdez*] disaster, or discounting the possibility of unpredictable long-term impacts on some of the affected species, I would suggest that the long-term damage to aquatic ecosystems is far less than what is being done by the zebra mussel as it continues to spread through the Great Lakes and the Mississippi Basin. There will never be any recovery of 'normal species composition, diversity and functional organization' in this huge area of natural aquatic resources in the heartland of North America."[3]

Whereas oil spills were episodic events that caused phenomenal harm, invasive species were like a slow moving, never-ending wildfire. To put it another way, zebra mussels and other invasive species were like herpes for the Great Lakes: a disease that never went away and was easily spread. The devastating, chronic illness called biological pollution was one legacy of the St. Lawrence Seaway, an engineering marvel that opened the floodgates to foreign species when it opened the Great Lakes to transoceanic ships. Just a few of the species ocean freighters imported have caused profound ecological changes in the world's largest freshwater ecosystem. The Seaway was supposed to create an economic bonanza in the industrial and agricultural heartland of North America. It certainly increased the amount of commerce moving in and out of the region, but at what cost? For the lakes' ecosystems, every ocean freighter that entered via the St. Lawrence Seaway was a potential Trojan horse: a gift of commerce that could carry in its ballast water tanks a load of potentially harmful invasive species and pathogens from distant waters. The results were disastrous—ecologically, economically, and aesthetically.

THE SEAWAY WAS SUPPOSED TO MAKE THE GREAT LAKES GREATER, AT LEAST in economic terms. The idea was to build a series of seven locks, six canals, and three dams that gave ocean freighters from around the world safe access to the Great Lakes by creating a navigational route through the previously unruly St. Lawrence River. The $1 billion project opened to ships in April 1959, but wasn't officially dedicated until June 26, 1959. Its completion ushered in a new era of global commerce for the eight U.S. states and two Canadian provinces that surround the Great Lakes. Coupled with locks on the Welland Canal and St. Marys River, the Seaway allowed ships up to 740 feet long and 78 feet wide to travel 2,340 miles inland from the North Atlantic Ocean to the western end of Lake Superior. It was a monumental feat of engineering that linked freshwater ports in places like Duluth, Minnesota, and Hamilton, Ontario, with ports around the globe.

The Seaway initially gave 80 percent of the world's merchant shipping fleet access to 60 percent of North America's population and 65 percent of its manufacturing capacity. In doing so, the Seaway planted the seeds of a biological New World Order in the Great Lakes. The *Toronto Globe and Mail* pondered the significance of the Seaway the day it officially opened. "It has moved the ocean a thousand miles inland," the newspaper said in an editorial. "The effects of this cannot as yet be estimated, but we can be certain that they will be very great."[4] The newspaper was referring, of course, to the projected economic benefits for the region.

At a dedication ceremony in Duluth on July 11, 1959, then Minnesota governor Orville Freeman said the port at the western end of Lake Superior, 2,340 miles from the Atlantic Ocean, would become "the center of trade and commerce at the crossroads of the continent. Now that our window to the world is opened, it can never be closed. We are tied to the national and world economy. Our profits are linked to developments on the other side of the world. Our peace is dependent on the peace of the world."[5] No one anticipated how profoundly ocean ships that sailed the Seaway would affect the Great Lakes.

It is important to keep the opening of the Seaway in historical context. The 1950s were an era of big dreams, mind-boggling feats of engineering, and incredible social change. A decade after World War II ended, the United States was living large. The baby boom was in full swing, the country's first McDonald's opened, and work began on the interstate highway system. Flush with pride and money, Americans built houses, bought cars, and made babies at a phenomenal rate. It was during this era of prosperity that the space race between the United States and Soviet Union began. In 1957, the Soviets launched the world's first satellite, Sputnik. The Americans responded a year later by launching the Explorer I satellite. NASA, the National Aeronautics and Space Administration, was established, and the space race was on.

There seemed to be no limit to America's technological capabilities in the 1950s. Joining Canada to build a series of canals, locks, and dams that opened the Great Lakes to ocean commerce was, at the time, a no-brainer. Building the Seaway gave the United States and Canada a strategic military asset in the event of another global conflict. The Great Lakes region was the bosom of America's military-industrial complex and the hub of North America's manufacturing empire. By linking the lakes to the international shipping network, Canada and the United States could enhance the flow of global commerce and strengthen their ties to allies in western European nations. That was an important consideration as the United States and the Soviet Union began an increasingly tense standoff known as the cold war. Though it made sense to build the Seaway, from economic and military standpoints, the project caused phenomenal collateral damage to the Great Lakes environment. Those effects were never anticipated.

The Seaway opened a pipeline through which foreign species of fish, plankton, invertebrates, and pathogens could sneak into North America in the ballast water tanks of transoceanic freighters. Ships that visited the Great Lakes from Europe and elsewhere carried cargo and a dirty secret that could best be characterized as a type of plague—a plague of biological pollutants. Transoceanic freighters, those engines of commerce that dumped ballast water teeming with foreign species into the lakes, turned these freshwater seas into one of the world's largest unplanned biological experiments. It was an experiment no one anticipated or could control once the genie was out of the bottle. Much like the storied opening of Pandora's Box unleashed all the evils known to humanity, opening the Great Lakes to transoceanic freighters inadvertently set loose a form of ecological evil in the lakes.

The environmental consequences of opening the Seaway were to the Great Lakes what Dutch Elm Disease was to one of North America's most beloved tree species. Dutch Elm Disease was imported to the United States from Europe in the 1930s, delivered in wooden crates made with wood from infected elm trees. The disease spread quickly across North America, wiping out millions of stately elms that created tunnels of trees on countless residential streets. It eventually killed more than 20 million trees in the United States and Canada; only a small percentage survived. The loss was huge in both nations, aesthetically and economically. Mature elm trees not only looked nice and provided shade, they increased residential property values. One estimate put the value of all the elm trees in the United States and Canada felled by the disease at $20 billion. No one knew it at the time, but Dutch Elm Disease was, in a way, a blueprint for the spread of aquatic invasive species in the Great Lakes. Like the ocean ships that transported infected elm wood to North America in the 1930s, ocean ships that the St. Lawrence Seaway invited into the Great Lakes carried unwanted imports with the cargo they transported.

Invasive species imported by ocean freighters were not responsible for all of the ecological harm foreign organisms caused in the lakes. But transoceanic ships—and the canals that gave them access to the Great Lakes—were responsible for many of the worst invaders, including zebra and quagga mussels, round gobies, the sea lamprey, and alewife. The engineers who designed the Seaway did not set out to plunge the lakes into ecological chaos. Theirs was an error of omission, not commission. They failed to consider the ecological risks of opening these freshwater seas to oceangoing freighters, even though it was obvious by the early 1950s—before construction began on the Seaway—that the lakes were vulnerable to invasions by foreign species. The sea lamprey's infestation of Lakes Erie, Huron, Michigan, and Superior in the 1940s and '50s clearly illustrated that tinkering with the lakes was a risky proposition. Still, engineers forged ahead with one of North America's most ambitious public-works projects. In the Seaway, supporters only envisioned the promise of economic growth.

For the lakes, the Seaway was the ecological equivalent of Pandora's Box. It released a tidal wave of unintended consequences—destructive changes that may never be reversed. "Unfortunately, it is usually true that invasion is forever. Biological invasions are the least reversible form of pollution. Chemical pollutants . . . do not reproduce; species do," said University of Notre Dame biology professor David M. Lodge.[6]

The Seaway and its nineteenth-century predecessors, the Erie and Welland canals, triggered an ecological domino effect in the lakes that still raged out of control in 2009, fifty years after the Seaway opened. Invasive species displaced the lakes' top fish predator, lake trout, and gave rise to the alewife infestation. The alewife invasion led to the creation of a wildly popular, and artificial, salmon fishery. The salmon fishery became the backdrop of one of the world's most bizarre cases of intentional food poisoning. In the 1980s and '90s, a surge of invaders transformed the chemistry and biology of Lakes Michigan, Huron, Erie, and Ontario so quickly, and dramatically, that scientists who studied the lakes could not keep pace with the changes. Fifty years after the Seaway opened, the lakes' ecosystems bore only a slight resemblance to those pristine, healthy waters—teeming with giant sturgeon and lake trout—that Native Americans discovered when they settled the region more than 2,000 years ago.

The plague of invasive species was not unique to the Great Lakes; it was a global problem. There were more than 50,000 nonnative plant and animal species in the United States, according to scientists at Cornell University. Even though some of those species were intentionally imported, invasive species caused $137 billion of damage annually in the United States by 2005.[7] Some Great Lakes species have been exported to other parts of the world, wreaking havoc on ecosystems in parts of Europe and Asia. Still, invasives have taken a particularly heavy toll on the Great Lakes, an assemblage of relatively young and biologically vulnerable ecosystems. In a sense, the vast and seemingly indestructible lakes were like human infants who lacked a fully developed immune system. Because of their relative youth, the lakes lacked the suite of native species necessary to ward off or control invaders. As a result, the lakes were an aquatic frontier easily conquered by freshwater creatures from other parts of the globe. A 2004 study of invasive species in North America reached this ominous conclusion: "The Great Lakes and San Francisco Bay . . . are among the most invaded ecosystems on earth."[8] That was a deplorable distinction for the largest source of fresh surface water on the planet.

THE BIOLOGICAL INVASION OF THE GREAT LAKES THAT BEGAN IN THE mid-1980s and continued into the first decade of the twenty-first century was unprecedented. At no time in recorded history had so many destructive foreign species invaded the lakes. A few perceptive researchers saw it coming. Scientists warned the U.S. and Canadian governments in the early 1980s—before many of the worst invaders arrived—that ocean ships entering the Seaway carried thousands of viable foreign organisms from distant ports in their ballast water tanks. But those warnings were ignored. After the zebra mussel was discovered in 1988, the U.S. and Canadian governments got serious about addressing the invasive-species problem. But the resulting regulations were riddled with loopholes and failed to close the door on new species sneaking into the lakes. Two decades after zebra mussels were discovered, and with a new invasive species taking hold in the lakes every seven months, the U.S. and Canadian governments still had not enacted the stringent regulations needed to stem the tide of invaders entering via ocean freighters. Government records showed that the United States and Canada had botched two golden opportunities to prevent zebra mussels and a host of other destructive species from invading the lakes. The U.S. government missed two more chances to dramatically slow a rising tide of invaders overwhelming the lakes.

For their part, shipping-industry officials dragged their collective feet and slowed efforts to develop ballast water standards and treatment systems. Consider the following comment by the director of the Great Lakes' busiest shipping port. "I don't think our industry took the ballast water problem really seriously until about 2003," said Adolph Ojard, executive director of the Duluth Seaway Port Authority. He made those comments to me during a 2007 interview. That same year, the director of the U.S. St. Lawrence Seaway Development Corp. offered a similar assessment. In a February 2007 speech to the environmental group Save the River, Collister (Terry) Johnson Jr. said: "Ballast water is a major vector for invasive species. It has been the source of invasive species in the past and, unless resolved, threatens to be so in the future. The shipping industry has been very slow to understand the seriousness of the problem, but is now playing catch up and realizes that business as usual is not acceptable."

By 2007, a growing number of frustrated environmental activists, scientists, and politicians were calling for a ban on ocean freighters entering the lakes until the ships were equipped with treatment systems that could kill all foreign organisms and pathogens in ballast water tanks. Reeves, who managed ballast water regulations for five years while in the Coast Guard, said closing the Seaway was the only sure way to keep ocean ships from importing more foreign species. "The only way to get a 100 percent level of protection against ballast water is to shut down the Seaway. That is not going to happen," Reeves said. "The practical question to consider, therefore, is how we can move up the level of protection . . . as soon as possible. Time is a critical factor in the calculus here."[9] He made those remarks in 1999. Transoceanic freighters continued to enter the Great Lakes for another decade without first sanitizing their ballast tanks.

Ocean freighters were required in the 1990s to exchange ballast water on the open ocean before entering the Great Lakes—the theory being that salt water would purge or kill any freshwater organisms lingering in ballast water taken on in European or Asian ports. However, that didn't solve the problem. The U.S. and Canadian governments knew the regulations weren't 100 percent effective. But neither country could muster the political will to force ships to treat ballast water with filters, heat, or chemicals that could have stopped or slowed the great invasion. The international shipping industry professed concern about the problem, but for years delayed serious efforts to deal with it. The result: regulatory and political gridlock in the United States and Canada that allowed potential solutions to remain on the shelf for more than a decade as the number of invasive species colonizing the lakes escalated.

What follows is the story of how ineffectual regulatory agencies, ignorant politicians, and an intransigent shipping industry allowed the insidious byproducts of ocean shipping to eviscerate and reconfigure the world's largest freshwater ecosystem. It is a tale of how a grand engineering project profoundly changed the Great Lakes and how people relate to these sweetwater seas. My understanding of this issue was acquired during two decades of writing newspaper articles about the Great Lakes. But it wasn't until 2005, after legendary conservationist Lee Botts questioned the wisdom of building the Seaway, that I began to ask myself whether this technological marvel was ultimately good or bad for the lakes. "Was it worth it?" Botts said in a newspaper interview. That profound question, coupled with an experience my wife and I had in 2007 while visiting the Sleeping Bear Dunes National Lakeshore in Michigan, solidified my decision to write this book.

LABOR DAY 2007 WAS A GLORIOUS SUMMER DAY IN NORTHWEST MICHIGAN. Temperatures hovered around 80 degrees, and there wasn't a cloud in the azure sky. My wife and I parked at the end of a dirt road that led to a secluded bay in the Sleeping Bear Dunes National Lakeshore, about 30 miles west of Traverse City. There, we kicked off our shoes and began the quarter-mile walk through low-slung sand dunes that separated the Lake Michigan beach from an adjacent forest. As we strode eagerly toward the lake, we encountered a young couple in beach attire. They walked at a brisk pace away from the deserted shoreline. Disappointment was etched on their faces. That was odd. It was a glorious afternoon, and there were few better places on the Great Lakes to spend a day at the beach. The remote stretch of sand afforded stunning panoramic views of South Manitou Island and dunes that rose 450 feet above the glistening waters of Lake Michigan. How could anyone walk away from such beauty so obviously disgusted? I was seized by curiosity.

"How's the water?" I asked.

"Oh, we're not that brave," the woman replied.

It was apparent from the scowl on her face and their quick pace that the couple was in no mood for small talk. I let it go, assuming the water must have been too cold for swimming. Perhaps an East wind had caused an upwelling, a phenomenon that pushed warm water offshore and brought deeper, colder water to the surface. Lake turnover, as it was known, could drop water temperatures in Lake Michigan 20 degrees in a matter of hours.

We were not going to be deterred by cold water. After all, this site was more than a beach to us. It was a refuge from the grind of life, a place we visited to clear our heads and recharge our batteries by soaking up the sublime beauty and abundant solitude. We had made a pilgrimage to this beach nearly every year for more than a decade. Hidden at the end of a winding dirt road that headed west off M-22, the Platte Bay Beach was a favorite among people who were intimately familiar with the 35-mile-long coastal paradise known simply as Sleeping Bear Dunes. So isolated was this beach, you could walk a mile on sugar-like sand without seeing another human being. At night, the beach afforded spectacular views of the stars. It was the place my children and I first experienced aurora borealis, the northern lights. An evening of storytelling around a beach fire was interrupted late one summer night by shimmering curtains of pale green, white, and pink light that danced across the northern sky. We stared at nature's amazing light show for so long, our necks cramped.

The drive to this beach was always filled with anticipation: Would we have the place to ourselves? Would we see raptors circling high above the dunes? Would we hear the haunting call of a common loon? That childlike sense of excitement never faded. A late summer visit to this beach on a sun-drenched day would help sustain us through the long, dark winter that was mere weeks away. This was to be a memorable Labor Day, to be sure—but for the wrong reasons.

Cresting the dune, our mouths fell open. We could see why the couple we passed on the path seemed so perturbed. A wide strip of dark green algae that looked like cooked spinach and had a faint odor of sewage blanketed the shoreline as far as the eye could see. A dead chinook salmon, its white belly glinting in the sun, rolled back and forth in the dark green, molasses waves. Several dead birds lay scattered across the miles-wide beach, eyes plucked out by scavengers. Thousands of dead *Dreissena* mussels—zebra and quagga mussels native to the Caspian Sea in eastern Europe—littered the beach. Clumps of mussels were interspersed among the bloated carcasses of round gobies, an ugly little fish with bulging eyes that made it look like a creature from a low-budget horror movie. Gobies also were native to the waters of eastern Europe. Out of nowhere, the invasive fish and native birds began washing up dead on scenic beaches that were far removed from the environmental insults of humans. It was a carnage we had never experienced in our previous visits to this wondrous lakeshore.

The park's biologist later explained how the problem began in August 2006 and quickly mushroomed. Dead gulls, mergansers, and common loons—the icon of the North—started washing up on the beach along Platte Bay that year. It began with a few birds, but the epidemic claimed 2,900 birds over a three-month period. Hundreds more birds died the following year. The killer was neither man nor beast, but a bacterium. Botulism, better known as food poisoning, killed the fish and birds. It seemed the foreign mussels and gobies had assembled an unnatural, toxic food chain that was claiming thousands of native birds and fish around the Great Lakes. The list of casualties included bald eagles, piping plovers, and massive lake sturgeon. Something was terribly wrong; our beloved lake was sick. That was evident in the dead birds and fish the lake regurgitated onto the beach. Sadly, the sickening scene was not unique to Lake Michigan. An identical strain of botulism had been killing birds for several years along the shores of Lakes Erie, Huron, and Ontario. Only Lake Superior was spared the bird kills that became an annual event in parts of the Great Lakes during the first decade of the twenty-first century.

Our plan for a relaxing day at the beach turned out to be something far different: an encounter with icons of a mounting ecological disaster. That was the moment I resolved to write this book. In my estimation, this was a story that had to be spread far and wide, its causes and consequences explained in vivid detail. To gain an intimate understanding of how the St. Lawrence Seaway and the ocean ships it lured here changed the Great Lakes, I spent three months traveling around these inland seas. Mine was a 10,000-mile journey of discovery that took me to places of breathtaking natural beauty and sickening ecological devastation. I traveled nearly the entire length of the Seaway by land, from Montreal to Duluth, during a series of trips. Along the way, I interviewed scientists, shipping industry officials, anglers, and lake lovers. I fished with charter-boat captains and tribal fishermen, toured Duluth

Harbor with foreign food brokers, trawled for bottom-dwelling fish with government biologists, meditated on the beauty and ecological significance of Niagara Falls, spent hours watching ships pass through the Seaway's massive locks and canals, and visited shoreline-property owners whose beaches were fouled by waves of zebra mussels, noxious algae blooms, and fish kills. Those experiences strengthened my emotional bond to these magnificent waters and steeled my resolve to educate the masses about the problems foreign species were causing in these greatest of lakes.

Many previous books have documented the political and economic events that led to construction of the Seaway. Most have focused on how the artificial shipping channel improved the region's economy and benefited port communities. None detailed how the structure and the ships it allowed into the heart of North America affected Great Lakes ecosystems, as well as people who relied on the lakes for work and pleasure. What better time than the Seaway's 50th anniversary to examine this transformative project from a new perspective? This narrative, unlike previous books, speaks for the lakes and all who rely on them for recreation, inspiration, and sustenance—economic, spiritual, or otherwise.

I have chosen to not answer Lee Botts's vexing question about whether the Seaway's economic benefits outweighed its environmental costs. That is a question best left to scientists, economists, and you, the reader. My goal was to shed light on the Seaway's role as the purveyor of a costly, albeit unintended, environmental calamity. Hopefully, this work adds to the general understanding of how invasive species have radically altered the Great Lakes ecosystem, and how those changes have affected the region's 40 million residents. For some people, the cost of invasive species has been the destruction of a favorite beach or a treasured fishing spot that no longer supports native fish. For others, species like zebra mussels have caused taste and odor problems in drinking water or increased the cost of water and electricity as utilities have spent millions of dollars to cope with the invaders. Amid the economic suffering and mounting ecological chaos, the U.S. and Canadian governments fiddled as the Great Lakes ecosystem burned. For too long, ignorance was a convenient excuse for the shipping industry's abuse of the lakes and the government's failure to protect these priceless waters. What follows is my attempt to explain what precipitated the great biological invasion of the lakes, how it could have been prevented—or at least slowed—and why it wasn't. These magnificent lakes, which contain 18 percent of all fresh surface water on the planet, are worthy of the most lavish odes. An honest review of how the St. Lawrence Seaway transformed the Great Lakes ecosystem demands an elegy.

TIMELINE

Important dates in the development of the St. Lawrence Seaway and invasive species that have entered the Great Lakes through the Erie Canal, Welland Canal, and the Seaway.

~ 14,000–9,000 B.C.E.	Glaciers from the last Ice Age retreated from the Great Lakes, leaving behind the world's largest source of fresh surface water. The five lakes drain into the St. Lawrence River, which flows into the North Atlantic Ocean.
~ 5,000 B.C.E.	Native Americans began to settle the Great Lakes region.

1535 French explorer Jacques Cartier sailed up the St. Lawrence River in search of a passage to the Orient.

1680 Dollier de Casson, a religious leader, tried to build a five-foot-deep canal around the Lachine Rapids in the St. Lawrence River, near Montreal. Others completed the canal in 1824.

1779 A series of three locks and shallow canals were built in the St. Lawrence River, near Montreal. The structures, North America's first canals and locks, were deep enough to allow fur-trading vessels to circumvent treacherous rapids in the river.

1825 The Erie Canal opened in New York, linking Lake Erie to the Hudson River. The 340-mile-long system of canals allowed boats from the Atlantic Ocean to skirt Niagara Falls and reach Lakes Erie, Huron, and Michigan via the Hudson River.

1829 The first Welland Canal opened in the Canadian province of Ontario. It allowed ships to bypass Niagara Falls and travel between Lake Ontario and the four Great Lakes above the falls.

1855 The first Soo Locks were built in the St. Marys River, near the cities of Sault Ste. Marie in Michigan and Ontario. The locks bypassed rapids in the river that linked lakes Superior and Huron.

1873 The alewife, a fish species native to the Atlantic Ocean, was discovered in Lake Ontario.

1895 The U.S.-Canadian Deep Waterways Commission was formed to study the feasibility of building the St. Lawrence Seaway.

1909 The United States and Canada signed the Boundary Waters Treaty to settle transboundary water issues. The agreement created the International Joint Commission to monitor the Great Lakes and other border issues.

1919 The Welland Canal was expanded, allowing Lake Eric to provide all the water that flows through its channel. The expansion also allowed the sea lamprey, an eel-like invasive fish, to spread from Lake Ontario to the other Great Lakes.

1921 Sea lampreys discovered in Lake Erie. By 1946, the blood-sucking parasites were in all the Great Lakes.

1950s After sea lampreys decimated the lake trout fisheries in Lakes Michigan, Huron, and Superior, alewives became the most abundant fish species.

1954 Construction began on the St. Lawrence Seaway.

1959 The St. Lawrence Seaway opened. Its canals and locks allowed large ocean freighters to reach the Great Lakes for the first time. Its massive hydroelectric dams generated three times more electricity than the Hoover Dam. Cost to build the Seaway was $470 million. Building the three associated dams cost another $530 million.

1966 The state of Michigan began stocking Pacific salmon in Lake Michigan to control the alewife population and to create a sport fishery.

1972 The United States and Canada signed the first Great Lakes Water Quality Agreement, in which both nations pledged to reduce pollution in the lakes.

1972 The U.S. Congress approved amendments to the Federal Water Pollution Control Act, which later became known as the Clean Water Act.

1973 The U.S. Environmental Protection Agency made ballast water discharges from freighters exempt from the Clean Water Act.

1981 A Canadian study discovered numerous viable organisms, including zebra mussel larvae, in the ballast water of ocean freighters entering the Great Lakes. The report warned that zebra mussels and other invasive species were entering the lakes and could colonize the massive bodies of water. The U.S. and Canadian coast guards shelved the report and did nothing.

1982 A species of zooplankton native to European waters called *Bythotrephes*, or spiny water flea, discovered in Lake Ontario.

1985 Eurasian ruffe, an invasive European fish not to be confused with orange roughy, was discovered in Lake Superior's Duluth Harbor.

1988 Zebra mussels were found in Lake St. Clair. Scientists later discovered the mussels had invaded Lake Erie in 1986.

1989 Zebra and quagga mussels discovered in Lake Erie.

1989 Canada enacted voluntary guidelines asking transoceanic freighters to exchange ballast water at sea before entering the St. Lawrence Seaway and the Great Lakes.

1990 Tubenose gobies discovered in the St. Clair River.

1990 Round gobies discovered in the St. Clair River and Lake Erie.

1990 The U.S. Congress passed the world's first mandatory ballast water regulations, the Nonindigenous Aquatic Nuisance Prevention and Control Act (NANPCA), to control the spread of zebra mussels and ruffe and prevent infestations by other foreign species.

1993 The United States began requiring all ocean freighters to exchange ballast water at sea before entering the Great Lakes. The U.S. Coast Guard's interpretation of the law made it applicable only to ships with pumpable ballast water on board. The result: Up to 90 percent of ocean freighters entering the Seaway weren't required to flush ballast tanks with sea water, because the vessels were loaded with cargo but had no pumpable ballast water on board. Those ships, called NoBOBs because they reported No Ballast on Board, usually carried tons of muddy slop teeming with foreign organisms in their ballast tanks. That muddy water was routinely dumped in the lakes as cargo was loaded and unloaded.

1995 Canada reorganized its Department of Transport, complicating its efforts to regulate ballast water.

1996 Canadian officials discovered that 90 percent of ocean freighters entering the lakes were NoBOBs and therefore exempt from ballast water exchange regulations.

1996 The U.S. Congress reauthorized NANPCA as the National Invasive Species Act. The law required all transoceanic ships to flush ballast tanks with seawater before entering U.S. ports. The Coast Guard still did not require NoBOBs to flush ballast tanks with seawater before entering the Great Lakes.

1998 Fish-hook water fleas discovered in Lake Ontario.

1999 Several environmental, conservation and fishing groups filed a petition asking the U.S. EPA to strike down the 1973 exemption for ballast water discharges under the Clean Water Act. The EPA rejected the petition, prompting the groups to challenge the exemption in federal court.

2002 The U.S. Congress failed to consider legislation that would have required all ocean freighters to disinfect ballast tanks before entering the Great Lakes.

2004 The International Maritime Organization, a branch of the United Nations, established the first international ballast water treatment guidelines. The guidelines did not take effect immediately and were not enacted as of 2008.

2005 Viral hemorrhagic septicemia, an often-fatal fish virus, killed hundreds of thousands of fish in Lakes Erie, Ontario, and Lake St. Clair. The source was never confirmed, but biologists suspected it was imported in an ocean freighter's ballast water.

2005 A federal court ruled that the U.S. EPA acted illegally when it made ballast water discharges exempt from the Clean Water Act in 1973. The court in 2006 ordered the EPA to develop ballast water discharge standards by September 2008.

2006 Bloody-red mysis, a foreign species of freshwater shrimp, discovered in a channel to Lake Michigan at Muskegon, Mich. It was the 183rd exotic species in the Great Lakes.

2006 Canada closed the so-called "NoBOB loophole" by requiring all transoceanic freighters destined for Canadian ports on the Great Lakes to flush ballast tanks with seawater before entering the St. Lawrence Seaway. The rule didn't apply to ocean ships headed for U.S. ports.

2007 Quagga mussels discovered in Nevada's Lake Mead and in California reservoirs. Two scuba divers died in California while removing quagga mussels from a municipal water pumping facility.

2008 U.S. researchers discovered that flushing ballast tanks with ocean water reduced the number of live organisms in ballast water by at least 95 percent.

2008 The United States followed Canada's lead and closed the NoBOB loophole. The new rule required all transoceanic freighters destined for U.S. ports to flush ballast tanks with ocean water before entering the St. Lawrence Seaway.

PANDORA'S LOCKS

When you see loons dying in large numbers it's very sad. You're losing that voice and symbol of the wilderness.

—Ward Stone, New York state wildlife pathologist

PROLOGUE

DEATH OF AN ICON

The sun was sinking toward the horizon over Lake Ontario, its fading light painting the autumn sky orange and lavender, when the icon of the North filled the air with its haunting yodel. The bird broadcasting the call was not visible from the beach at Henderson Harbor in northern New York. Still, there was no mistaking that a common loon, *Gavia immer*, was nearby. The loon had one of the most distinctive calls in the community of water birds. Its memorable call was matched only by its beauty and affection for its offspring. With monochrome feathers, crimson eyes, a black-and-white-barred necklace and checkered pattern on its back, the common loon struck a distinctive pose as it cruised low in the water. The gorgeous, symbolic birds were known to carry their chicks on their backs to protect them from predators. They were anything but

The common loon, with its distinctive markings and haunting calls, is an icon of wilderness in the Great Lakes region.

Invasive mussels and fish, imported to the Great Lakes by transoceanic freighters, passed a strain of botulism up the food chain, killing tens of thousands of loons and other water birds. This bird was one of many that washed up at Presque Isle State Park in Erie, Pennsylvania.

common. Among water birds of the North, the common loon had no equal, in my estimation.

So revered was the common loon, its likeness adorned the Canadian one-dollar coin, the so-called "loonie." It was Minnesota's state bird and a symbol of the wilderness that still reigned supreme in northern boreal forests. Its plaintive wail and laughing tremolo were the sounds by which many who visited the northern Great Lakes and Canada measured their wilderness experience. "Loons . . . provide much of the magic of the northern wilderness. Their haunting calls become the raw material for our own dreams and imaginations," said writer Judith W. McIntyre, who spent 20 years studying the birds.[1] Hearing a loon was a reliable indication you had escaped the cacophony of city life and found a measure of solitude. It was the beauty and symbolism of these birds that made it so disturbing to see them washed up, dead, on Great Lakes beaches.

In November 2007, I went looking for birds I did not want to see. Dead loons were washing up on the eastern shores of Lake Ontario. It wasn't an isolated problem. More than 8,000 dead loons washed up on Great Lakes shorelines between 1999 and 2007, their carcasses casting a pall over hundreds of miles of otherwise scenic beaches along Lakes Erie, Ontario, Huron, and Michigan. Sadly, the shoreline of Lake Ontario's Henderson Bay, about 15 miles southwest of Watertown, New York, was but one of a growing list of beaches that became death traps for loons in the first decade of the twenty-first century.

Irene Mazzochi, a biologist for the New York Department of Environmental Conservation, allowed me to tag along on her weekly check for dead birds on the coast of Henderson Bay. The picturesque beach fronted the state's Black Pond Wildlife Management Area and El Dorado Shores, a Nature Conservancy preserve that featured some of the most spectacular sand dunes in the eastern Great Lakes. It certainly wasn't the kind of place one would expect to find bird carcasses littering the shoreline. Mazzochi was driving a state-issued pickup truck along the beach, a death patrol carried out 30 feet from the water's edge, when one of her assistants sitting in the back pounded on the rear window.

"Bird," he yelled. Mazzochi hit the brakes and turned off the truck's engine. Everyone piled out and walked somberly toward the water. "Good eye," she told her colleague. He had spotted a dead loon rolling on its chest in the tiny waves that lapped at the shoreline.

Everyone knew the drill: Two workers in rubber boots and latex gloves pulled out a large, clear plastic bag; a third put on a pair of latex gloves and picked up the dead bird by its webbed feet and dropped it in the bag. The bag was placed in the back of the truck for transport back to a field station. From there, it would be sent to a laboratory to confirm, or dismiss, the presence of avian botulism. Then the carcass would be incinerated. The bird was the third of four dead loons and five deceased long-tailed ducks we found on the beach that raw November day. Our haul was a pittance compared to what Mazzochi's crew had found the previous week. In the second week of November 2007, the documenters of death turned up 66 dead loons and 41 other dead birds on that same stretch of beach. "When we drove out on the beach, everywhere you looked you could see white spots—those were the loons," Mazzochi said. The dead loons often were found lying on their backs, wings clutched tight to their bodies, their snowy chests facing the heavens.

A spirited Italian woman working in the male-dominated field of wildlife management, Mazzochi said she considered removing dead birds from the beach one of the worst parts of her job. Still, it was important work that had to be done. Leaving dead birds on the beach increased the risk of raccoons and other scavengers ingesting the deadly botulism bacteria. On an emotional level, the sight of dead birds on a beach was depressing. Physically, it was nauseating.

"People don't like to go out to the beach and see dead birds," Mazzochi said. "And if we can keep other animals from ingesting botulism it's a good thing. But this is pretty sad for me, picking up all these dead birds. It's not a glamorous part of my job. I'd rather be monitoring birds in a wetland or doing environmental education for students. I usually spend a lot of time in a canoe in wetlands. I feel like I almost have to get a cold heart when I go to pick up these dead birds. You can't look at every dead bird you pick up and say, 'Oh,'" she said, her words fading to an awkward silence.[2]

Mazzochi's cell phone rang as we talked about the deaths of glorious loons. A colleague called to report several dead loons along the rocky shoreline at Cape Vincent, the town situated where Lake Ontario flows into the St. Lawrence River. "You might want to go up there if you need more pictures," Mazzochi told me. It wasn't too long a drive, and Cape Vincent afforded a chance to see where nearly all water from the Great Lakes funneled into the St. Lawrence River.

I drove north up Highway 12E, a rolling two-lane road that passed through bucolic farmland and the curiously named villages of Limerick and Lyme. Arriving in Cape Vincent, I headed west along the St. Lawrence River. Within minutes, I spotted dead loons. The birds' brilliant white chests stood out like soldiers waving the white flag of surrender. I pulled off the side of the winding road and walked down the rocky bank to get a closer look.

While snapping photographs, I heard a faint, rhythmic sound approaching from the east: Thump, thump, thump, thump, thump, thump. The deep, pounding sound grew a bit louder with each cycle. A minute or so passed before an ocean freighter headed for Lake Ontario emerged from behind a small peninsula. The timing was uncanny. It was freighters like that one, ocean ships from Europe and Asia, that infected the Great Lakes with the foreign mussels and fish believed to be responsible for a chain of events that ultimately killed the loon lying at my feet.

The dead loons on the shores of Lake Ontario and the St. Lawrence River were eerily reminiscent of the scenes Rachel Carson described in her landmark 1962 book, *Silent Spring*. Carson wrote about how the careless, rampant use of the pesticide DDT in the 1950s killed countless robins and other songbirds. The chemical also thinned the eggs of bald eagles, greatly reducing the number of offspring and pushing the spectacular birds to the brink of extinction in the Great Lakes region. Carson's book, which led to a nationwide ban on DDT and launched the modern environmental movement, questioned the wisdom, the arrogance, of humans who decided which parts of the natural world could be sacrificed for the sake of economic gain. She wrote: "Who has made the decision that sets in motion these chains of poisonings, this ever widening wave of death that spreads out, like ripples when a pebble is dropped into a still pond? . . . The decision is that of the authoritarian temporarily entrusted with power; he has made it during a moment of inattention by millions to whom beauty and the ordered world of nature still have a meaning that is deep and imperative."[3]

Fifty years after the St. Lawrence Seaway opened, I wondered if a scenario similar to the DDT tragedy was playing out in the Great Lakes. Was the decision to open the lakes to ocean freighters for the sake of commerce carried out by a powerful few with no regard for how it might affect the nature of these freshwater seas? Were the deaths of 70,000 water birds that washed up on beaches over a 10-year period an ecological crisis, or simply part of the cost of foreign freighters doing business in the Great Lakes? Looking at a dead loon lying face up on the shore of Lake Ontario, its haunting call silenced, I wondered if this iconic species was the canary of the lakes. There was a time when coal miners took caged canaries with them deep into mines to gauge air quality. If the birds stopped singing, it was a warning that methane levels were dangerously high and might explode.

Common loons that spent the majority of their lives on remote lakes in northern Canada were an indicator of water quality, fish populations, and serenity. Lakes too polluted to support healthy fish populations, or too heavily trafficked by powerboats or personal watercraft, rarely supported loons. As top predators, loons were a barometer of ecosystem health. Problems in the food chain, caused by toxins or a lack of nutrients, were often reflected in the health of fish, birds, and mammals at the top of the food chain.

When thousands of loons fell silent on Great Lakes beaches, it said something about an ecosystem that provided the birds sustenance during their fall migration. Were these dead icons of the North a warning, much like the dead canaries that foretold dangerous conditions in coal mines? If so, what were the silenced loons telling us about the health of lakes that the St. Lawrence Seaway linked to the global shipping network and all the world's oceans 50 years ago? Let's find out.

And God said, Let us make man in our image, after our likeness: and let them have dominion over the fish of the sea, and over the fowl of the air, and over the cattle, and over all the earth, and over every creeping thing that creepeth upon the earth.

—Genesis 1.26

There is no trifling with nature; it is always true, grave, and severe; it is always in the right, and the faults and errors fall to our share.

—Johann Wolfgang von Goethe

01 DOMINION

Here the great Erie Canal has defied nature, and used it like a toy.

—**Caroline Gilman, American poet, 1838**

This [Welland] canal is intended to connect Lakes Erie and Ontario, and thereby remove the natural barrier caused by the wonderful and well known falls of Niagara.

—**William Hamilton Merritt, Canadian businessman, 1828**

01 CONQUERING NATURE

Niagara Falls is one of the most breathtaking and commercialized natural wonders on the planet. Its massive liquid curtains, created by water from four of the five Great Lakes spilling over the shale and dolomite edge of the Niagara Escarpment, measure a half-mile wide and 175 feet high. Words cannot adequately convey the magnificence of the Horseshoe Falls on the Canadian side of the Niagara gorge, and the American and Bridal Veil falls on the United States side of the border. The trinity of waterfalls is the second largest on Earth, but it has no equal. The beauty and fury of 750,000 gallons of Great Lakes water pouring over the American and Canadian falls each second—a thundering white wall of water that crashes onto a rocky gorge and explodes into a tower of mist 50 stories tall—is at once mesmerizing and terrifying.

U.S. Environmental Protection Agency

Niagara Falls was a natural barrier that kept ocean species and ships that reached Lake Ontario via the St. Lawrence River from reaching the four Great Lakes above the falls.

The tourist havens that line both sides of the river below the falls are in sharp contrast to Niagara's natural beauty. The cities of Niagara Falls, New York, and Niagara Falls, Ontario, are gaudy, loud, and visually confrontational. High-rise hotels, observation towers, casinos, and countless souvenir shops line the river gorge. Those manmade structures create a menagerie of concrete, glass, and bright neon lights above the natural canal that the raging Niagara River carved out of stone over the past 12,000 years. Gift shops dominate many street corners in both cities, all hawking a remarkable quantity and variety of trinkets bearing images of the glorious falls—hats, shirts, cups, spoons, pencils, ashtrays, lighters, and place mats. Some merchants even peddle tiny vials of colored water that supposedly "survived going over the falls." Just $3.98.

Perhaps it is the combination of natural beauty and commercial glitz that lures 50,000 newlywed couples and 12 million visitors to the falls each year. Tourism promoters boast that the falls attract more visitors each year than Disney World or the Grand Canyon. I choose to believe that people are drawn to the falls by their beauty, raw power, and mystery. Surely, some are attracted to the brink of the falls by a morbid desire to see where daredevils have gone over the edge in barrels and other contraptions. Whatever the attraction, visiting the falls seems to be a sort of required pilgrimage for people from many countries and cultures. On any given summer day, it is common to see Jews and Muslims, Africans and Asians, Germans and Brits standing shoulder to shoulder, all eyes locked on the churning, turquoise Niagara River as its water races over the precipice and bursts into a falling sheet of mist.

The falls, for my money, are the best place to grasp the immensity of the Great Lakes. It is one thing to read about the lakes' vital statistics. Carved into the continent roughly 10,000 years ago by glaciers, which then filled the massive basins as the mile-thick layer of ice melted, the lakes contain 18 percent of all fresh surface water on the planet. The five lakes hold six quadrillion gallons of water spread across 95,000 square miles of surface area. If water from the lakes were spread evenly across the continental United States, it would create a pond 9.5 feet deep from coast to coast. Those are impressive facts, to be sure. Now conjure this: Nearly all the water from Lakes Superior, Michigan, Huron, and Erie that doesn't evaporate eventually spills over the brink of Niagara Falls. It takes decades for water from the lakes to reach the falls. Lake Superior has a retention time of 191 years, meaning its water is completely replaced every two centuries. Lakes Michigan and Huron have a 99-year retention time. The water from those three lakes flows quickly through Lake Erie, spending about three years in the smallest and shallowest of the Great Lakes. By the time water from the four upper lakes flows out of Erie and feeds the raging Niagara River, some of the liquid assets have been moving toward the falls for roughly three centuries.

A volume of water slightly greater than that found in an Olympic-sized swimming pool spills over the falls every second of every day, every day of the year. That amounts to about 65 trillion gallons of water each day, enough water to supply about 21 cities the size of Chicago. Only a phenomenally huge, deep body of water could lose that much volume each day without running dry. If that doesn't drive home the immensity of the lakes, consider this: These freshwater seas are visible from outer space.

PALEO-INDIANS WERE THE FIRST HUMANS TO LAY EYES ON THE FALLS, thousands of years before European explorers began trickling into the region in the seventeenth century. They were followed by Woodland Indians and, later, by five distinct bands of the Iroquois Indian Nation. When Europeans finally reached the falls, they found the area occupied by warring factions of the Iroquois Nation. The Iroquois called the mighty river Onguiaahra, meaning "the strait." European immigrants, determined to make peace with the feuding Iroquois tribes, settled on a neutral Indian term for the falls, one that meant "thundering water." It was a fitting moniker. Standing on a rock outcropping at the side of the falls, the cascading water roars past, creating a constant rumble as it pounds the rocks below. Talking to a person next to you requires a strong set of lungs. Conversations are all but impossible.

Father Louis Hennepin, a French missionary who claimed to be the first white man to see Niagara Falls in 1678, described a scene of sheer terror. "When one looks down into this most dreadful Gulph [*sic*], one is seized with horror." The falls, he said, were "a vast and prodigious cadence of water which falls down after a surprising and astonishing manner, insomuch that the universe does not afford its parallel . . . the waters which fall from this horrible precipice do foam and boil after the most hideous manner imaginable, making an outrageous noise, more terrible than that of thunder."[1] Hennepin was known to exaggerate.

Canadians took the first steps toward making the falls a tourist attraction, building Portage Road along the river gorge in 1790. The first inn near the falls opened the next year, and the race was on to squeeze every possible dollar of tourism revenue from the natural spectacle. A visitor to the area in 1847 bemoaned the commercialization of the falls. "Now the neighborhood of the great wonder is overrun with every species of abominable fungus—the growth of bad taste, with equal luxuriance on the English and American sides—Chinese pagoda, menagerie, camera obscura, museum watch tower, wooden monument and odd curiosity shops."[2] In the late 1800s, the falls became the backdrop for an increasingly circus-like atmosphere. Daredevils flocked to Niagara to test their mettle against the mighty falls. One of the most bizarre spectacles took place in 1859, when a famous Canadian tightrope walker named Jean Francois Gravelet, aka "The Great Blondin," walked across a rope spanning the 1,100-foot wide gorge near the falls. He accomplished the amazing feat 21 times, including one journey with his manager sitting on his shoulders.[3]

A widowed schoolteacher named Annie Edson Taylor became the first person to survive going over the falls in a barrel. Taylor was a destitute teacher from Bay City, Michigan, who was hoping to cash in on her feat, provided she lived to tell about it. She figured such an experience would bring instant fame and fortune. On the afternoon of October 24, 1901—her 46th birthday—Taylor climbed into a 160-pound oak barrel and went over the Horseshoe Falls. Inside the barrel was a 100-pound anvil to keep it upright in the river. Taylor lived to tell about the terrifying ordeal, but vowed never to go over the falls again. "If it was with my dying breath, I would caution anyone against attempting the feat. I will never go over the falls again," she said. "I would sooner walk up to the mouth of a cannon knowing it was going to blow me to pieces than make another trip over the falls."[4] Several other people duplicated Taylor's feat in subsequent years; many others died trying.

I found it striking how Niagara Falls was a magnet for one form of commerce (tourism), while another (shipping) avoided the falls at all costs. For nineteenth-century entrepreneurs eager to extend the reach of global shipping into the heart of North America, the falls were a deadly obstacle that could not be overcome. Business and political leaders in the United States and Canada knew they had to find a way around the falls, and create a safe passage around its deadly wall of water, if they hoped to expand their empires west of the burgeoning cities of New York and Montreal.

THE NIAGARA FALLS REGION OF WESTERN NEW YORK WAS A RUGGED WILDERness in 1800. The uncharted frontier, occupied by just a few thousand Indians who didn't believe in the concept of owning land, was ripe for the taking. There were just 16 states in the United States at the time. But there was a strong desire among politicians to expand the young nation's sphere of influence to the vast expanse of land west of the Appalachian Mountains. The Appalachian range, which extended from Quebec to what is now northern Alabama, were as formidable an obstacle to western expansion as was Niagara Falls. Efforts to extend the western boundary of the United States began shortly after George Washington signed the *Declaration of Independence* in 1776. "If nothing were done, the young United States would be left squeezed between the mountains and the sea, a constricted minor league nation compared with the growth and power developing on the other side of the mountains," historian Peter L. Bernstein wrote in his book *Wedding of the Waters: The Erie Canal and the Making of a Great Nation*.[5] Washington organized the Patowmack Company with a mind toward converting the Potomac River into a canal extending from the Atlantic Coast at Alexandria, Virginia, to the Appalachian Mountains. Construction stalled shortly after it began, and the company fell into bankruptcy after Washington died in 1799. It was during that era of expansionist desire, when the French government in Canada was trying to monopolize trade with Native Americans in the western frontier, that New York became a focal point for efforts to build a shipping canal through the Appalachian Mountains and into the promised land of the boundless frontier. A fearless leader was needed to shepherd the construction of a canal across New York and through the stone wall of the Appalachian range. DeWitt Clinton, New York's governor in 1810, enthusiastically took up the cause. Clinton made construction of a shipping canal linking Lake Erie to the Atlantic Ocean, via the Hudson River, one of his top priorities. He was undeterred by the magnitude of the project or opposition from powerful federal leaders such as Thomas Jefferson, who derided the project as "little short of madness."[6]

The result of Clinton's tireless effort was the Erie Canal, a 363-mile-long ditch with 18 aqueducts and 83 locks that formed a shipping channel four feet deep and forty feet wide. The canal bypassed Niagara Falls by connecting Lake Erie to the Hudson River. Side channels connected the Erie Canal to Lake Ontario, which lay below Niagara Falls. The canal allowed ocean ships from the Atlantic Ocean to reach the Great Lakes by traveling north up the Hudson River and heading west into the Erie Canal a few miles north of Albany. No longer would ships from Europe have to travel to the St. Lawrence River, north of Maine, to gain access to rivers that led into the heart of North America. The Erie Canal, which could handle low-draft boats carrying up to 30 tons of freight or human cargo, provided a time-saving shortcut into New York and the unsettled territory that surrounded the massive lakes above Niagara Falls: Erie, Huron, Michigan, and Superior.

Building the Erie Canal was a stroke of technological and economic genius, one that expanded business markets and secured vast areas of the frontier for a burgeoning world power. Some called the canal the eighth wonder of the world. It reduced travel time from the East Coast to Lake Erie by half, and cut shipping costs by 90 percent. But there was a tradeoff. Clinton's Ditch was the first of many canals that neutralized the Great Lakes' natural defense mechanism by linking the lakes, hydrologically and biologically, to the Atlantic Ocean. In their natural form, the lakes had a natural barrier that shielded the freshwater ecosystems against invaders from the sea. Niagara Falls kept the four lakes above it biologically isolated from the Atlantic Ocean. The impenetrable falls, and the fact that the lakes were above sea level, made it impossible for ocean species to enter the freshwater seas without human assistance. If the powerful St. Lawrence River wasn't enough to ward off potential invaders from the North Atlantic, Niagara Falls was. The Erie Canal changed that. Its locks enabled boats from New York City and Europe to travel uphill, essentially, to Lake Erie and the vast waters of Lakes Huron, Michigan, and Superior.

Completion of the canal was a momentous occasion, economically, technologically, and socially. To celebrate the canal's historic opening, Governor Clinton and other dignitaries gathered in Buffalo on October 26, 1825, to fill a keg with Lake Erie water before traveling down the canal to New York City. They were greeted by cheering crowds, bonfires, and cannon salutes nine days later as their flotilla entered New York Harbor at 7 A.M. on November 4. Standing aboard his boat, the *Seneca Chief*, Clinton poured the barrel of Lake Erie water into the Atlantic Ocean and declared the "Wedding of the Waters" to be complete. "May the God of the heavens and Earth smile most propitiously on this work and render it subservient to the best interests of the human race," Clinton said.[7] A similar ceremony was held in Buffalo, with a keg of Atlantic Ocean water emptied into Lake Erie.

The Erie Canal was justifiably heralded as an engineering marvel and agent for social change. It gave European immigrants eager to claim their piece of the American dream easy access to what is now the Midwest. The canal opened the nation's heartland to commerce and made New York the trading capital of North America. Its influence on the settlement of the Great Lakes region cannot be overstated. The combined population of Illinois, Indiana, Michigan, and Ohio increased fifteen-fold between 1810 and 1850, swelling from 273,000 people to 4,217,000. Many of those settlers were European immigrants who began their American odyssey in New York City before boarding boats that took them north up the Hudson, west along the Erie Canal, and, finally, into Lake Erie. From there, the possibilities were endless.[8]

No one realized at the time that creating an unnatural link between Lake Erie and the Atlantic Ocean also opened a conduit through which ocean species, not just European immigrants on boats, could stream into the world's largest freshwater ecosystem. For all its benefits, and there were many, the

Erie Canal was the first in a series of Pandora's Locks that opened the doors for invasive species now wreaking havoc on the Great Lakes. The Erie Canal lit the fuse on a biological time bomb that exploded in the lakes a century later. The world's largest freshwater seas would never be the same again.

THE ERIE CANAL WASN'T THE FIRST CANAL TO MODIFY THE GREAT LAKES and its connecting rivers. Nor was the dream of opening the lakes to ocean ships born in the 1800s. French explorer Jacques Cartier, the first European known to traverse the St. Lawrence River, discovered in 1535 that the river was impassable near Montreal. Fierce rapids made paddling or sailing up that steep stretch of the river impossible. Sailing down the stair-step rapids was, at best, dangerous. At worst, the rapids could be deadly. That roiling stretch of river, which became known as Lachine Rapids, kept French explorers from proceeding further up the river for a century after Cartier arrived in Montreal. It wasn't until 1680 that a religious leader named Dollier de Casson began work on a five-foot-deep canal around the rapids. His efforts were slowed by endless amounts of seemingly impenetrable rock, and later by Iroquois Indians who opposed the canals. The Iroquois harassed workers to the point that the project was finally scrapped.

The British Royal Army Engineers began work in 1779 on four small canals on the north shore of the St. Lawrence, near Montreal. The canals were designed to connect Lake St. Louis to Lake St. Francis. Workers finished the job four years later. Measuring 2.5 feet deep, 6 feet wide, and 40 feet long, the primitive stone canals were believed to be the first locks built on the North American continent. The St. Lawrence River locks were large enough to allow fur-trading boats to bypass the river's treacherous rapids. Though modest in size, the structures were hugely significant. The canal and locks proved that humans could conquer natural obstacles and probe further into what was then an untamed wilderness.[9]

The first St. Lawrence River canal allowed shallow-draft boats to travel safely between Lake Ontario and the Atlantic Ocean. Still, Niagara Falls blocked all vessels from venturing further into the North American continent. It was simply impossible to navigate the treacherous Class V rapids downstream of the falls. Even if a sailor could navigate up the Niagara River, there was no way to get over the falls. That obstacle would soon be conquered as well.

In 1816, a year before work began on the Erie Canal, an ambitious Canadian businessman put forth a plan to keep pace with his canal-building rivals in the United States. William Hamilton Merritt proposed building a canal across the Niagara Peninsula, a 28-mile-wide band of land in southern Ontario that divided Lakes Erie and Ontario. The Niagara River, which flowed between Lakes Erie and Ontario, cut a natural path across the peninsula. But its menacing falls and raging rapids made the river impassable. Merritt, who was motivated by business interests and national pride, set out to make an end run around Niagara Falls.

Building the Welland Canal was Merritt's chance to show up the Americans, against whom he harbored some ill will. His father, Thomas Merritt, was a member of the Queen's Rangers, who had battled upstart rebels during the American Revolution 40 years earlier. The younger Merritt also fought the Americans during the War of 1812, a conflict between the United States and Great Britain that was waged on land and in the Great Lakes. As a Canadian soldier loyal to Britain, Merritt participated in the battle of Queenston Heights as a militia soldier. He fought many battles against the Americans until he was captured in 1814 and imprisoned in Cheshire, Massachusetts. Merritt was released when the war ended in 1815 and promptly headed back to St. Catherines, Ontario, to begin work as a merchant. There, his dream of building the Welland Canal was born.[10]

Merritt purchased a small mill on Twelve Mile Creek in 1816, which he expanded into a grist mill, salt spring, and distillery. There was a problem, though. He needed more water in the creek to achieve his entrepreneurial dreams. He initially proposed building a canal between the Welland River and Twelve Mile Creek. Those plans quickly evolved into something far more ambitious. Merritt formed a company to promote the construction of a canal that would allow ships to bypass Niagara Falls and sail between Lake Ontario and Lake Erie. His dream had blossomed into a fantastic effort to open all five Great Lakes to ships from all the world's oceans.

Construction of the first Welland Canal began in 1824, the year before the Erie Canal opened. With the project safely under way, Merritt unveiled his most ambitious plan: a project to improve the locks and canals in the St. Lawrence River, creating a standardized system of locks and canals that would allow ocean ships to travel 2,300 miles deep into the heart of North America via the Great Lakes. Together, the locks on the Welland and St. Lawrence rivers would transform the freshwater oceans into a new seacoast. Merritt explained his vision in an 1828 essay: "Lake Erie will be connected with the ocean by canals of only seventy six miles in length—sixteen to Ontario, and sixty on the St. Lawrence, which will render this extensive lake coast a sea coast, to all intents and purposes."[11]

Merritt noted somewhat jealously in an 1829 speech that the completion of the Welland Canal, not the Erie Canal on the American side of the border, marked the true marriage of Great Lakes and ocean commerce. "The artificial wedding of the Great Lakes of the West and North with the waters of Ontario and eventually with the St. Lawrence and the ocean," he said, "was complete."[12] Dismissed by critics as an "extravagant theorist," Merritt's dream ultimately became reality when the St. Lawrence Seaway opened 130 years after the first Welland Canal was completed.

History would prove the Welland Canal to be far more integral than the Erie Canal to the cause of opening the Great Lakes to global commerce. Without the Welland Canal, the St. Lawrence Seaway might not have been built. The Welland was expanded four times over the course of a century so it could accommodate ever-larger ships that moved iron ore, salt, and grain between ports on all five Great Lakes. The Erie Canal was expanded only slightly after it opened. By the time the Seaway opened in 1959, the Erie Canal was obsolete, a relic that could no longer meet the needs of a modern shipping industry centered around freighters that were as long as two football fields. The Welland Canal, on the other hand, was poised to welcome most of the world's ships into the upper Great Lakes.

A century of engineering and backbreaking labor went into developing the Welland Canal into its current form. The first Welland Canal, which cost $8 million to build, was modest by today's standards: it was just shy of eight feet deep and used a series of 40 wooden locks, each of which were at least 109 feet long and 22 feet wide, to lift and lower ships over the Niagara Escarpment. A second, larger Welland Canal opened in 1845 to accommodate bigger ships.

The bottom of the first and second Welland canals had a slight hump near the town of Welland, a site that became known as the Deep Cut. That was a problem. Water from Lake Erie didn't flow over the hump at the Deep Cut and into Lake Ontario. So engineers diverted the nearby Grand River to add flow to the canal. The Grand River was made into a 25-mile-long feeder canal that funneled water into the Welland Canal at the Deep Cut. The Grand River

U.S. Environmental Protection Agency

This lock is one of eight in the Welland Canal that allows ships to bypass Niagara Falls and to travel safely between Lake Ontario and Lake Erie. The canal also allowed sea lampreys to spread from Lake Ontario to the other four Great Lakes.

Jeff Alexander

An ocean freighter traveling from Lake Ontario to Lake Erie enters a lock in the Welland Canal. The locks lift and lower ships 327 feet, allowing vessels to climb the Niagara Escarpment and avoid Niagara Falls.

Jeff Alexander

After the ship is securely in the lock, water is allowed to flow into the structure, lifting the ship.

Jeff Alexander

The ship is released from the lock after it is lifted to a level equal with the water upstream.

diversion allowed ships safe passage through the length of the Welland Canal, but it did not change this fact: the manmade channel remained divided, with water in the southern section flowing toward Lake Erie and water from the northern section flowing into Lake Ontario. The phenomenon kept Lake Ontario fish from swimming the length of the Welland Canal and into Lake Erie. Author William Ashworth explained the phenomenon in his book *The Late Great Lakes:*

"A fish from Lake Ontario, swimming upstream through the canal in its instinct-driven search for the headwaters, would suddenly—at Welland—find itself swimming downstream instead. Instinct would complain vociferously; the fish would turn aside, passing through the feeder canal into the Grand River and never reaching Lake Erie at all."[13]

During a third expansion of the canal, completed in 1887, engineers carved the Deep Cut to a depth that allowed a portion of Lake Erie's water to flow directly into Lake Ontario via the canal. Though it technically was not needed, the Grand River diversion remained in place until 1919. A fourth expansion of the Welland Canal cut off the Grand River diversion, creating a direct pathway between Lakes Erie and Ontario that was 14 feet deep and 45 feet wide. With the Grand River diversion eliminated in 1919, fish that previously swam up the canal and were diverted into the river at the Deep Cut had nowhere to go but further up the canal and into Lake Erie.

One of the species known to inhabit Lake Ontario and swim up the Welland Canal prior to 1881 was the sea lamprey—a hideous, parasitic fish that looked like an eel and fed like a vampire. The sea lamprey literally sucked blood and other bodily fluids from its prey. Sea lampreys entered Lake Ontario in the early 1800s, using a side channel to the Erie Canal. If the lampreys could just get past Niagara Falls, they would be free to roam the other four Great Lakes. That moment came in 1919, when the last modification to the Welland Canal created a continuous flow of water from Lake Erie into Lake Ontario. That was the chink in the Great Lakes' natural armor that the lamprey needed to invade the world's largest freshwater fishery. Biological mayhem was about to rain down on Lakes Huron, Erie, Michigan, and Superior.

The first lamprey in Lake Erie was discovered in 1921. The unsightly creatures must have felt like kids in a candy store—Erie and its sister lakes teemed with lake trout and whitefish. By 1940, one of the world's worst fish massacres was underway. The assassin: a foreign fish species that humans allowed into the bountiful waters of the Great Lakes by building the Erie and Welland canals.

I visited the Welland Canal in the fall of 2007, nearly a century after the artificial channel allowed sea lampreys to invade the lakes above Niagara Falls. Truth be told, I expected to feel anger and revulsion at the sight of the massive canal that had plunged the lakes into biological chaos. Instead, I was awed by the size and brilliant simplicity of a series of stair-step locks that gently lifted and lowered 650-foot-long freighters over the Niagara Escarpment. Standing a stone's throw from the locks, it was easy to forget that the Welland Canal opened the Great Lakes to more than international shipping. There was no question that the Welland Canal was, and remains, an engineering masterpiece that helped transform the economies of the eastern United States and Canada. In bypassing the falls, engineers and politicians opened a new frontier for global commerce. But foreign species, just like European immigrants, soon flowed into the upper Great Lakes. There was no disputing the fact that the Welland and its American counterpart, the Erie Canal, visited biological mayhem upon the Great Lakes. The canals nullified one of Niagara Falls' most important roles, that of a biological barrier that kept ocean species from reaching Lakes Erie, Huron, Michigan, and Superior. Granted, the sea lamprey invasion was an unintended consequence of building canals that linked freshwater and saltwater ecosystems. The grim reality was that the Erie and Welland canals were like stray threads in a sweater, slowly unraveling the finely woven biological tapestry of all five Great Lakes.

Sea lamprey are probably the most destructive invasive species ever to hit the Great Lakes.

—Doug Jensen, Minnesota Sea Grant

02 VAMPIRES OF THE DEEP

On a cool September night in 1954, Marilyn Bell stood on a retaining wall in Youngstown, New York, and tried to overcome her worst fear as she prepared for an unprecedented feat. The 16-year-old Canadian girl was about to attempt a swim across Lake Ontario, a 32-mile journey across dark, mysterious waters in the middle of the night. No one had ever completed the grueling swim from Youngstown to Toronto. Bell knew it would take every ounce of strength and endurance she could muster. But it wasn't the distance, fear of strong currents pushing her off course, or the reality that she would be swimming through polluted water that worried Bell. Her greatest fear was a hideous fish that looked like an eel and sucked the blood of its victims. She was terrified at the prospect of a sea lamprey clinging to her body, boring a hole, and drawing blood. "If I find an eel on me," Bell told her coach before diving into the lake, "I'll scream."[1] Ever since sea lamprey invaded Lake Ontario in 1835, scientists assured the public that the unsightly fish would not attack humans. Lampreys were only attracted to cold-blooded fish, not warm-blooded humans, according to biologists. Those assurances carried little weight with Bell. She was gripped by fear.

Bell dove into the water at 11:08 P.M. on September 8 and began swimming north across the lake's ebony water, toward the bright lights of Toronto. It wasn't long before she felt a gnawing sensation on her abdomen; something was tugging at her bathing suit. Suddenly, the girl's worst nightmare was a terrifying reality. A sea lamprey—the horrid fish that used its disc-shaped, tooth-lined mouth to cling to fish and drain their blood—was attacking Bell. With the menacing fish grabbing her midsection, Bell did what her coach advised. "I struck hard at it with my hand and my blow knocked it off," Bell said later.[2]

Bell was one of three women who attempted the swim across Lake Ontario the night of September 8, 1954. Florence Chadwick, the 34-year-old American swimmer who was the first woman to swim the English Channel, also was attempting to swim across the lake. The third swimmer was Winnie Roach Leuszler, a 28-year-old champion swimmer from Ontario. Chadwick's motive was money: the Canadian National Exhibition, a huge Toronto festival, had offered the legendary American swimmer $10,000 if she swam across Lake Ontario. Bell and Leuszler were motivated by personal and national pride. Both wanted to compete against the American swimmer, even if it meant the only prize would be a crowning sense of accomplishment. Leuszler gave up her quest shortly after it began. Chadwick quit seven hours into the swim, a victim of rolling waves that caused her to vomit repeatedly.

By the morning of September 9, the 16-year-old Bell was engaged in a singular battle against the perils of Lake Ontario. Strong currents from the mighty Niagara River surged through the lake, pushing Bell off course, and sea lamprey repeatedly struck at her body. Near the end of Bell's 21-hour swim, a lamprey latched onto her thigh. She pulled if off and tossed it aside. Three more attacked her as she battled waves and mounting fatigue.[3] When she finally reached the Toronto shoreline, a crowd of 250,000 people gave her a hero's welcome. The unassuming teenager who had bested an American champion and a fierce Great Lake swam her way into the hearts of Canadians. Bell became a national hero and was honored with parades, cash, and gifts. A

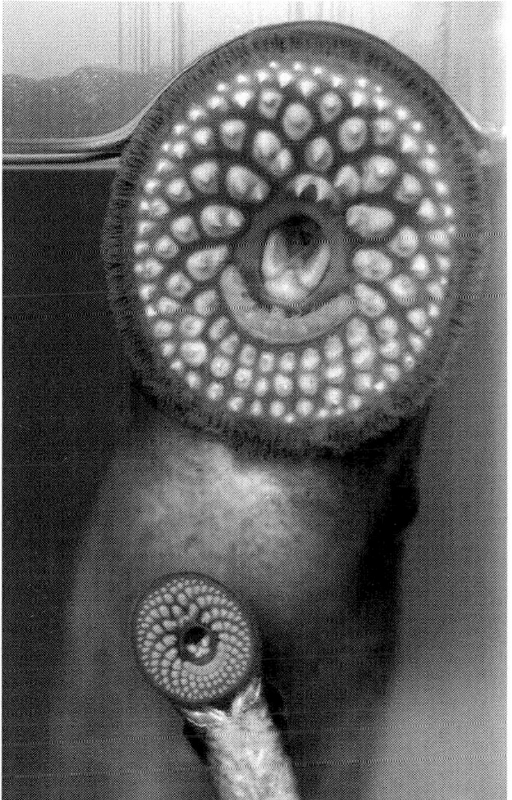

Dave Brenner, Michigan Sea Grant

Construction of the Erie Canal in 1825 allowed sea lampreys in the Atlantic Ocean to migrate into Lake Ontario. An expansion of Canada's Welland Canal in 1919 allowed sea lampreys to spread from Lake Ontario to Lakes Erie, Huron, Michigan, and Superior. The eel-like fish are parasites, which cling to fish and drain their blood and other bodily fluids.

Sea lampreys are shown clinging to a lake trout. These parasites drove lake trout, the Great Lakes' top native fish predator, to the brink of elimination before a control program reduced their number.

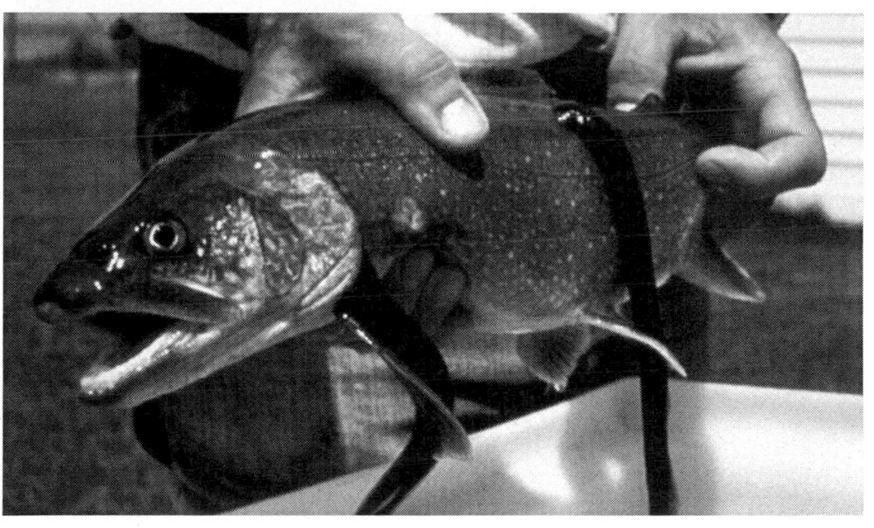

Great Lakes Fishery Commission

park on the Toronto shoreline bears her name. Hers is a heartwarming story of incredible personal achievement against tremendous odds.

The fact that Bell completed the unprecedented swim while doing battle with bloodsucking sea lampreys reads like the script of a low-budget horror movie. Her experience is one of the few documented cases of sea lampreys attacking humans. It was unlikely that lampreys would have pierced Bell's skin and drawn blood. Still, who wouldn't have been terrified by the prospect of a snake-like fish clinging to their body? Fortunately, Bell had the strength and good sense to fight off the swimming vampires before they had a chance to sample her flesh. Great Lakes fish, on the other hand, were no match for the ravenous parasites.

SEA LAMPREYS ARE, AS THE NAME IMPLIES, NATIVE TO SALTWATER ENVIRONments—the Atlantic Ocean, to be precise. Though the snake-like beasts look like eels, sea lampreys belong to a species of jawless fish that have roamed the world's oceans for 250 million years. That sea lamprey survived whatever cataclysmic events eliminated dinosaurs 65 million years ago says something about the species' durability.

Sea lamprey is one of about 50 species of lamprey. Several are native to the Great Lakes basin, including the northern brook, chestnut, and silver lamprey. Some native species of lamprey cling to fish and drain bodily fluids, just as the sea lamprey does. But the sea lamprey is the largest and most ferocious lamprey to colonize the Great Lakes, and rivers that flow into them. The fish grows to nearly three feet long. Full-grown males weigh about two pounds; females can reach five pounds. Its disc-shaped mouth, lined with rings of sharp teeth and a piercing tongue, allow the lamprey to cling to the most powerful of fish. Coupled with its ability to zip through water like a snake on speed, the sea lamprey is a highly effective aquatic assassin.

The sea lamprey's appearance is almost as grotesque as its dietary habits. The fish has slimy skin and a mottled appearance that ranges from pale green and yellow-brown to dark brown on its back and pale gray on its belly. It has two dorsal fins and a single ventral fin, all of which are reddish brown along the edges. Behind the eyes, on both sides of the tubular body, are lines of seven round gill openings. Sea lamprey is easily the ugliest of all fish species in the Great Lakes.

Sea lamprey weren't known to live in the Great Lakes prior to the opening of the Erie Canal in 1825. The first breeding population was discovered in 1835 in Duffins Creek, which flows into Lake Ontario east of Toronto. Scientists theorized that the parasites snuck into Lake Ontario through a side channel of the Erie Canal. Because lampreys are anadromous fish, like salmon, they move out of oceans and into rivers to spawn before dying. It was only natural for sea lamprey that lived in the Atlantic Ocean to expand their range into the Erie Canal and Lake Ontario after humans, in building the Erie Canal, opened a new freshwater frontier. The 1921 expansion of the Welland Canal allowed sea lamprey to migrate into the four Great Lakes above Niagara Falls. There, the ravenous parasites found the world's largest freshwater fish market.

Scientists discovered lamprey in Lake Erie in 1921, in Lake Michigan in 1936, in Lake Huron in 1937, and a year later in Lake Superior. For sea lamprey, the Great Lakes were a dietary nirvana. There was an abundant supply of fish—lake trout, whitefish, and burbot—and thousands of miles of streams where they could spawn. Most importantly, there was no natural predator to control the lamprey population. The parasites quickly established their species as king of the lakes, dethroning the native ruler: lake trout. Sea lampreys feasted on lake trout and whitefish, causing those fisheries to collapse with stunning speed. By 1949, lake trout were on the verge of being eliminated from Lake Huron, and the population in Lake Michigan was dropping like a lead balloon.

Lake trout were a staple of the Great Lakes commercial fishery before sea lamprey invaded. Anglers harvested some 15 million pounds of lake trout each year in Lakes Michigan, Huron, and Superior. By the early 1960s, the lake trout catch dropped to 300,000 pounds. The lake trout harvest in Lake Huron dropped from 3.4 million pounds in 1935 to 1,000 pounds in 1949. The catch in Lake Michigan dropped from 5.5 million pounds in 1946 to 402 pounds in 1953. In Lake Superior, the catch dropped from an average of 4.5 million pounds annually to 368,000 pounds in 1961. An unsightly fish with no commercial value was making quick work of the most valuable sector of the Great Lakes commercial fishery, lake trout.[4]

Michigan Department of Natural Resources biologist Vernon Applegate summarized the sea lamprey's conquest of the lakes in a 1947 report. He wrote: "The problem of the sea lamprey and its potential threat to the commercial fisheries of our state, especially the lake trout, has grown alarmingly in magnitude within the past decade."[5] Applegate concluded in a 1951 report that sea lamprey, not excessive fishing by humans, was responsible for the collapse of the lake trout fishery in the Great Lakes. "The most careful consideration of available evidence nevertheless permits only one conclusion—namely, that the sea lamprey was the major, perhaps the only significant, cause of the decrease of the lake trout in both lakes [Michigan and Huron]," he said in an article in *Scientific Monthly* magazine.[6]

The sea lamprey accomplished in three decades what thousands of anglers and hundreds of commercial fishing operations could not: the virtual elimination of lake trout from all but one of the lakes—Superior.

Bob King, a third-generation tribal fisherman who harvested whitefish in northern Lake Michigan, had vivid memories of the sea lamprey's biological rein of terror. "When I was a young boy, in the late 1940s, the sea lamprey came in and just about wiped out the fishery; it wiped out pretty much everything," King said.

Paul Jensen, a commercial fisherman in Muskegon, Michigan, said lamprey were the primary reason commercial fishing waned in Lake Michigan. "Every port along southern Michigan had some commercial fishermen. South Haven, St. Joseph, Ludington, Manistee. . . . There aren't any [now], they're gone. And one of the main reasons they're gone, I think, can be traced back to sea lamprey."[7]

Dracula had nothing on these ruthless killers. After a lamprey bored into a fish, it injected an anticoagulant, lamphredin, that allowed it to suck blood and other fluids until the predator was sated or its prey died. Adult lamprey spent up to 18 months feeding in the lakes. Each lamprey killed up to 40 pounds of fish before exiting the lakes to migrate upstream and spawn another generation before dying.

Lamprey are as efficient at reproducing as they are at hunting. The average female lays 61,000 eggs, but can produce as many as 100,000 eggs. After spawning, the adults drift downstream and die. Their offspring burrow into the soft silt on river bottoms after hatching, where they remain for as long as six years before heading out into the lakes to feast on fish. It is an elaborate reproductive cycle that has survived for 250 million years. Imagine trying to drive that kind of durable, evolved invader out of an ecosystem that is just 10,000 years old.

In the 1940s and '50s, at the peak of the lamprey infestation, up to 85 percent of Great Lakes fish that anglers caught had scars from lamprey attacks.[8]

There was no mistaking a lamprey scar—they were round, about the size of a half dollar, and ranged from pale red marks on the scales to open wounds that exposed flesh or internal organs. The fish often died of infection if they didn't bleed to death.

Faced with the very real prospect of losing all lake trout, whitefish, walleye, and burbot in the Great Lakes, the U.S. Fish and Wildlife Service in the 1940s declared war on sea lamprey. The agency installed mechanical and electric barriers in dozens of rivers that flowed into Lakes Michigan and Superior, hoping to keep lamprey from migrating upstream to spawn. Government scientists worked with researchers from chemical manufacturing corporations for four years, testing more than 6,000 chemicals in their desperate quest for a magic bullet that would kill lamprey without decimating other fish species. Intense research at the Hammond Bay Biological Station, on the Michigan side of Lake Huron, led scientists to a chemical called TFM, 3-trifluoromethyl-4-nitrophenol. TFM was a chlorinated compound that turned out to be highly effective at killing juvenile lamprey in streams, while causing minimal collateral damage to native species. "We had some super bright people working on the program at the Hammond Bay Biological Station [in the 1950s]. Otherwise, we would have developed another DDT and we'd all be in trouble," said Jim Seelye, who supervised the lamprey research program from 1982 to 1995.[9] TFM applications occasionally killed native fish and salamanders, but scientists concluded that the benefits of using the chemical far outweighed the negative side effects. Without TFM, there might not be any lake trout, whitefish, burbot, chubs, or salmon in the Great Lakes today.

The U.S. and Canadian governments in 1955 established the Great Lakes Fishery Commission to manage the war on sea lamprey. The development of TFM as a lampricide led some government officials to predict, incorrectly, that lamprey would be eradicated from the Great Lakes. The Mosquito River, which flowed into Lake Superior in northern Michigan, was the first stream treated with the lampricide, in 1958. Sea lampreys had already been in the lake and its tributaries for 20 years. Reducing their numbers would take a monumental effort and cost millions of dollars. The Great Lakes Fishery Commission spent $915,000 to kill 122,275 lampreys in Lakes Superior and Michigan in the first year of the program—$7.48 per lamprey.[10] Despite the daunting nature of the task and the astronomical cost, some biologists were convinced they could eliminate sea lamprey from the 95,000 square miles of Great Lakes and thousands of tributaries that flowed into the freshwater seas. "We will beat the eel," said James W. Moffett, head of the U.S. Fish and Wildlife Service's lamprey control campaign, in a 1956 newspaper interview. "But victory isn't around the corner. It will be a long, hard fight but the issue is not in doubt, for we have the means of winning."[11] It was a bold prediction, to be sure—one born more of wishful thinking than reality.

The lamprey's steady march across the lakes had profound ramifications for several species of Great Lakes fish. Its decimation of lake trout and other native species left a massive void in the fish community, which was promptly filled by another invader: the alewife. By 1969, when lamprey attacks on lake trout and other fish surged anew, scientists accepted the reality that they would likely never see the elimination of sea lamprey. "We thought we had them under control. But we do not," said Charles Lloyd, chief of Wisconsin's fish-management division, in a 1969 newspaper interview. It took government fish biologists another 20 years to reduce the lamprey population by 90 percent, a level that would permit the recovery of whitefish and other desirable species. Sea lamprey reproduced in 433 of the 5,747 streams and tributaries that flowed into the Great Lakes. Of those 433 streams, 250 were treated regularly with TFM; about 70 streams were treated each year to keep the aquatic monsters in check. The war on sea lamprey was raging on in 2008 and will likely continue in perpetuity.[12]

Eager to reduce the use of TFM, the Great Lakes Fishery Commission in the 1990s developed new weapons for its war on sea lamprey. Biologists began collecting thousands of male sea lamprey each year and sterilizing them at the Hammond Bay Biological Station. The fish were fed through a complex machine that destroyed their sperm. The males were then released back into the wild to trick unsuspecting female lamprey, the eggs from which would go unfertilized and perish. The commission also dispatched a helicopter each spring to drop granular Bayluscide, a new lamprey-killing chemical, into the St. Marys River. The St. Marys, which linked Lake Superior to Lake Huron, was a large, fast-flowing river and the mother of all lamprey breeding grounds in the Great Lakes.

The Great Lakes Fishery Commission spent $318 million in the first 50 years of its war on sea lamprey. But the battle was an eternal one. Lamprey numbers swelled every time the parasites found a hole in the chemical safety net government biologists cast over the lakes. The scorned fish demonstrated remarkable resilience in the late 1990s, finding their way through fissures in a crumbling dam in northern Michigan. Lamprey migrated through the dam and infested a stretch of the Manistique River thought to be inaccessible to the invaders. By 2000, the number of Lake Michigan fish with lamprey wounds was on the rise. Scientists were baffled. They initially thought the lamprey population had increased in the St. Marys River. Biologists pumped more TFM into the St. Marys to beat back the lamprey, but the number of lamprey wounds on Lake Michigan fish continued to rise. Biologists in 2002 estimated the number of spawning lamprey in Lake Michigan at 200,000—the highest number recorded since control efforts began four decades earlier.

"It is obvious there are more lampreys in Lake Michigan than at any time since the initiation of control efforts," said Mark Holey, project leader at the U.S. Fish and Wildlife Service's office in Green Bay, Wisconsin. "It just goes to show how fragile the fishery that anglers enjoy and rely on really is to lamprey predation, one of the original invasive species."[13]

Alarmed by the prospect of lampreys regaining control of the Lake Michigan fishery, scientists fanned out to check dozens of streams around the lake where the parasitic fish were known to breed. They discovered thousands of lamprey breeding in the Manistique River, one of the largest rivers in Michigan's Upper Peninsula. The Fish and Wildlife Service treated the Manistique with TFM in 2003 and 2004. The benefits of those treatments became evident in 2006, when fewer fish showed signs of lamprey attacks.

The lamprey suppression effort, though imperfect, was the most successful invasive-species control program in the Great Lakes. And yet, if the Great Lakes Fishery Commission didn't spend $19 million annually to control the predator, its population would explode. Populations of lake trout, salmon, whitefish, and walleye would most certainly crash. Seelye believed that the Great Lakes Fishery Commission could eradicate sea lamprey from the lakes if it had sufficient financial resources. "The original plan was to eradicate sea lamprey, then it became a control and manage program," Seelye said. "We could eradicate the lamprey but it would cost a lot of money. Once it was done, it would be done for good."

There was always the possibility, however, that more lampreys would enter the lakes through the St. Lawrence Seaway or the Erie Canal. The Great Lakes were defenseless against invaders from the sea. Artificial canals that connected Great Lakes ports to the global shipping network were like open sores on a human body that never healed—the victim was always vulnerable to new infections. The vexing nature of the struggle highlighted the paramount importance of preventing new invasive species from gaining a foothold in the Great Lakes.

THE SEA LAMPREY SAGA RAISED A SIMPLE QUESTION IN MY MIND: IF scientists knew in the 1940s that man-made canals allowed the ferocious ocean fish into the Great Lakes, did anyone bother to ask whether building the St. Lawrence Seaway might allow more invaders into the lakes? Apparently not. There was no record of fisheries biologists or engineers involved in the design of the Seaway pondering such a scenario. Similarly, there was no evidence of anyone questioning whether ocean freighters might import foreign species along with the cargo they would deliver to the lakes. The engineers who designed the Seaway obviously weren't students of popular culture. For in September 1954, the same month construction began on the Seaway, a 16-year-old girl swam across Lake Ontario and into the annals of Canadian history.

I asked Marc Gaden, a Great Lakes Fishery Commission official who knew the lamprey story as well as anyone, whether anyone questioned the ecological wisdom of building the Seaway before construction began. Gaden's response came in the form of a story. "Only one other person has asked me that question over the years that I've been speaking about this," Gaden said. "He was a fifth-grader who sat and listened to my sea lamprey talk politely and then asked, 'So wait, if the lampreys were so bad, why did they build the St. Lawrence Seaway?' My only reply was that I wished someone as smart as he would have asked that question in 1959. So the answer is no, I have never come across any literature that suggests invasive species were on people's minds when they proposed and built the Seaway."[14]

Gaden had his own theories about why no one connected the sea lamprey disaster to the tidal wave of invasive species that might flood the Great Lakes once the Seaway allowed ocean freighters into the lakes. He said few people, other than fisheries biologists, were aware of invasive species in the lakes in the 1950s. And since sea lamprey swam into the lakes through canals that were built long before the Seaway, why would a few more artificial canals matter? Ballast water wasn't known to be a vector for the introduction of invasive species

50 years ago—it wasn't on scientists' radar. And, perhaps most significantly, scientists in the 1950s knew far less about the lakes' ecosystems than they did a half century later. The learning curve would be steep and costly.

Engineers who designed the Seaway and politicians who lobbied for its completion blithely assumed that opening the Great Lakes to ocean freighters could have only positive consequences for the U.S. and Canadian economies. They didn't question whether opening these freshwater seas to ocean freighters might be detrimental to the lakes. They lacked the scientific expertise or intuition to recognize that the sea lamprey's conquest of the lakes was a shot across the bow of all who dared tinker with the natural order of the lakes. For them, ignorance was bliss.

Along the steel-blue St. Lawrence River, seaward outlet of the world's busiest inland waterway, a century-old dream is coming true.

—*Time*, 1955

03 SALT IN THE WOUND

Dense fog blanketed a section of the St. Lawrence River from Montreal to Massena, New York, on a summer morning in 1959. That day was supposed to be the brightest moment in the shared maritime history of Canada and the United States. After nearly five years of construction that followed six decades of political bickering, the time had come to dedicate the St. Lawrence Seaway. It was, at the time, one of the world's largest public-works projects. The momentous occasion brought a queen and a president to the banks of the St. Lawrence River. England's Queen Elizabeth II and U.S. President Dwight Eisenhower arrived at St. Lambert lock, near Montreal, on the morning of June 26, 1959. They came together to dedicate a $1 billion project that opened the Great Lakes to intercontinental shipping by taming one of North America's mightiest rivers. The enormous St. Lawrence, which flowed

Getty Images

U.S. President Dwight Eisenhower and British Queen Elizabeth II celebrated the official opening of the St. Lawrence Seaway on June 26, 1959. They traveled along a portion of the Seaway aboard the queen's royal yacht, the *Britannia.*

out of Lake Ontario, carried nearly all the water from the Great Lakes along its nearly 800-mile long course to the Atlantic Ocean. Conquering the river was the final step in a 130-year campaign to link Great Lakes shipping to all the world's ports. The project entailed building six canals and seven locks in a section of river that dropped 226 feet between Lake Ontario and Montreal. When combined with the existing Welland Canal and the Soo Locks, situated between Lakes Huron and Superior, the Seaway gave freighters from around the world safe passage into all five Great Lakes. It was a historic project that would alter the course of economic and environmental history across a vast expanse of North America.

Lavish as it was, the dedication of the Seaway was anticlimactic—ocean ships had begun using its canals and locks three months earlier. The icebreaker *D'Iberville* was the first ship to pass through the Seaway, traveling from Montreal to Lake Ontario. Officials delayed the grand opening until they were sure all technical bugs had been resolved before sending a royal yacht up the river. On that celebratory day in June, the queen and president were scheduled to

U.S. Army Corps of Engineers

Seven locks in the St. Lawrence Seaway, combined with eight locks in the Welland Canal and parallel locks at the Soo Locks, allow ocean ships in the North Atlantic to travel 2,340 miles into the heart of North America, to Duluth, Minnesota.

board the royal yacht *Britannia* in Montreal and travel about 30 miles up the St. Lawrence River, passing through the Seaway's canals and giant locks en route to the Eisenhower Lock in Massena, New York. Unbowed by the gloomy weather, the queen and president smiled and waved to the crowd as the sleek blue and white yacht moved slowly through the St. Lambert Lock. Guns fired a salute, trumpets sounded, rockets burst in the air, and a huge throng of dignitaries and commoners cheered as the royal yacht inched through the first of seven locks—each measuring 766 feet long, 80 feet wide, and 30 feet deep. The royal journey, which was delayed three hours by fog, had to be cut short when the weather refused to yield to the gala celebration. The *Britannia* could only transport the two world leaders as far as Snell Lock, in New York, because the cotton-like fog reduced visibility in the river to near zero. Perhaps it was an omen of the Seaway's cloudy future.

When the queen and president arrived by car at the giant Moses-Saunders hydroelectric dam near Massena, the queen unveiled a plaque commemorating the Seaway as a symbol of cooperation between two nations that had gone to war twice in the previous two centuries. The British monarch said she hoped the project would be a model of cooperation for the rest of the world, one that would become "infectious."

Warships and freighters traveled to several Great Lakes ports during the ensuing 10 days to celebrate the Seaway's opening. Large crowds welcomed the ships as symbols of economic prosperity and military security. One of the largest celebrations was in Duluth, Minnesota, at the western end of the 2,340-mile-long Seaway. A crowd of 100,000 people turned out for the celebration. The Navy destroyer *Forrest Sherman* led the procession into Duluth Harbor, followed by the submarine *Corsair* and a birch-bark canoe that replicated the main source of transportation for Native Americans and the first European settlers in the Great Lakes. There were sailors decked out in dress whites, and voyageurs clothed in animal skins. The television hero Wyatt Earp, portrayed by actor Hugh O'Brien, rode a horse that led a parade through downtown Duluth. Fighter jets screamed past, flying low and in tight formation.

After a series of speeches, then Minnesota governor Orville Freeman presided over a ceremonial wedding of the waters. He poured water that had been collected from the world's seven seas into Duluth harbor. Former Wisconsin governor Gaylord Nelson, who founded Earth Day 11 years after the Seaway opened, conducted a similar ceremony on the other side of the bay, in Superior, Wisconsin. For the second time since the Erie Canal had opened 134 years earlier, Great Lakes waters were mixed with those of the world's oceans. The second wedding of the waters—marking the first time large freighters from around the world could enter all five Great Lakes—ultimately proved more ecologically disastrous than the first. The Seaway certainly became a global model, just not the kind Queen Elizabeth, President Eisenhower, or other ardent supporters of the project ever anticipated.

THAT THE SEAWAY WAS BUILT AT ALL WAS A SMALL MIRACLE, GIVEN ITS controversial history. The campaign to build it lasted 60 years before the first spade of dirt was turned over. In 1895, the United States and Canada formed the Deep Waterways Commission to study the feasibility of building a shipping canal that would open the lakes to commerce from all the world's seaports. Midwest farmers viewed the Seaway as a way to ship more grain, at lower cost, to overseas customers. Quebec mining firms wanted the Seaway built so they could ship iron ore from the Labrador range to steel mills in the Midwest, which were churning out the raw materials for a thriving auto industry.

The Great Lakes were not landlocked prior to the Seaway opening. For decades, small ships sailed in and out of the lakes by using low-draft canals near Montreal that bypassed rapids in the St. Lawrence River. Paddlewheel ships that ran the staircase Lachine Rapids in the late 1800s and early 1900s were a major tourist attraction before the Seaway was built. Though passable to smaller ships that could tolerate a rough ride, freighters carrying 25,000 tons of cargo could not navigate the river's rapids or be assured of safe passage through the varied depths of its channel. The Seaway was designed to allow 80 percent of the world shipping fleet to enter the lakes by completing a shipping channel with a uniform depth of 27 feet that stretched 2,340 miles from the Gulf of St. Lawrence to Duluth, Minnesota.

The Seaway also had strategic value for the Canadian and U.S. militaries. Both nations viewed the project as a military asset, one that could be used to transport ships and weapons to the Atlantic Ocean during times of conflict. Moreover, harnessing the power of the St. Lawrence River with hydroelectric dams would give Canada and the United States access to cheap, reliable electricity, which was essential to the production of tanks, guns, and other instruments of war. It's important to remember that the Seaway was built a decade after World War II ended, just as the cold war between the United States and the Soviet Union was heating up. The military significance of the proposed shipping channel was made clear at a 1940 Seaway conference, when a high-ranking U.S. official read a message from President Franklin D. Roosevelt:

> The United States needs the St. Lawrence Seaway for defense. The United States needs this great landlocked sea as a secure haven in which it will always be able to build ships and more ships in order to protect our trade and our shores. The United States needs, tremendously needs, the [hydroelectric] power project which will form a link in the Seaway in the International Rapids Section of the St. Lawrence River to produce aluminum and more aluminum for the airplane program which will assure command of the air. . . . Shipyards on the Great Lakes, with access to the ocean, yet close to the sources of supply of labor, raw and finished materials, further removed from possible attack, may be a vital factor in successful defense of this continent.[1]

Federal politicians in the United States had waged a rhetorical war against the Seaway since the idea was first floated. Though every U.S. president who served from 1900 to 1954 supported building the Seaway, every Congress during that period torpedoed funding for the project. Members of Congress were swayed by railroad lobbyists, who feared the Seaway would break their industry's stranglehold on cargo shipments in the eastern United States. Those fears were justified. Freighters using the Seaway could transport cargo from the Midwest to the Atlantic Ocean for about $1.70 per ton, far less than the $13 per ton it cost to ship cargo the same distance on trains.

Frustrated by decades of political wrangling, Canadian officials began threatening in the 1940s to build the Seaway on their own. The steady rise of cold-war tensions between the United States and the Soviet Union increased support for building the Seaway, but not enough to get Congress to fund the project. The stalemate wasn't broken until 1954, three years after Canadian leaders announced that their nation would build the Seaway on its own, if necessary. That created a frightening scenario for the United States. If Canada built and solely owned the Seaway, it would control the major access route to the Great Lakes from the Atlantic Ocean. That possibility, Congress decided, was simply unacceptable. In January 1954, a young United States senator named John Fitzgerald Kennedy made a speech on the floor of the Senate in favor of the project. His colleagues who had opposed the project soon followed his lead and jumped on the bandwagon. Kennedy, after all, was the most powerful member of Congress from the East to support the Seaway. If Kennedy could justify the project, then surely other lawmakers could add their support with less risk of committing political suicide. His speech broke the political logjam that had prevented construction of the Seaway for a half century. "Both nations now need the St. Lawrence Seaway for security as well as for economic reasons," Kennedy told the Senate in his January 14, 1954, speech. "Mr. President, our ownership and control of a vital strategic international waterway along our own border would be lost without passage of this bill."[2]

The U.S. House of Representatives and Senate passed legislation four months later supporting construction of the Seaway. Work began August 10, 1954—four centuries after French explorer Jacques Cartier paddled up the St. Lawrence River with dreams of finding a safe passage to the Orient. A sense of euphoria swept the Great Lakes region after construction began.

Port communities anticipating a flood of cargo from overseas built new docks and grain elevators that would store Midwestern grain destined for overseas markets. The newspaper in Massena, New York, expressed the hopes of many port cities in 1954 in a special series of articles on the Seaway. "Massena is today on the brink of a great future that may one day turn our quiet, well-loved community into a teeming "Pittsburgh of the North. . . . What lies ahead for Massena? Our community may turn into a large metropolis with booming industries. It may one day mushroom far beyond its present village limits. It may someday become a center of tourist travel as visitors come from all over the world to view the completed project. But only time will tell."[3]

Great Lakes communities large and small believed the Seaway would usher in a golden age of prosperity. The Lake Michigan port city of Muskegon, Michigan, renamed its main thoroughfare "Seaway Drive." The excitement was understandable: The Seaway added 8,000 miles of Great Lakes coastline to the global shipping system. It effectively transformed dozens of coastal cities in the United States and Canada—including Buffalo and Detroit, Hamilton and Toronto—into international ports. So revered was the project, *National Geographic* made it the cover story of the magazine's March 1959 edition. The magazine devoted 40 pages to the Seaway, tracing the long history of efforts to remake the St. Lawrence River, and explaining the technical aspects of the project in vivid detail. The article's headline said it all: "New St. Lawrence Seaway Opens the Great Lakes to the World." The magazine concluded that the Seaway would usher in a new era when the first ocean ship steamed through its locks the following month.

"Officially, the passage of this ship will celebrate the opening of the St. Lawrence Seaway—with its linked power development one of the most incredible engineering and construction jobs men have ever attempted, and in some ways the hardest," the magazine said. "Unofficially, the day will mark much more: It will begin a major reshaping not only of America's but of the world's geography, and of its economy as well."[4]

The Seaway was the biggest construction project in Canada's history. An American engineer working on the project called it the biggest engineering project in the world at the time. It was a colossal endeavor, one that changed the mighty St. Lawrence and the lives of 6,500 Ontario residents forced to relocate to make way for the reengineered river. As residents were relocated, an army of construction workers equipped with dynamite and fleets of tractors carved 33 miles of canals out of the St. Lawrence's stone riverbed. They then built seven huge concrete locks that would utilize nothing more than the force of gravity and water to gently lift and lower freighters over the steepest sections of the St. Lawrence River. Workers also built three large dams in the river, each of which had a distinct and important role in the St. Lawrence Seaway. The Iroquois Dam controlled water levels in Lake Ontario; the Long Sault Dam regulated the flow of the St. Lawrence and tamed the violent Long Sault Rapids; and the Moses-Saunders dam transformed the river and the Great Lakes water it carried into one of the world's most productive power plants. Spanning a half-mile-wide section of the river, the 32 massive turbines in the Moses-Saunders Dam generate a phenomenal two million kilowatts of cheap electricity—enough to power two cities the size of Washington, D.C.

The Lost Villages

Submerged beneath the St. Lawrence River near Cornwall, Ontario, visible only to pilots and scuba divers, are the haunting remnants of villages that were swallowed by the reshaped river. There are roads but no vehicles, foundations but no houses, a trestle that once carried trains over the river that later flowed over the wooden structure. Invisible to the eye, but seared into the minds of thousands of Canadians, are the memories of life in the charming villages of Mille Roches and Moulinette, Wales and Dickinson's Landing, Farran's Point and Aultsville. All are gone—ghost towns drowned in 1958 by the St. Lawrence Seaway construction project.

Turning the St. Lawrence River into a liquid highway for cargo involved more than digging miles of canals and building massive locks and dams. Replumbing the river and flooding 100 square miles of land required relocating 6,500 residents in waterfront communities in Ontario. Several tiny hamlets were uprooted and relocated to make way for the deep reservoirs that formed behind the Long Sault, Iroquois, and Moses-Saunders dams. The dams transformed miles of North America's most powerful river into a massive artificial lake that submerged 40,000 acres of farmland, commercial areas and neighborhoods, schools and cemeteries. When all was said and done, crews relocated 530 homes, 225 farms, 17 churches, and 18 cemeteries (containing 2,000 bodies) that stood in the path of progress. All were moved to higher ground. Dozens of older homes were demolished or burned, along with all the trees.

The Lost Villages Historical Society

A church is moved from the village of Iroquois, Ontario, in the early 1950s to make way for the St. Lawrence Seaway project.

Six homes in Aultsville were torched in the name of science—researchers studied the blazes to determine how fire spread in residential structures.

The old village of Iroquois was moved a mile and a half north, to an area called New Iroquois. Cottage dwellers on the islands of Sheek, Barnhart, and Croil's also were forced out by dams, which drowned rocky islands that once rose above the river's surface. Many of the dislocated residents moved to new towns that sprouted near Cornwall: Ingleside and Long Sault. New communities rose along the shores of Lake St. Lawrence, an artificially bloated section of the once-mighty river. Beneath its deep blue waters lay what came to be known as the Lost Villages.

A group of people who lived in communities that were sacrificed to the St. Lawrence Seaway formed the Lost Villages Historical Society in 1977 to preserve their shared heritage. The society's museum at Ault Village, in Ingleside, features a functional general store and a church that hosts weddings. The buildings and the artifacts displayed within are bittersweet reminders of the riverside communities that fell to the hammer and shovel of progress.

On July 1, 2008, the Lost Villages Historical Society marked the 50th anniversary of dynamite blasts that destroyed temporary coffer dams on the St. Lawrence River. The explosions allowed the river to fill in the valley behind the three dams, creating the deep, calm water that freighters needed to traverse the once unruly river. "A half century has not faded the memories of life in the Lost Villages of Mille Roches, Moulinette, Wales, Dickinson's Landing, Farran's Point and Aultsville and the lost hamlets of Woodlands, Santa Cruz and Maple Grove," the historical society said in a press release promoting the event.

Fifty years after they saw their homes, schools, and churches relocated or demolished, residents of the Lost Villages gathered to mourn all they had lost and celebrate what they had gained when the St. Lawrence Seaway transformed their lives. It was a bittersweet event. Jane Craig, who was president of the Lost Villages Historical Society in 2008, expressed the views of many Lost Village residents in an interview with Canada's National Post:

"I have two children and I have grandchildren now. And I would give anything to walk down my old sidewalk and say, 'I went to school here, I swam here, I used to skate here, I got in trouble here once,'" Craig told the newspaper. "That's gone, I can't do it."

The upside, Craig said, was the creation of sparkling new towns—Long Sault and Ingleside—along the altered river. There remained, however, a lingering anguish in the hearts of people who were driven from their homes, their towns. Time could not erase the stinging images of grown men crying as their homes were demolished, their families uprooted. For some members of the Lost Villages Society, the memories of watching the river creep up deserted streets until it buried a part of Canada's past were as vivid in 2008 as they had been five decades earlier. The passage of 50 years could not heal all the physical and emotional wounds the Seaway left in its wake.

Remarkably, the 25,000 workers assembled to build the Seaway completed the mammoth job in just four years, nine months. The St. Lawrence was transformed into a working river with a uniform shipping channel that was 27 feet deep, which was essential to the international shipping industry. Of equal significance, the river was suddenly capable of churning out tremendous amounts of inexpensive electricity for the growing nations. The Seaway was an extraordinary accomplishment—a monument to brilliant engineering, advanced technology, fierce determination, and relentless hard work by thousands of laborers. After four centuries of failed attempts, humans had finally conquered the powerful St. Lawrence River.

The successful completion of the Seaway was the climax of a technological feat that Jacques Cartier could not have imagined four centuries earlier when he encountered an impassable river at the menacing rapids near Montreal. The Seaway and its predecessors—the Welland Canal and Soo Locks, near the eastern end of Lake Superior—were finally a uniform network of shipping canals. Comprised of six sets of canals that spanned 60 miles and featured 19 locks in three distinct regions of the Great Lakes and the St. Lawrence River, the Seaway allowed ocean ships to penetrate deep into the heart of North America and reach the highest point in the Great Lakes—Lake Superior, 600 feet above sea level. The Seaway was an aquatic stairway to the economic heaven of North America, a region rich with grain and iron ore that could feed much of the world and equip nations with the raw materials of boundless commerce. No longer were the lakes isolated, ecologically or economically, from the world's oceans.

"The Great Lakes–St. Lawrence Seaway was open. Man had redesigned the continent," author Carleton Mabee said in his book *The Seaway Story*.[5] "He had given it an interior sea coast with a more productive hinterland than any of its other coasts, on the Atlantic, Pacific, Arctic or Gulf of Mexico." The question was whether the Seaway could live up to all the hype.

ENGINEERS DESIGNED THE SEAWAY'S LOCKS TO MATCH THOSE IN THE Welland Canal, the 28-mile-long channel between Lake Erie and Lake Ontario that allowed ships to bypass Niagara Falls. The Welland Canal's eight locks, last updated in 1932, were 766 feet long, 80 feet wide, and 30 feet deep. Those locks could handle ships up to 740 feet long and 78 feet wide. With the global shipping industry evolving, and ships getting larger during the 1940s and '50s, Seaway engineers fit the locks to design standards of the past instead of the future. It was a fateful decision—driven largely by political pressure from rail and shipping interests on the East Coast of the United States—that kept the world's largest ships out of the Great Lakes. It was a strategic error that proved financially devastating to the Seaway. The Panama Canal, which was built in 1914 and competed with the Seaway for shipping traffic, featured locks that were 1,000 feet long and 110 feet wide. As journalist Dan Egan observed in 2005, "The Panama Canal was built for the future; the Seaway was built for the past."[6]

The Seaway was essentially obsolete the day it opened. It was accessible to 80 percent of the world's merchant ships in 1959, but that figure plunged in subsequent years as the shipping industry evolved. Shipping companies in the 1950s began moving away from break-bulk carriers—smaller ships that stored cargo loaded in bags, barrels, or on pallets—and started building container ships that could hold hundreds of standardized metal boxes. Those containers could be transferred easily among ships, trains, and trucks. The first modern container vessel was launched in 1956, three years before the Seaway opened. Container ships dominated the global shipping industry by the 1970s. By 2005, the Seaway could only handle 2 percent of the cargo-carrying capacity of the world's shipping fleet, and only 5 percent of the world's container fleet.[7]

Davis Helberg, the longtime director of the Port Authority of Duluth—the busiest port on the Great Lakes—acknowledged in 1993 that the changing nature of ocean commerce had largely bypassed the Seaway and the Great Lakes ports it served. By 1989, only 28 percent of the world's dry-bulk carriers could fit through the Seaway. The vast majority of ships were too wide. Helberg was one of many shipping officials who lobbied for enlarging the Seaway less than 35 years after it was built.

"By the 1980s, the Seaway's size limitations began to take their toll on Great Lakes ports," Helberg said in a 1993 essay. "I refuse to believe that the Seaway's founders—who opened world markets for thousands of farmers and manufacturers in the North America interior—intended to create a system that would outlive its physical practicality in a few decades. I firmly believe we owe it to future generations to modernize the Seaway and keep it—and its users—globally competitive."[8]

Efforts to expand the Seaway were derailed by environmentalists, including 1960s radical Abbie Hoffman. Critics claimed that enlarging the Seaway could not be justified economically and might prove to be environmentally disastrous for the Great Lakes and the St. Lawrence River. The critics ultimately prevailed. A quarter century after its completion, the Seaway was undersized, outdated, and incapable of competing with the Panama Canal. The Seaway couldn't even handle the largest ships working the Great Lakes, the so-called "lakers" that hauled coal, iron ore, salt, concrete, and gravel around the lakes. The massive lake freighters, measuring up to 1,000-feet long, could only move cargo within Lakes Superior, Huron, Michigan, and Erie. The largest lakers couldn't fit in the Welland Canal or the Seaway's locks between Lake Ontario and Montreal.

New York native Daniel J. McConville, who grew up near the Seaway and worked on the construction project, reflected on its plight in a 1995 essay. He called the project he helped build a "commercial debacle."

"The St. Lawrence Seaway was an important part of my life for at least 40 years, from the early 1930s to 1975. As a native and longtime resident of Ogdensburg, New York—Seaway country—I watched it develop from a hotly disputed geopolitical concept to a massive construction enterprise to a vital international trade route. And then I watched it decline from a world-renowned engineering masterpiece to a half-forgotten artifact of the 1950s, a victim of global economic and logistics changes and its own design shortcomings," McConville said. "Though I scarcely realized it at the time, I was a witness to what is still one of the most complicated public works projects ever brought off, and the largest construction undertaking of its kind ever done in a populated area. Yet, among world-class civil-engineering wonders, the Seaway holds the dubious distinction of being less famous now than before it was built."[9]

Helberg said in a 2005 newspaper interview that the Seaway was plagued from the outset by ridiculously high expectations. "The expectations were so high that anything short of the streets being paved with gold bricks was going to be considered a huge disappointment," he told the *Milwaukee Journal-Sentinel*.[10]

Despite its many shortcomings, the Seaway remained an impressive feat of engineering 50 years after it opened. Though it failed to meet initial economic projections, the Seaway did increase the volume of cargo moving between the ports on the Great Lakes and those in Europe, Asia, and South America. That had tremendous economic benefits for the northeast United States and eastern Canada. Since the Seaway opened, ships had moved two billion metric tons of cargo, valued at $400 billion, on the Great Lakes and Seaway.[11]

The Seaway's inaugural year turned out to be one of its busiest, in terms of the number of ships—7,452—that passed through its locks in the St. Lawrence River. Freighters carried 18 million tons of cargo in the Montreal section of the Seaway, from the lower St. Lawrence River to Lake Ontario. Less than one-third of that cargo, 5 million tons, was transcontinental freight shipped between ports on the Great Lakes and those overseas. The majority was freight hauled between ports on the St. Lawrence River and the Great Lakes. Initial predictions that freighters would carry at least 49 million metric tons of cargo annually through the Montreal section of the Seaway were achieved just seven times in the first 50 years of operation. An average of 40 million metric tons of cargo was shipped annually on the Seaway during its first 34 years, a figure that dropped to 35 million metric tons annually between 1998 and 2007.[12] Most of that cargo was shipped between U.S. and Canadian ports, not to other continents.

Had the Seaway locks been built larger to compete with the Panama Canal, and the Welland Canal enlarged to match the Seaway, the project might have lived up to the initial hype and kept pace with an evolving shipping industry. Instead, the Seaway's role as a global economic force diminished over time. Most of the world's largest ships avoided the Seaway by 1980, accessing North America via the Panama Canal and depositing cargo containers at ports on the East and West coasts and the Mississippi River. The Seaway's financial performance was so poor, the U.S. and Canadian agencies formed to build it never paid off the construction debt. Tolls paid by ships that used the Seaway were supposed to make the project economically self-sufficient. That didn't happen. Both nations subsidized the Seaway to keep it afloat.[13]

In 2007, ocean ships accounted for just 4 percent of all cargo moved on the Great Lakes and through the Seaway. The lakers moved the vast majority of cargo. The Seaway was simply too small to accommodate the largest ships moving cargo on the Great Lakes or the world's oceans. Instead of becoming a beloved public-works project, like New York's Brooklyn Bridge or the Gateway Arch in St. Louis, the Seaway became notorious for a most dubious distinction. The foreign freighters the Seaway invited into the economic heart of North America became the primary vector for aquatic invasive species that disfigured the world's largest freshwater ecosystem.

I visited the Eisenhower Lock near Massena in 2007 to watch ocean ships pass through the mammoth structure in the river. As a freighter entered the lock from the Wiley-Dondero Canal, my mind drifted back to 1959, when a president and queen visited this site to commemorate a project that wed the Great Lakes to the global shipping industry. It struck me how the Seaway had fallen far short of initial economic projections and, worse, had become an agent of harmful environmental change. Massena, a gritty community of 10,000 people located on the banks of the St. Lawrence River, was a case in

St. Lawrence Seaway Development Corp.

A freighter heads out of the Eisenhower Lock in Massena, New York, and steams toward Lake Ontario.

point. The city that billed itself as the gateway to America's fourth coast never became the Pittsburgh of the North, as the editors of the local newspaper envisioned in 1954. Massena, like many industrial-based communities around the Great Lakes in 2007, struggled as U.S. companies shipped manufacturing jobs overseas. The Seaway, like many of the rust-belt cities it served, became a sparsely visited tourist attraction.

I joined a crowd of about 30 ship-lovers on the elevated observation deck at the Eisenhower Lock on an autumn day to watch ocean freighters crawl through the awe-inspiring contraption. Among the ships was an ocean freighter capable of carrying oil, gasoline, or other hazardous chemicals. As the massive gates of the lock opened, the ship slowly entered from the man-made Wiley-Dondero Canal. With the ship securely in the lock, the gates closed behind it and the lock was flooded with 24 million gallons of water from the St. Lawrence River. It took about seven minutes to fill the lock with water and, in the process, lift the massive ship 30 feet. The lock raised the ship to the same water level as the river upstream. The entire process of moving the ship through the lock took about 45 minutes. Amazingly, no pumps were used to fill or drain the locks—only the power of gravity. The only electricity used was that which opened and closed the huge gates at either end of the lock.

A warning sign painted in large white letters on an equipment housing on the ship's deck came into view as the freighter rose in the lock. It read: "Warning: Dangerous Cargo." The sign referred, of course, to explosive chemicals in the ship's cargo holds. But the warning was symbolic for an altogether different reason. Ocean freighters that entered the Great Lakes through the St. Lawrence Seaway were like dirty syringes, each one capable of infecting the lakes with foreign species of fish, mussels, and potentially deadly pathogens. Though I marveled at the technology of the ship and the locks that gave it safe entry into the world's largest source of fresh surface water, a sense of unease crept into my mind. Would this ship contribute to the hordes of foreign species trampling the ecosystems of these freshwater seas? Would the next ocean freighter through this lock deliver another invader, one capable of harming the lakes or the creatures that lived therein? It was entirely possible, given the lax environmental regulations on the books in 2007.

After the water level in the Eisenhower Lock reached equilibrium with the river, the forward gate opened and the freighter slowly accelerated into the open waters of the St. Lawrence River. The ship-lovers who snapped photographs from the observation deck slowly dispersed as the ship steamed toward Lake Ontario. I stayed behind to watch the freighter as it vanished into the glare of the setting sun. It was during that time of silent reflection that I grasped the significance of the transoceanic freighter passing through the Seaway lock, en route to the Great Lakes. I had witnessed an event that was, for the lakes, the ecological equivalent of Russian roulette.

I never saw such a mess in my life.

—George F. Liddle, city manager of Muskegon, Mich., following an alewife die-off in 1966

04 ALEWIFE INVASION

For the better part of three centuries, anglers and fish-eating birds have converged on rivers in Canada's Maritime Provinces each spring to stalk a small fish known to locals as gaspereau. It is an anadromous ocean fish, a type of herring known in the United States as alewife, river herring, or sawbelly. Technically, its name is *Alosa pseudoharengus*. But the common names afforded it are far more interesting. Some anglers call it sawbelly due to the row of scales along its belly that resemble a saw blade. Many Americans refer to it as alewife, a name some historians attributed to an early angler's barmaid wife. Others claimed the fish was named alewife because the belly protruding from its slight frame resembled a beer gut. In the fishing villages of eastern Canada, the fish is known simply as gaspereau. It is a beloved species.

Every May or June, depending on water temperatures in the rivers, alewives vacate their saltwater havens in the Atlantic Ocean and headed for spawning grounds in freshwater streams. Measuring up to a foot long, the fish—silver, with a dark green stripe along their backs—swarmed into the rivers and streams that slice through New Brunswick, Nova Scotia, Prince Edward Island, and Newfoundland. Each female deposits as many as 100,000 eggs on river bottoms before heading back to sea for another year. The fish are an important bait species for larger ocean fish. They also provided an income for commercial anglers in the Maritime Provinces. Canadian anglers sold millions of dollars worth of gaspereau every year. Some became dinner for humans, but the majority were sold as fish bait and cat food.[1] The fish's annual spawning ritual occurred up and down the Atlantic seaboard, from Newfoundland to South Carolina. It was an event widely celebrated for its ecological significance, as well as its symbolic renewal of life on the heels of winter.

There is little mystery surrounding the arrival of spring in the quaint Maine fishing village of Damariscotta. Migrating alewives usher in the season, attracting hundreds of visitors to the quiet coastal community of 2,000 residents. Locals and tourists flock to the banks of the Damariscotta River to watch massive schools of alewives migrate through the river's estuary, into the Great Salt Bay and up a fish ladder that lifts the creatures into Damariscotta Lake. Thousands of fish congregate at various points in the journey, their masses forming moving clouds of fish beneath the water's surface. Ospreys, herons, and bald eagles flock to the river for an easy meal. Birders converge on the river to see the majestic birds in action. It was fitting that alewives were celebrated in Damariscotta. The village name was a crude interpretation of the Algonquin word Madamescontee—a place of abundant alewives.

The beloved ocean fish were not afforded a warm welcome when they arrived in the upper Great Lakes in the 1940s. They were greeted with all the ceremony afforded a disease, which was precisely what alewives were to the lakes. Through no fault of its own, the marine species found its way into the world's largest freshwater ecosystem. And it was good, at least for alewife. The species was the beneficiary of human tinkering that opened the lakes to ocean species and knocked the freshwater ecosystems wildly off kilter. The chaos created dysfunctional predator-prey relationships that allowed alewife to become the dominant species in three of the five Great Lakes. That made the alewife a scapegoat for humans who built the canals that allowed the foreign fish to conquer the lakes.

Sea lamprey and alewife were the first ecologically significant fish species to invade the Great Lakes. Both arrived through the Welland Canal, the predecessor to the Seaway. The lamprey's devastating effects on lake trout and other native fish populations were evident before the Seaway opened. The alewife's effects remained largely unknown until the 1960s, when millions of the fish died and washed up on beaches. Despite their differences, sea lamprey and alewife were intricately related in their adopted home. Lamprey created a biological void in the lakes, which alewives promptly filled. To use a street-fighting analogy, the lamprey was like the new kid on the block who moved in and knocked out the resident tough guy. It then allowed its smaller, weaker friend—alewives in this case—to take over the day-to-day operations of the neighborhood.

No one sounded the alarm when alewives were discovered in Lake Erie in 1931. After all, the species had been in Lake Ontario since at least 1873. Some biologists believed alewives were native to Lake Ontario and had lived in the lake for thousands of years. Others insisted that the fish was an ocean species until the Erie Canal created a pipeline through which alewives could swim from the Atlantic into Lake Ontario. This much was universally accepted: alewife could not reach the other four Great Lakes above Niagara Falls until the Welland Canal was enlarged in the 1920s. The canal expansion, which established a constant flow of water from Lake Erie into the canal and downstream into Lake Ontario, allowed sea lamprey to invade lakes Erie, Huron, Michigan, and Superior. Schools of alewives soon followed, invading Erie in 1931, Huron in 1933, Michigan in 1949, and Superior in 1954. With the Erie and Welland canals in place, alewives and other ocean species hard-wired to spawn in fresh waters could swim up the Hudson River from the Atlantic Ocean, through the Erie Canal, across Lake Ontario, and through the Welland Canal—the gateway to the upper Great Lakes.

The implications of the alewife invasion didn't come into focus until the late 1950s, after sea lamprey had wiped out lake trout. With no natural predators to keep their numbers in check, alewives enjoyed a population explosion in Lakes Michigan, Huron, and Erie that would make rabbits blush. The result was a biological disaster of nearly biblical proportions.

David Jude, University of Michigan, School of Natural Resources and Environment

The alewife invaded the upper Great Lakes after the Welland Canal was expanded. These tiny ocean fish were kept in check by lake trout until sea lampreys reduced the number of lake trout.

THE ALEWIFE OCCUPIES A PLACE NEAR THE BOTTOM OF THE FISH FOOD chain in its native environment. It is a prey-fish species, essentially food for larger fish and fish-eating birds. It carries out that role well, producing phenomenal numbers of offspring each year. Alewives can live as long as seven years; they begin spawning at two years of age. A healthy female can produce up to a half-million eggs during her lifetime, assuming she lives seven years, spawns for five of those years, and escapes anglers' nets and predatory birds. Not bad for a fish that maxes out at one foot long (in the ocean) and weighs less than a pound.

Alewives tend to run smaller in their adopted freshwater environment, growing up to six inches long. They also produce fewer offspring than their ocean kin. Female alewives in the Great Lakes lay between 17,000 and 38,000 eggs annually. Whether in its native range or the freshwater seas, the alewife is an unassuming creature. It certainly isn't the type of fish one would think capable of taking over a body of water the size of Lake Michigan, which spans nearly 23,000 square miles of surface area. Alewives lack the sharp teeth or demeanor of, say, piranha, which seized control of tropical rivers through fiercely aggressive behavior. Yet, alewives achieved a measure of ecological dominance in the 1960s that was worthy of piranhas and other top predators. That was due in large part to the fact that Lake Michigan in 1960 was an ecosystem devoid of top natural predators in the fish community. Lake trout had been essentially eliminated, and sea lamprey didn't attack alewives, which were too small for the bloodsucking parasites.

Absent predators to keep its numbers in check, the alewife became the dominant fish species in Lake Michigan and Lake Huron in the 1960s. The tiny

David Jude, University of Michigan, School of Natural Resources and Environment

With few lake trout to feed on alewives, the invader's population increased dramatically in the 1950s. Dead alewives washed up on Great Lakes beaches each spring until the late 1960s, when Pacific salmon were imported to reduce the alewife population.

fish were so numerous, they accounted for 90 percent of biomass (the weight of all aquatic life) in the lakes.[2] Scientists in 1967 estimated that there were 167 *billion* alewives in Lake Michigan alone.[3] That meant there were, on average, 7,260,869 alewives in every one of Lake Michigan's 23,000 square miles of surface area. It was a phenomenal number of fish. Think of it this way: If all of those fish were laid end to end, and each one was just four inches long, the alewives in the lake that year would have circled the Earth 423 times.

That alewives were able to thrive in all but the frigid waters of Lake Superior wasn't surprising. Water temperatures in Lakes Ontario, Erie, Huron, and Michigan were only slighter colder than the ocean temperatures where the fish originated. Lake Superior, the coldest of the Great Lakes by far, was too cold to support larger numbers of alewives. The only natural controls on the booming alewife population, absent lake trout, were the availability of food and the water temperatures. It turned out that lake water temperatures were most volatile in the late spring and early summer, the same time alewives were spawning along Great Lakes coasts or migrating up rivers. Quick-moving storms that pushed warm water offshore, causing an upwelling of cold water in nearshore areas where alewives spawned, routinely caused massive kills in the 1960s that created slicks of dead fish several miles long. Despite the periodic fish kills, alewives ruled Lakes Ontario, Michigan, and Huron for much of the 1960s. Theirs was a biologically destructive tyranny.

Alewives in their native environment fed largely on plankton. When the fish ran low on plankton in the Great Lakes, they went after smaller fish species and the eggs of other fish. The results were devastating: Alewives depressed native fish species—perch, whitefish, and lake herring—through sheer numbers and by hogging the plankton that many native species needed for sustenance. The invaders also feasted on newborn lake trout, which further decimated a species already driven by sea lamprey to the brink of elimination. Adult lake trout ate alewives, but the tiny invaders were the worst kind of junk food. The flesh of alewives contained an enzyme called thiaminase, which destroyed thiamine (Vitamin B1) in lake trout. When lake trout consumed alewives, the thiaminase reduced the level of thiamine in their bodies, killing or causing birth defects in many of their offspring.

Devastating as they were to native fish stocks, alewives became best known for massive die-offs that turned vast stretches of Great Lakes beaches into fish graveyards and breeding grounds for flies, maggots, and botulism. The die-offs routinely clogged municipal water intakes and smothered miles of beaches with dead, smelly fish. Beaches that had been popular a decade earlier were deserted in the 1960s. The economic effects were devastating: Dead alewives caused a $100 million loss to Lake Michigan's recreational economy in 1967 alone, according to federal officials.[4] *Time* magazine declared in its July 7, 1967, edition that Lake Michigan was suffering from an "alewife explosion." The article described the sickening scene in the lake:

"From the Chicago waterfront to the Mackinaw Bridge, the shores of Lake Michigan were taken over last month by dead alewives. The fish . . . washed ashore on every incoming wave, piling up on the beaches faster than bulldozers and tractors could clear them away. They filled the air with the odor of decay and drew swarms of mosquitoes and flies. Chicago's municipal water-supply inlets and those of industries that draw water directly from the lake became clogged time and again with the little alewives. Off Benton Harbor, Michigan, an aerial photographer reported a ribbon of dead fish 50-feet wide and 40-miles long floating on the surface of the lake."[5]

Dead alewives blanketed beaches in Michigan, Illinois, Indiana, New York, and Ontario. Lake Michigan was hit hardest, because it had the greatest density of alewives. The annual alewife die-off whipped residents of some coastal communities into a frenzy of disgust and outrage. A beach manager in Muskegon, Michigan, described the scene of dead alewives floating on Lake Michigan in a 1967 newspaper article. "The water was white with them," Claude Williams told the *Muskegon Chronicle*.[6] Some communities dispatched teams of workers to beaches to rake up tons of the dead fish. Others used tractors to collect and remove the fish, which were then hauled to landfills. Federal officials in 1968 estimated it would cost as much as $12,000 per mile to remove masses of dead fish that reached densities of one ton per 10 feet of beach.[7] Federal agencies paid commercial fishing crews hundreds of thousands of dollars to skim dead alewives off the surface of Lake Michigan before their carcasses washed up on beaches. Commercial anglers, with the blessing of state and federal agencies, began harvesting alewives for sale as fishmeal, fish oil, and chicken and cat food. But they were no match for the army of invasive fish that dominated the lakes. In 1967, for instance, commercial anglers harvested 40 million pounds of alewife from Lake Michigan alone. That haul was a drop in the bucket.[8] Some cities joined the fight, deploying more fishing boats to skim dead alewives off the lakes. It was an ambitious but unsuccessful campaign.

There was little doubt by the mid-1960s that humans could not reduce the alewife population on their own. They needed a new weapon in the war on alewives, something that would fight the fish on its own turf.

The chosen solution, ironically, was a controversial biological experiment that involved the introduction of yet another foreign fish species. It was a desperate effort born of desperate times. The overriding goal of many coastal communities was to shrink the alewife population. People were willing to try almost anything, and pay any cost, to achieve that goal. It was against that backdrop of near hysteria that a biologist came to Michigan, like a knight on a white horse, to orchestrate one of the most daring, successful, and controversial fish experiments in modern history.

People love to come to Lake Michigan and fish for salmon.
Some don't realize salmon are an introduced species.

—Michael Chiarappa, Environmental Historian, Western Michigan University

05
A KING IS BORN

The early 1960s were the environmental Dark Ages in the Great Lakes. Industries and cities with inadequate sewage-treatment plants used the lakes as toilets for all manner of chemical and biological wastes. Decades of excessive fishing, coupled with the sea lamprey invasion, had virtually eliminated the Great Lakes' top fish predator, the lake trout. Alewives that snuck into the lakes after lamprey multiplied at a frenetic pace and then died by the billions, fouling beaches. Water quality wasn't much better: Pollution tainted fish and wildlife and fueled summer algae blooms that turned Lake Erie and shallow bays the color of pea soup. The immense waterways were so profoundly disrupted by human activities that relatively few people bothered to swim, fish, or go boating in the filthy waters. So dire was the situation, many individuals and communities simply turned their backs on the lakes, figuring they were hopelessly polluted. William Taylor, a Michigan State University fisheries expert who grew up near Lake Ontario in the 1960s, remembered avoiding beaches that were a stone's throw from his house in Rochester, New York. "We didn't go to the lake to swim because it was polluted and the beach was covered with dead alewives," Taylor said. His parents, like many of their neighbors, devised their own solution to the pollution that fouled the lake. They built a swimming pool. "That was the technological solution of the day," Taylor said. "Rather than clean up your lake, you built a swimming pool."[1]

The multitude of human abuses left the world's largest freshwater ecosystem gasping for life. The Great Lakes needed a savior—a bold leader who would think outside the box, challenge conventional wisdom, and push for a radical solution. That person was one Howard Tanner. He was working as a fisheries biologist in Colorado in 1964 when his home state, Michigan, came calling. Ralph A. MacMullen, then director of the Michigan Department of Conservation (now the Department of Natural Resources), asked Tanner to take over as director of the DNR's fisheries division. It was a dream job for Tanner, who grew up fishing the idyllic lakes and streams of the state's northern Lower Peninsula. MacMullen, a conservation legend in Michigan, gave Tanner a bold directive: Fix the ailing Lake Michigan fishery. Quickly.

"I want you to do something very visible and if you can make it spectacular, please do so," MacMullen told Tanner.[2] For a young biologist eager to make his mark, MacMullen's directive was the challenge of a lifetime. Tanner was given carte blanche to try anything to restore biological order in Lake Michigan, the largest lake entirely within U.S. borders. His solution: Stock coho and chinook salmon native to the Pacific Ocean in the fresh waters of Lake Michigan. It was a radical proposal that rewrote the rules of fisheries management. Tanner wanted to place a new predator at the top of the fish food chain—a foreign species, no less. Salmon, not lake trout, would reign supreme in Lake Michigan if Tanner had his way. On the surface, the notion of staking the future of Lake Michigan's fishery on an ocean species seemed like sheer lunacy. No one had ever succeeded at creating an ocean fishery in so large a freshwater ecosystem. But if the gamble paid off, the 23,000-square-mile lake would become the world's largest salmon fish farm, and a magnet for anglers from near and far.

SALMON WEREN'T ENTIRELY FOREIGN TO THE GREAT LAKES. ATLANTIC salmon were native to Lake Ontario and abundant until the late 1800s. The fish, which entered Lake Ontario through the St. Lawrence River, thrived in the lake and its tributaries. Atlantic salmon never spread to the other Great Lakes because Niagara Falls blocked the only aquatic pathway from Lake Ontario to Lakes Erie, Huron, Michigan, and Superior. The abundance of Atlantic salmon in Lake Ontario and the rivers that flowed into it in the 1800s was almost incomprehensible. A State University of New York history of the species reported that one angler caught 400 Atlantic salmon in New York's Salmon River in one day in the 1800s. Farmers used pitchforks in the 1860s to toss salmon onto the banks of the Seneca River. But the good times couldn't last. By 1898, excessive fishing, deforestation, the construction of dams, and pollution eliminated Atlantic salmon from Lake Ontario and its tributaries. Writer Richard Follett lamented the loss of the majestic fish in a poignant 1932 essay: "Today, very few people know Lake Ontario was once the greatest salmon lake in the world, or that salmon even existed in these waters, or that the St. Lawrence was once the greatest salmon river of the world, considering all its tributaries."[3] Six decades after Atlantic salmon vanished from Lake Ontario, Tanner set out to create a new Great Lakes salmon fishery in Lake Michigan.

Tanner decided early on to go with coho salmon from the Pacific Ocean because it was relatively easy and cheap to raise the fish in hatcheries. Moreover, the fish were virtual eating machines that were capable of consuming huge quantities of alewives. From the outset, he dismissed the notion of restoring native lake trout. "We concluded that the lake trout was not a very interesting fish to catch. . . . It was our opinion that a truly exciting fishery was essential" to attract anglers to Lake Michigan, Tanner wrote in a 2002 essay.[4] If coho thrived in the lake, Tanner planned to add an even larger, more powerful species of Pacific salmon: chinook. Among anglers, chinook—an incredibly powerful fish that could grow to 45 pounds—were known simply as kings. The title was entirely deserved.

If you've never had the thrill of landing a chinook salmon, it would be difficult to fathom its incredible power. Reeling in a moderately sized chinook, in the 15-pound range, is an arduous task. The fish are essentially swimming muscles. I've caught salmon with deep-sea rods on boats in Lake Michigan, and with fly rods in rivers that flow into the lake. All were memorable experiences. When a chinook struck, it would rip the fishing rod out of your hands if you weren't paying attention. Reeling in one of the maniacal fighters was like playing tug of war with a pit bull. This is not an exaggeration. Landing a chinook would test the strength and durability of the strongest human.

Tanner, himself an avid fisherman, knew anglers would flock to the long-beleaguered Lake Michigan if he could create a thriving salmon fishery. The salmon odyssey got off to a rocky start. In fact, his first salmon-stocking experiment was a bust. Tanner sent a crew of Michigan biologists to Colorado in 1964 to obtain one million kokanee salmon eggs. The eggs were reared in a Michigan hatchery and released in 1965, with great fanfare, into a tributary of Torch Lake, a glorious lake in Tanner's home county. Michigan governor George Romney was on hand to pour the kokanee salmon fingerlings into the lake from a gold-plated bucket. Thousands more salmon were planted in Higgins Lake, another deep, cold, and clean inland lake in northern Michigan. But the project fizzled. Some of the fish reproduced, but the results, in Tanner's view, were "insignificant."

Tanner needed a strain of Pacific salmon that would reproduce in Michigan rivers and survive their three-to-four-year life spans growing in Lake Michigan. Though Pacific salmon were ocean fish, they hatched in rivers and returned to those same streams to spawn before dying. If the fish could mature in the Great Lakes the way they did in the Pacific Ocean, Tanner's plan would work. He would be considered a fisheries genius.

"The challenge in adapting the coho to the freshwater environment of the Great Lakes is an intriguing one," Tanner and his assistant, Wayne Tody, wrote in a 1966 report. "Nowhere in the world has the species been permanently established outside its native range in the north Pacific coastal area. The ultimate aim is to convert an estimated annual production of 200 million pounds of low value fishes—mainly alewives—that now teem in the upper Great Lakes into an abundance of sport fishes for the recreational fisherman. Secondly, we hope to restore the depressed commercial fisheries to a productive and economically viable industry."[5]

In an interview three decades after launching his grand experiment, Tanner said he never doubted that salmon could thrive in the Great Lakes. "The notion that coho salmon could not survive in a totally freshwater environment was largely a mental barrier," he said. "We knew better. . . . Salmon have to come to freshwater to spawn, but there is no obligatory reason for them to go to saltwater except in search of food."[6]

Creating a salmon fishery was not as simple as acquiring eggs, growing the fish in a hatchery, and planting them in rivers that flowed into Lake Michigan. There were many political battles to be fought before the first batch of fish was released into a Great Lakes tributary. Tanner needed to marginalize Michigan's commercial fishing operations, which posed a direct threat to the planned salmon fishery. He also had to ward off opposition from federal officials at the Bureau of Commercial Fisheries, which was pushing for restoration of lake trout and the commercial harvest of alewives.

The Bureau of Commercial Fisheries was determined to restore lake trout to its throne atop the Great Lakes food chain. Tanner's proposal to plant millions of salmon in the lakes, and manage the lakes for sport fishing over commercial fishing, was nothing short of an attempt to carry out a biological and political coup. The two sides went to war in the arena of public opinion. That battle climaxed in 1965 when Fenton Carbine, director of the Bureau of Commercial Fisheries, wrote a letter to Michigan governor George Romney demanding that he scrap the salmon-stocking program. Carbine, who supported the commercial harvest of alewife, issued his diatribe after Michigan had already acquired coho eggs and was raising them in a hatchery. Michigan officials leaked the letter to the media; it caused an uproar among sport anglers desperate for a solution to the oppressive alewife population. The war of words was a classic power struggle for control of one of the world's largest freshwater fisheries, half of which was controlled by the state of Michigan. Tanner eventually prevailed. The Bureau of Commercial Fisheries was moved out of Michigan and merged with the U.S. Fish and Wildlife Service. Carbine was transferred out of the state. Tanner was given free rein to create the sport fishery of his dreams.

At the same time he was doing battle with federal officials, Tanner was working to reduce the size of Michigan's commercial fishing fleet. Commercial anglers relied on gill nets to catch whitefish, lake trout, and perch; their nets posed a direct threat to the salmon fishery Tanner and Tody hoped to create. The Michigan DNR persuaded the state legislature to revise the state's commercial fishing laws. Laws passed in the mid-1960s banned the commercial harvest of salmon and lake trout, outlawed the use of gill nets, and restricted the number of commercial fishing licenses that could be issued. The state provided grants to help commercial fishing operations switch to trap nets—which were less harmful to salmon—but the regulations drove many commercial anglers out of business. In 1966, there were more than 1,000 active commercial fishing licenses in Michigan. That number dropped below 200 by the mid-1970s.[7]

Having shoved aside federal officials and commercial anglers, Tanner launched his revolutionary fisheries management plan. The first act took place on April 2, 1966, along a tiny stream near the town of Honor, in northwest Michigan. Tanner took a golden bucket filled with two-year-old coho salmon and emptied it into the Platte River. A total of 650,000 coho salmon were released into the Platte River and Bear Creek, which flow into Lake Michigan. "The salmon fisheries that would develop throughout the Great Lakes had begun," Tanner said in a 2002 essay.[8]

The coho planted in the Platte River and Bear Creek were expected to swim 20 miles downstream to Lake Michigan, where they would feast on alewife for two to three years before returning to their adopted streams to spawn before dying. Tanner hoped salmon would become the foundation of a new sport fishery and rein in the troublesome alewife. No one, not even the hyper-optimistic Tanner, anticipated how well the experiment would work. Or how quickly. The first run of immature coho up the Platte River and Bear Creek took place five months after the fish were planted. Thousands of rambunctious young cohos, called "jacks," migrated up the streams. Anglers caught 1,500 of the fish, which ranged from two pounds to seven pounds. The fish weren't huge, but the fact that any were caught triggered a buzz in the angling community. Educated anglers knew that a solid return of jacks usually meant a huge run of mature, spawning salmon the following year. The fuse was lit on salmon fever. It spread like wildfire the next year, with anglers from several states flocking to Lake Michigan to fish for coho. Men, women, and children stood shoulder to shoulder in northern Michigan rivers in the autumn of 1967 to cast for one of the largest, most powerful fish ever to be taken from Great Lakes waters. "The 1967 run of coho salmon electrified not only sport fishermen, but also the general population of Michigan, and it was hailed as top news from coast to coast of the United States," Tanner said.[9]

The Michigan DNR added fuel to the burgeoning salmon craze in the spring of 1967 when it stocked chinook salmon smolts in several streams that flow into Lake Michigan. Mature chinook were larger than coho and fought harder. Rivers of anglers, fishing boats in tow, flowed into northwest

Muskegon Chronicle

The state of Michigan imported Pacific salmon to Lake Michigan in 1966 to control the alewife population and create a sport fishery. The biological experiment created a $1 billion fishery.

Jeff Alexander

The artificial salmon fishery transformed Lake Michigan from a biological wasteland to a valuable sport fishery that attracted anglers from near and far. The fishery pumped millions of dollars of revenue into coastal communities around the Great Lakes.

Michigan in the fall of 1967. Michigan author Jerry Dennis, who grew up near Lake Michigan and participated in the early years of salmon fishing in the lake, described the euphoria surrounding the artificial fishery in his book *The Living Great Lakes*.

"A kind of gold rush mentality prevailed," Dennis said. "Anglers drove straight through from Pennsylvania and Tennessee and North Dakota. Most had no idea how to fish for salmon. Many had never before visited the Great Lakes and were surprised they couldn't see from one shore to the other. . . . It was terrific fun. A carnival atmosphere. . . . Salmon 'porpoised' from the water, leaping so high they sometimes landed in the laps of surprised anglers. No one had seen anything like it."[10]

Dave Borgeson, a young DNR biologist working for Tanner and Tody when they launched the salmon fishery, said the fish triggered a frenzy every fall when they migrated up rivers to spawn a new generation of offspring. "It was unprecedented in the annals of Michigan sport fishing," Borgeson said. "On Labor Day weekend of 1967, people caught thousands of fish, cohos weighing 18, 19, 20 pounds. They would come to the docks wagging their heads saying, 'I don't believe this.' To say that it was a howling success is an understatement. It's a great story. None better in conservation."[11]

In little more than a year, Lake Michigan went from being viewed as a biological wasteland—dominated by alewife that owned the water while alive and ruined beaches when they died—to one of the world's most productive, albeit artificial, salmon fisheries. Small towns along the Lake Michigan coast were transformed from struggling rust-belt communities into hotbeds of recreational tourism. Sagging economies were resurrected as anglers gleefully shelled out money in gas stations, tackle shops, bars, restaurants, and hotels. Within a decade, the Lake Michigan salmon fishery was valued at $1 billion.

The economic side effects of the salmon fishery were enormous and widespread. Sales of Chrysler fishing boats increased 40 percent by 1968. Officials at Evinrude Motor Corp. said the Lake Michigan coho fishery was one of the "greatest stimulants to motor sales in the history of outboard manufacture."[12] Other states saw the phenomenal economic success of Michigan's salmon fishery and quickly jumped on the bandwagon. By the mid-1970s, salmon were being stocked in all five Great Lakes. Communities like Pulaski, New York, where anglers caught Atlantic salmon in the 1800s, were reinvigorated by a Pacific salmon experiment that began in Michigan and soon spread to Lakes Huron, Superior, Erie, and Ontario.

"The economic benefits of the salmon industry cannot be overemphasized," said Taylor, who grew up near Lake Ontario and eventually became a professor and chairman of the Department of Fisheries and Wildlife at Michigan State University. "Salmon controlled the alewife, which cleaned up beaches, created a billion dollar fishery, revitalized coastal communities and gave people a reason to go to the lakes."[13]

Tanner had pulled off one of the world's most successful fisheries experiments. That his ploy worked, and became the cornerstone of a resurgent Great Lakes fishery now valued at $7 billion, was nothing short of miraculous. Biologists had stocked salmon at various sites in the Great Lakes on at least 35 occasions prior to 1966. None paid the huge dividends that Lake Michigan's new fishery provided.[14]

"Everybody wanted to know where I got the idea to stock coho," Tanner said. "I did not get the idea. I do not claim to be the originator of the idea. That idea had obviously been around for a long time. What I do take credit for was I took the responsibility for doing it. Whenever somebody wants to criticize it, they criticize me. When everybody wants to compliment, they can just as well compliment me, because I took the responsibility. I was allowed to make the decision to stock the coho salmon. . . . I took the idea and I sold it." Pulling off such a radical biological experiment would be impossible under the current regulatory environment, said Michael Chiarappa, a Western Michigan University history professor. Chiarappa and his wife, Kristin Szylvian, co-authored *Fish for All*, an oral history of Michigan's divisive salmon fishery. "The Michigan Department of Conservation altered the ecosystem of one of the world's largest resources of freshwater without holding public meetings and hearings, conducting exhaustive scientific studies and writing environmental impact statements," they said in the book.[15]

Tanner was praised by sport anglers as the "father of the Michigan coho," and demonized by commercial anglers and federal officials who viewed him as a reckless cowboy. Critics resented Michigan's fish manager placing sport fishing ahead of commercial fishing. But the salmon fishery had an ancillary benefit that commercial fishing could not offer: sport fishing put people in contact with Lake Michigan. Anglers could catch salmon from canoes, small boats, and yachts, or from any number of piers that lined the Lake Michigan coast. The salmon also took a huge bite out of the alewife population, greatly reducing the number of dead fish that littered Great Lakes beaches. The result: People returned to beaches in droves. Lost in all the excitement was the irony that an ocean fish species helped Great Lakes residents reconnect with their magnificent lakes.

"The salmon fishery put a lot more people on the lakes and I think the result was a greater appreciation of the lakes," said Gary Fahnenstiel, an ecologist at the National Oceanic and Atmospheric Administration's Lake Michigan Field Station in Muskegon, Michigan. "You can't help but fall in love with the lakes when you're out on them all the time. That's the thing I appreciate most about the salmon fishery—it restored some of the glory to the Great Lakes."[16]

Despite the success of the salmon fishery, critics wondered whether introducing another foreign species was ecologically wise. "If you look at this from a sport fishing standpoint, the salmon introduction was successful. It has attracted a lot of people and it is legendary," Chiarappa said. "But are the lakes healthier in an ecological sense? That's a harder question. The lakes are still a mess, exotics remain a huge problem. It's great that Lake Michigan can hold salmon, but we've basically got a fish pond. The introduction of salmon was a benchmark moment in the history of the lakes but to say it was great across the board for the lakes would be a pretty simplistic assessment of the salmon fishery."[17]

Tanner, for his part, never second-guessed his bold experiment. He explained his rationale for changing the course of Great Lakes fishery management in a 1991 interview:

> I think we have a totally altered system and we have to live with what we have. Not only that, it is still changing. People often say, "What have you done to the Great Lakes fishery?" And I say, "All right, which native species do you want? The brook trout is not native. The rainbow or steelhead is not native. The native lake trout of Lake Michigan is gone. Which native species do you want me to manage for? Tell me and I will try. But I don't have a native species . . ." If it had been a stable, natural system, that is one thing. But we have been disrupting the Great Lakes system and fish population for a hundred years or more.[18]

The success of the salmon program drove fisheries management in the Great Lakes after 1966. But there was a price to be paid for turning the lakes over to an imported species. Maintaining the lucrative fishery required keeping a large enough alewife population in the lake to support salmon. Alewife was essential to the survival of salmon in the lakes. Salmon showed little interest in native prey-fish species. Biologists charged with managing the salmon fishery had to find a balance between salmon and alewife numbers in the lakes. If there were too few salmon, alewives would die by the millions and foul beaches. If there were too many salmon and not enough alewife, salmon could starve or become susceptible to disease. Native fish species—perch, lake trout, and walleye—would bear the ecological cost of sustaining the alewife population. The alewife was detrimental to those species, but such concerns were lost in the jubilation over the salmon fishery. Sport anglers were hauling in an astonishing 9.3 million pounds of chinook in the mid-1980s. For salmon anglers, life on Lake Michigan was good indeed. And then the bubble burst.

Anglers who set out on southern Lake Michigan in the spring of 1988 encountered hundreds, sometimes thousands, of dead salmon floating on the lake's surface. Many washed up on beaches, prompting concern that some chemical pollutant was killing the beloved fish. It was a stunning turn of events: in the span of seven months, from the fall of 1987 to the spring of 1988, Lake Michigan's salmon fishery went from boom to bust. "The fall of the chinook salmon fishery was even more dramatic than its growth," according to a 1998 report by state and federal scientists. "In the spring of 1988, an estimated 7,000 to 8,000 dead or dying chinook were found floating near the surface or washed up on beaches, primarily in the southern end of the lake. Chinook salmon mortalities were subsequently observed each spring through 1992."

Salmon catch rates plunged, and anglers wondered if it was safe to eat the fish. The number of chinook caught per hour of angler effort in 1988 dropped 40 percent from record-high rates in the previous three years. The vaunted fishery seemed to be in free fall. By 1993, the chinook salmon catch rate in Lake Michigan—the number of fish caught per hour of angler effort—dropped 85 percent from the peak recorded in 1986.

Scientists determined early in the crisis that a bacterial disease that attacked fish kidneys was killing salmon. But scientists and anglers were alarmed by the fact that no one could pinpoint why the ailment, bacterial kidney disease (BKD), started killing salmon in 1988. Biologists and fish pathologists concluded in 1998, after years of research, that the chinook weren't getting enough to eat. That made them vulnerable to disease. In other words, the alewife that Tanner and his team sought to control in 1966 were suddenly too sparse. The notion that a lack of food could endanger fish health would return to haunt the treasured salmon fishery two decades after the BKD outbreak.

Fish biologists in 1993 were able to control the BKD outbreak by discarding the eggs of infected fish that were captured when they swam upstream to spawn. Still, the disease remained in Lake Michigan chinook, affecting between 10 percent and 20 percent of fish in any given year. The number of chinook anglers caught each year after the BKD outbreak remained well below the record catches recorded in 1986. Though salmon slowly rebounded in the 1990s, a dark cloud of uncertainty hung over the future of the man-made fishery. "The future stability of the Lake Michigan chinook salmon fishery is now uncertain after the spring [BKD outbreaks] and the decline in chinook salmon abundance observed from 1988 to 1992," according to a 1998 report by the U.S. Fish and Wildlife Service and the states of Michigan and Wisconsin. "Chinook salmon is the most popular salmonid species in Lake Michigan and the demand to restore the fishery to its former abundance is high. However, because Pacific salmon species have an apparently higher susceptibility to BKD relative to other salmonids, managers probably cannot guarantee anglers a return to the record chinook salmon catches of the 1980s. The economic development of the sport fishery in Lake Michigan can be attributed to the popularity of the chinook salmon fishery. Now the question arises, can the sport fishery withstand significant fluctuations in harvest and effort and still survive?"[19]

The rise of bacterial kidney disease was further evidence that importing foreign species into the Great Lakes often resulted in unforeseen ecological consequences. Worse, it demonstrated how little control humans exerted over the heavily manipulated Lake Michigan ecosystem, one teeming with dozens of foreign species.

Salmon also caused subtle changes to the ecosystems of rivers that flowed into the Great Lakes. When thousands of salmon migrated up a river and dug redds in which to deposit their eggs, the fish displaced invertebrates and disrupted habitat for native fish species. Some of the most dramatic such effects were documented in the Platte River, birthplace of Michigan's salmon fishery. The abundance and condition of twelve species of invertebrates in the river declined after salmon were imported. "As a result, food supplies for native fishes in this region could be affected negatively through the spawning activities of introduced salmon," according to Charles Krueger, science director at the Great Lakes Fishery Commission.[20]

That the imported fish would change the Great Lakes ecosystem only made sense, given the tremendous number of salmon stocked in the lakes and their tributaries. The eight Great Lakes states and the province of Ontario planted more than 745 million salmon in the lakes between 1966 and 1998. That amounted to an average of more than 66,000 fish stocked in the lakes and their tributaries every day for 33 years. More than 90 percent of those fish were stocked in the lakes by fish managers on the U.S. side of the lakes, according to Canadian biologist Steven Crawford.[21]

Salmon also became an unintended distribution system for toxic chemicals in the lakes. Salmon and lake trout accumulated toxins—remnants of industrial pollution in the lakes—in their body fat. Unlike lake trout, which reproduced in the lakes, salmon spawned exclusively in streams. Spawning salmon effectively acted as carriers of chemical pollution as they migrated upstream. Those poisons were transferred to eagles, mink, and other animals that ate the fish. "Taken together, this body of evidence supports the conclusion that the ongoing introduction of nonnative salmonines poses an ecologically significant risk to the Great Lakes ecosystem and its native organisms and that the introductions should be terminated," Crawford said in a 2001 report.[22]

Toxic Gamble

Lake Michigan's billion-dollar salmon fishery became the backdrop for a bizarre criminal enterprise worthy of a novel. Robert Gehl owned a fish-processing firm in the small town of Hart, Michigan, which collected and sold surplus salmon for the Michigan Department of Natural Resources. The DNR trapped thousands of migrating salmon in rivers every autumn before the fish spawned. Biologists captured the fish to collect a volume of eggs and fish sperm sufficient to propagate a new generation of fish needed to sustain the artificial salmon fishery. The salmon-harvesting contract allowed Gehl to control Michigan's salmon market

Muskegon Chronicle

Michigan businessman Robert Gehl is shown leaving a New York court in 1993 after he and his company, Michigan-based Tempotech Industries, were indicted for selling salmon eggs laced with toxic chemicals as caviar. Gehl and his assistant were convicted of the charges and served time in federal prison.

for nearly two decades. But to Gehl, the fish were the means to a more profitable end. The big money was in the salmon eggs, or roe, which could be sold as caviar. It didn't take a financial genius to figure out that getting tons of salmon eggs each year, at almost no cost, and selling it as caviar was a recipe for potentially huge profits. Still, the lucrative deal didn't sate Gehl's thirst for money. He and his trusted aide, George Jackson, hatched an elaborate, illegal scheme to fatten their wallets.

Federal officials in 1993 indicted Tempotech, Gehl, and Jackson on charges of knowingly selling 500 tons of caviar laced with potentially dangerous concentrations of toxic chemicals. Authorities said Gehl, Jackson, and Tempotech earned $16 million by illegally obtaining contaminated roe from Lake Ontario salmon, shipping it to the firm's Michigan facility, and repackaging the toxic eggs as clean caviar from Lake Michigan salmon.

Authorities said Tempotech officials bought about 150,000 pounds of Lake Ontario salmon eggs annually from 1984 to 1992. The eggs were packaged as fish bait at Tempotech's facility in Pulaski, New York. Tempotech could legally sell the eggs as fish bait, despite the presence of chemical contaminants. That's where Gehl ran afoul of the law. Instead of selling the eggs to bait dealers, Tempotech employees shipped the roe to the company's fish processing facility in Michigan. There, the Lake Ontario eggs were blended with cleaner salmon eggs from Lake Michigan and repackaged as caviar. The contaminated caviar was then shipped to food brokers in New York, who sold the eggs to cruise lines, airlines, and restaurants along the eastern seaboard. Federal authorities said the food brokers paid Tempotech executives with shopping bags full of cash. The cash was then deposited in banks in multiple sums of less than $10,000 to avoid reporting the deposits to the U.S. Internal Revenue Service. U.S. District Judge Neal McCurn summarized the elaborate scheme in a 1994 court ruling:

> The evidence showed that over a period of more than eight years, Robert Gehl and employees of his companies, Tempotech and Gehl Productions, including George Jackson, had illegally processed and sold nearly one million pounds of salmon egg caviar in exchange for millions of dollars in cash, which they structured to avoid detection. They carried out this scheme by taking caviar from Lake Ontario—from which fish may not be sold, under New York law, except as bait—shipping it to their plant in Michigan as bait, repackaging it there as Lake Michigan caviar and then shipping it to New York for human consumption.

A jury convicted Tempotech Industries, Gehl Productions, Gehl, and Jackson of 25 federal counts of selling contaminated food and structuring cash deposits to avoid paying federal taxes. Judge McCurn sentenced Gehl to 7.2 years in jail and ordered him to pay a $250,000 fine. Jackson was ordered to serve 5.8 years in federal prison. Tempotech Industries and Gehl Productions were fined another $2.6 million. Michigan anglers, who had long contended that company officials were behaving badly, hailed the conviction of Tempotech and its top executives. Then, the unthinkable happened. Gehl slipped through a web of law-enforcement agencies, much like a fish squirming free from a net, and fled the country.

Gehl and Jackson were ordered to report to federal marshals in January 1996. Jackson showed up as directed, but Gehl was nowhere to be found. He jumped bail and fled the country, even though authorities had taken away his passport.

Police tracked down Gehl five months later in Costa Rica. He was extradited to the United States and served his federal prison sentence.

The convictions put Tempotech out of business and forced the state of Michigan to hire other firms to harvest surplus salmon. The case also closed one of the most bizarre chapters in the history of Michigan's celebrated salmon fishery. Said Judge McCurn in his 1994 ruling: "A great commotion has been caused by tiny salmon eggs."

The notion of halting salmon stocking in the Great Lakes would be considered heresy by many anglers in the region. Anyone promoting such an idea would likely be run out of town on a rail. Like it or not, Tanner's grand experiment was a huge economic success and placed salmon on a pedestal the fish would not easily relinquish. Salmon adapted well to the freshwater environment and became a naturalized species. Coho and chinook salmon would remain in the lakes even if government agencies stopped stocking the fish. About half the chinook in Lakes Michigan and Huron in 2006 were the result of natural reproduction in rivers that flowed into the lakes. The imported fish were churning out millions of offspring each year without human assistance. Salmon were here to stay, but theirs would be a challenging existence. Disease remained a constant threat, and other foreign species discovered in the late 1980s began dismantling the biological house of cards upon which the artificial salmon fishery was built.

The health and integrity of the Great Lakes basin ecosystem, including the forty million humans who live in the basin, are jeopardized by an immediate and growing problem: the rampant colonization by ship-borne exotic organisms.

—Great Lakes Fishery Commission and International Joint Commission, 1990

Here we have 20 percent of the world's fresh water in one little area and that Seaway, and what has happened because of it, has really screwed it up. That has to be one of the great ecological disasters of all time.

—David Schindler, University of Alberta freshwater ecologist, 2005

02 PLAGUE

Why have federal efforts to prevent new introductions of species by ships been so anemic?

—Allegra Cangelosi, Northeast-Midwest Institute policy analyst, 2002

06 FATAL ERROR

Among landmark environmental events, a disastrous incident on June 22, 1969, lived in infamy. That was the day the Cuyahoga River caught fire near Cleveland, Ohio. An oil slick on the river ignited—water turned to flame—and burned for 24 minutes. So polluted was the Cuyahoga with chemicals, oil, and grease, a spark was the only ingredient needed to trigger a fire. A train passing over the Cuyahoga produced the ingredient that set the filthy river ablaze. The 1969 fire wasn't the first time the Cuyahoga burned. There were much larger fires on the river in 1936 and 1952. The 1952 conflagration was the most destructive, causing more than $1 million damage to boats and a riverfront office building. But it was the 1969 blaze that captured national attention, a phenomenon due in part to an article in the August 1, 1969, edition of *Time* magazine.

> Some river! Chocolate-brown, oily, bubbling with subsurface gases, it oozes rather than flows." the article said. "'Anyone who falls into the Cuyahoga does not drown,' Cleveland's citizens joke grimly. 'He decays.' . . . The Federal Water Pollution Control Administration dryly notes: 'The lower Cuyahoga has no visible signs of life, not even low forms such as leeches and sludge worms that usually thrive on wastes.' It is also—literally—a fire hazard.

The magazine's portrayal of the river was harsh but, sadly, accurate. That a river could be so polluted as to burst into flame was more than the American public would tolerate. The Cuyahoga fire became an icon of the United States' abuse of its surface waters—a seminal moment that sparked public outrage. The timing of the fire, seven years after the publication of Rachel Carson's *Silent Spring,* added fuel to a growing environmental movement sweeping college campuses. Carson's book exposed the harmful consequences of rampant pesticide use. *Silent Spring* may have been the genesis of America's modern environmental consciousness, but the Cuyahoga River fire focused public outrage on the wanton fouling of land, air, and water. No longer were the masses content to sit by and watch industries and municipalities use lakes and streams as toilets for all manner of household and industrial wastes. Just as the Japanese attack on Pearl Harbor drew the United States into World War II, the Cuyahoga fire galvanized public opinion and forced politicians to address the abuse of America's surface waters.

The U.S. Congress responded with remarkable speed, passing the National Environmental Protection Act by the end of 1969. President Nixon signed the legislation—the first comprehensive federal law regulating pollutant discharges—on January 1, 1970. That was followed by a series of unprecedented actions: The first Earth Day was celebrated four months later, on April 22; Congress pressured Nixon into establishing the U.S. Environmental Protection Agency in 1970; and the first EPA administrator, William D. Ruckelshaus, threatened to sue Cleveland, Detroit, and Atlanta if those cities didn't reduce water pollution within four months. In 1972, Congress and President Nixon approved amendments to the Water Pollution Control Act of 1948. Those amendments, which led to the law that became known as the Clean Water Act in 1977, established the nation's first specific and enforceable limits on water pollution.

The Clean Water Act revolutionized water-pollution control by limiting the amount of waste that industries and cities could discharge into surface waters. Prior to 1972, state and federal agencies regulated pollution based on water quality. If a lake or stream became too filthy, government agencies ordered industries and cities to reduce discharges of chemical and biological wastes. The 1972 rules imposed the nation's first enforceable discharge limits on every company and city that sent waste into surface waters. The EPA assigned each city and industry a volume of waste that could be released into lakes and rivers. Every permittee had to meet its discharge limit, using the best available control technology, or face fines of up to $25,000 per day. The limits prompted the development of better treatment technology, which led to tighter discharge limits, which led to even better treatment systems. Improved treatment of wastewater was no longer an option—federal law required it.

The Clean Water Act forced cities and industries to invest billions of dollars in wastewater treatment equipment to meet their assigned discharge standards. Water quality and the health of aquatic life improved as a result. A 1999 report by the Congressional Research Service summed it up best: "The [Clean Water] Act has been termed a technology-forcing statute because of the rigorous demands placed on those who are regulated by it to achieve higher and higher levels of pollution abatement under deadlines specified in the law."[1]

The law initially required cities and industries to reduce discharges of conventional pollutants—suspended solids and bacteria-laden wastes. When that problem was largely under control, the EPA demanded reductions in discharges of toxic chemicals that harmed water quality and poisoned fish and wildlife. A key provision of the law was the establishment of the National Pollution Discharge Elimination System. The system amounted to a license to pollute, to be sure, but it marked the first time the federal government established limits on the amount of waste that could be discharged into lakes, streams, and oceans. The regulatory scheme was akin to police establishing speed limits on highways. Absent speed limits, authorities could tell motorists to avoid driving excessively fast—but such orders were entirely subjective, and largely meaningless, without enforceable standards. By setting specific limits on wastewater discharges, the NPDES process put the brakes on water pollution in the United States.

Cities and industries railed against the discharge limits, claiming that treatment technology to meet the standards did not exist. Critics also argued that the discharge limits would bankrupt countless companies and municipalities. It didn't happen. Instead, the vast majority of cities and businesses took the financial hit and bought the equipment necessary to meet the new wastewater discharge limits. America's lakes, streams, and oceans—and the ecosystems and economies they supported—were the beneficiaries of the tough new law. Proof that the Clean Water Act succeeded was found in surface waters across the nation: water quality improved, fish populations rebounded, and polluted lakes and streams became recreational magnets. The groundbreaking law was, and remains, an incredibly powerful tool for regulating water pollution. The era of industries and cities dumping unlimited quantities of chemical and biological wastes into the Great Lakes and other surface waters had finally come to an end. So it seemed.

Then, in 1973, the EPA quietly and unilaterally emasculated a key provision of the Clean Water Act. The agency, without consulting Congress, ruled ships that discharged ballast water into the nation's waters were exempt from the law. EPA officials claimed that the law didn't apply to discharges "incidental to the normal operation of a vessel." Moreover, the EPA concluded that ballast water discharges from ships "causes little pollution." The agency could have, and should have, applied the Clean Water regulations to ballast water. Ballast water, after all, brimmed with potentially deadly viruses, microscopic algae, tiny zooplankton, fish, and in some cases, crabs. Instead, the exemption laid the foundation for three decades of ineffective ballast water regulations. It was a bureaucratic blunder of the highest order. The misinterpretation of the law would haunt the nation's surface waters—particularly the Great Lakes—for decades to come. Under the exemption, the following discharges did not require a permit:

"Any discharge of sewage from vessels, effluent from properly functioning marine engines, laundry, shower and galley sink wastes, or any other discharge incidental to the normal operation of a vessel. This exclusion does not apply to rubbish, trash, garbage or other such materials discharged overboard; nor to other discharges when the vessel is operating in a capacity other than as a means of transportation, such as when used as an energy or mining facility, a storage facility or a seafood processing facility, or when secured to the bed

Jeff Alexander

A transoceanic freighter pumps out ballast water as the ship takes on a load of grain in the port Duluth-Superior, at the West end of Lake Superior.

of the ocean, contiguous zone or waters of the United States for the purpose of mineral or oil exploration or development."[2]

The exemption meant ships could dump millions of gallons of untreated ballast water, swarming with aquatic life, into the Great Lakes. But throwing a barrel of trash overboard was strictly prohibited and could draw a hefty fine. It was an absurd double standard that defied two key provisions: The Clean Water Act, as written, regulated ships as point sources of water pollution; and the law regulated biological pollutants, such as fish, aquatic vegetation, and pathogens. EPA officials viewed the law differently. They explained their rationale in the preamble of the ballast water exemption: "Most discharges from vessels to inland waters are now clearly excluded from the NPDES [National Pollution Discharge Elimination System] permit requirements. This type of discharge generally causes little pollution and exclusion of vessel wastes from the [Clean Water Act] permit requirements will reduce administrative costs dramatically."[3] Pause for a moment to let the significance of that last comment sink in. The EPA chose to cut costs by refusing to regulate ballast water discharges. It was a colossal, illegal gaffe that allowed transoceanic ships entering the Great Lakes via the St. Lawrence Seaway to dump billions of gallons of ballast water teeming with foreign organisms into the lakes for the next three decades. The result was a tidal wave of invasive organisms that infected the lakes, much as an electronic virus affects a computer. The invaders rewired entire ecosystems, causing billions of dollars in environmental and economic damage in the process.

Few people realized how critical a role the Clean Water Act—and the EPA's failure to apply it to ballast water discharges—played in the swarm of foreign species that spread across the Great Lakes. When zebra mussels were discovered in Lake St. Clair in 1988 and quickly colonized the Great Lakes, scientists and environmentalists promptly blamed the shipping industry. No one pointed out that the problem could have been prevented had the EPA enforced the Clean Water Act as the law and Congress intended. Who knew? Instead of politicians and environmentalists placing the blame where it belonged, the shipping industry became the fall guy for the EPA's misinterpretation of the Clean Water Act.

The ballast water exemption also allowed a cruel irony to play out in the Great Lakes. The Clean Water Act forged dramatic improvements in water quality, which made the lakes more hospitable to foreign species. Once here, the invaders spread like a stampede of land prospectors racing to claim new territory. The exemption was judged to be illegal, but not until 2005—32 years after it was approved. By then, dozens of invasive species were causing ecological mayhem in the lakes. How important was the ballast water loophole in the grand scheme of the invasive species issue? Consider this: In 1973, there were 126 invasive species in the Great Lakes. In 2006, when a federal court ordered the EPA to regulate ballast water discharges, there were 183. Ocean freighters imported 38 of the 57 new species that invaded the lakes between 1973 and 2006.[4] The shipborne invaders included many of the worst foreign species now common in the lakes: zebra mussels, quagga mussels, Eurasian ruffe, round goby, spiny water flea, bloody red mysis, Asiatic clam, New Zealand mud snail, blueback herring, and several types of fish viruses.

There is no way of knowing whether the Clean Water Act would have kept all of those species out of the Great Lakes if the EPA had begun regulating ballast water discharges in 1973. Some of those species might have snuck into the lakes by other means, by clinging to ships' anchor chains or fouling equipment other than ballast water tanks. But it isn't much of a stretch to suggest that regulating ballast water discharges in 1973 would have slowed the number and rate of invasions. Such regulations would have forced the development of treatment systems that could have shielded the lakes from some of the 57 foreign species that ocean freighters imported in the absence of rules. The burden for developing ballast treatment systems would have been placed squarely on the shipping industry. Instead, the EPA's exemption shifted the burden of developing ballast treatment systems to the government. By the time a federal court ruled that the ballast water exemption was illegal, and that the EPA had overstepped its statutory powers in creating the regulatory loophole, an uncontrollable biological plague was raging across the Great Lakes.

THE EPA'S BALLAST WATER EXEMPTION, BY DEFAULT, PUT THE U.S. COAST Guard in charge of regulating all environmental issues on ships. That was a glaring error. The Coast Guard did many things very well, such as rescuing stranded sailors, freeing ships trapped in ice, and intercepting drug runners on the high seas. Regulating ballast water was not one of the military agency's areas of expertise—Coast Guard officials knew little about regulating pollution discharges. Shippers argued, correctly, that the EPA also was clueless about regulating ballast water discharges in the early 1970s. The difference was that the EPA could have used the Clean Water Act to force shipping companies to develop ballast water treatment systems. The Coast Guard didn't have that luxury—it wasn't the agency charged with enforcing the law. The Coast Guard also was hamstrung by political factors. Until 2003, the Coast Guard was part of the U.S. Department of Transportation, the same agency that managed America's portion of the St. Lawrence Seaway. Asking the Coast Guard to police one of its sister agencies within the Department of Transportation was like asking a heroin addict to lead a war on illicit drugs.

Three decades of history demonstrated, with painful clarity, that the EPA's ballast water exemption was a catastrophic blunder on many fronts. It cheated the Great Lakes out of achieving the maximum environmental recovery from industrial pollution and, in fact, caused progress in restoring the lakes to backslide. The exemption also crippled future efforts to regulate ballast water discharges. Ocean freighters that entered the Great Lakes were virtually untouchable by the time the Coast Guard got around to working on ballast water discharge standards. Why? The Coast Guard was extraordinarily friendly with the shipping industry it was supposed to regulate, which allowed shipping companies to keep meaningful ballast water regulations at bay for three decades. To put the matter in baseball terms, the EPA's decision to abdicate its authority to regulate ballast water was strike one in the U.S. government's halfhearted battle to keep foreign species from mounting a hostile takeover of Great Lakes ecosystems. It was the first of several blown opportunities to shield the lakes from foreign organisms that arrived daily in the bowels of ocean freighters.

It is estimated that more than 10,000 marine species each day hitch rides around the globe in the ballast tanks of cargo ships.

—James Carlton, Williams College biologist, 1993

07 DANGEROUS CARGO

Ships are among the most alluring of all human inventions. A freighter moving through the high seas invokes a sense of wonder for observers on land. It is difficult for non-shipping types to comprehend the physics and technology that enable a vessel longer than two football fields to carry 50 million pounds of cargo, or more, and not sink like a stone. But the romanticism surrounding ships is more than a reverence for technology. Freighters entering ports from distant waters stoke the curiosity and imagination of ship lovers and landlubbers alike. Where has a given ship come from? What is it carrying in its cargo holds? Did its crew battle angry seas on their journey? And the most common query: How big is it? That sense of wonder lures thousands of self-declared boat nerds to Duluth, Minnesota, every year. Planted on the side of a rocky bluff overlooking the west end of Lake Superior, Duluth is the busiest of all Great Lakes ports. It was where I ended up on a mild autumn night in 2007, watching the Russian freighter *Grigoriy Aleksandrov* slip into Duluth Harbor under cover of darkness. The 605-foot-long bulk carrier came to the west end of the St. Lawrence Seaway—the heart of North America—to pick up a load of spring wheat it would haul back to Rotterdam. The *Aleksandrov* was not particularly notable. It wasn't the largest, fastest, or most historic ship to call on Duluth. Rather, it was a symbol of a grand, unplanned biological experiment born of the St. Lawrence Seaway.

The *Aleksandrov* announced its arrival with the traditional series of horn blasts: one long bass blast, followed by two shorter blasts that echoed off the nearby hillside. After passing under Duluth's famous lift bridge, the ship headed to one of the towering grain elevators that lined the harbor. There, crews filled the ship's cargo holds with thousands of tons of wheat, which would be distributed to milling operations across the Netherlands. As it took on cargo, the *Aleksandrov* left a part of Europe in Duluth. The ship pumped millions of gallons of untreated ballast water from distant oceans and freshwater ports into the St. Louis River estuary, which forms Duluth Harbor before flowing into Lake Superior. Just as civic leaders did at the dedication of the Erie Canal in 1825, the Welland Canal in 1829, and the St. Lawrence Seaway in 1959, the *Aleksandrov* performed a wedding of the waters. Was the Russian freighter's wedding of waters from around the globe with those of the Great Lakes the type of marriage that supporters of the St. Lawrence Seaway had in mind when they lobbied so forcefully for opening the lakes to global shipping? That was doubtful.

The wedding of waters forged by the Seaway was supposed to bolster the industrial economies of port cities around the lakes. It did. In the year 2000, ships hauled 192 million tons of cargo on the Seaway, the network of ports around Lakes Superior, Michigan, Huron, Erie, and Ontario, and the St. Lawrence River. About 44,000 jobs in the United States and Canada in 2000 were directly related to shipping activity on the Seaway. Another 108,000 jobs in the region were indirectly related to the Seaway. Industries connected to shipping on the Seaway generated $1.6 billion in direct wages and another $3.4 billion in revenue annually in the United States. Seaway-related industries also paid $1.3 billion in federal, state, and local taxes in 2000. Shipping on the Great Lakes was a huge industry but it wasn't the only industry.

Jeff Alexander

Transoceanic freighters from around the globe visit the Great Lakes each year. Pictured is a Russian freighter taking on grain at the port of Duluth-Superior. Foreign freighters discharge more ballast water in Duluth harbor than in any other Great Lakes port.

Smithsonian Environmental Research Center

The Chinese mitten crab was discovered in the St. Lawrence River in 1965. Imported to North America in the ballast tanks of ocean freighters, the foreign crabs have also been spotted in Lakes Erie and Ontario. The invader has not yet established a reproducing population in the Great Lakes.

The lakes' sport and commercial fisheries were valued at $7 billion in the first decade of the twenty-first century. Recreational boating also was a huge aspect of the Great Lakes economy. In 2007, the U.S. Army Corps of Engineers reported that the 4.3 million pleasure boats registered in the eight Great Lakes states on the American side of the border generated nearly $16 billion in annual spending on boats and related activities.

Though the Seaway increased global commerce on the lakes, domestic shipping remained the region's bread and butter. Transoceanic freighters accounted for a small fraction of all cargo moved in the Great Lakes and St. Lawrence Seaway. In 2000, ocean freighters hauled 8.8 million tons of cargo in and out of the Great Lakes—4 percent of all freight shipped across the lakes.[1] That figure increased to 12.3 million tons in 2003, which was 6.8 percent of all cargo shipped on the lakes that year. Lake freighters, those ships that remained in the Great Lakes, hauled the vast majority of cargo on the lakes—before and after the Seaway opened.[2]

Duluth was, by far, the Seaway's biggest beneficiary. More domestic and international ships visited Duluth each year than any other Great Lakes port. Freighters arrived daily during the nine-month shipping season to collect coal, taconite pellets, and grain, which was then distributed across the lakes and around the world. Domestic and foreign ships moved 36 million metric tons of cargo in and out of Duluth in 2001, generating about 2,000 jobs and $210 million in total economic impact.[3] Shipping also was a tourist attraction in Duluth. The city's Canal Park attracted hundreds of thousands of visitors each year—tourists who shopped in stores, slept in waterfront hotels, ate at the famed Mother's restaurant, and, of course, watched ships come and go from the city's pier. Adolph Ojard, executive director of the Port of Duluth, said people were drawn to Great Lakes ships for many reasons. "Shipping has a thin margin [of profit] and it's filled with risk. It's a vital service we're providing and ships are romantic as hell. There's something graceful about a ship moving through the water."[4]

Freighters that visited Duluth to tap the Midwest's huge stockpiles of iron ore and grain pumped millions of dollars into the local economy. But there was a tradeoff to being the Seaway's busiest port. Duluth also was the ballast-water-dumping capital of the Great Lakes. It was the port where freighters that inhaled ballast water at ports around the globe were most likely to disgorge the biologically rich water. For 30 years after the Seaway opened, transoceanic ships regurgitated tens of billions of gallons of untreated ballast water—and all that lived in it—into the Great Lakes. The practice was routine, unregulated, and incredibly risky—ecologically speaking.

SHIPS HAVE CARRIED BALLAST AS LONG AS HUMANS HAVE TRAVELED THE seas—it is essential to maintaining the stability of a ship. In the early days of shipping, sailors packed dirt and rocks into the bowels of ships to weigh down and maintain the stability of their vessels. Water became the preferred source of ballast with the advent of steel ships in the late 1800s. Water was readily available, easier to handle, and could be loaded and unloaded faster than dirt or rocks.

The importance of proper ballasting cannot be overstated—the ability to control the weight, trim, and stability of a ship is a matter of life and death. Carrying too much ballast, or not enough, makes a ship ride too low or too high in the water. The result could be catastrophic: an improperly ballasted ship runs the risk of capsizing or literally breaking in half. The M/V *Flare* was a tragic example of such a disaster. The 590-foot-long bulk carrier was headed to Montreal from Rotterdam in 1998 when it snapped in half and sank after being pounded by waves 50 feet high. The disaster killed 21 of the 25 men on board the Cyprus-flagged vessel. Canadian investigators concluded that the ship had too little ballast water in its forward tanks, which caused the *Flare*'s bow to bounce off waves as it struggled against the fierce North Atlantic.[5] The Cypriot owners of the ship blamed the accident on the *Flare*'s captain and extreme weather conditions.[6] Whatever the cause, the *Flare* accident illustrated the dangers inherent in moving freight across angry seas. Hauling ballast water from one continent to another posed another, hidden danger: the water was a convenient way for species to hopscotch oceans and colonize new frontiers.

I found it difficult to fathom that something as endearing as a ship could be the carrier of a type of plague. That was, until my focus went beyond a ship's outer appearance and focused on what was sloshing around in its ballast tanks. When it comes to transoceanic freighters, beauty is skin deep. Lurking in the bellies of those ships, in ballast water tanks, is an assortment of species from around the globe that could hurt Great Lakes fisheries, disrupt ecosystem function, and endanger human health. Scientists familiar with the inner workings of ships refer to them as biological islands—self-contained ecosystems that move around the world, expanding the range of species along the way. Unlike islands, which are largely biologically isolated from other land masses, ships move species between continents on a regular basis. The process is analogous to a supernatural force lifting part of an aquatic ecosystem from, say, Belgium, and dropping it in the Great Lakes. That notion of ships as traveling biological islands is no trivial matter: About 98 percent of world trade, more than 4.5 billion metric tons of cargo annually, and 80 percent of world commodities are shipped on the world's oceans. Most ocean freighters that traverse the Great Lakes visit Duluth at some point in their journey. About 70 percent of all the ballast water those ships hauled into the Great Lakes between 1981 and 2000 was dumped at ports on Lake Superior.[7]

Ballast water discharges weren't considered a threat to the Great Lakes prior to 1988—the issue wasn't on the public's radar. Few scientists were curious about what ocean freighters might be hauling into the Great Lakes in ballast tanks. Environmental groups were fixated on reducing toxic chemical discharges into the lakes. Government agencies in the United States and Canada were preoccupied with figuring out how to clean up industrial contaminants that poisoned fish, disfigured birds, and left pockets of toxic mud in 42 harbors around the lakes. Nobody, save one Canadian bureaucrat, was even remotely concerned about the millions of gallons of ballast water that ocean ships dumped in the Great Lakes on a daily basis.

JOE SCHORMANN WAS A SENIOR PROGRAM ENGINEER FOR ENVIRONMENT Canada in 1980 when he asked one of the most insightful "what if" questions of all time. Schormann wondered if the ballast water that transoceanic freighters carried into the lakes might be a vector for foreign species. Was it possible that the hundreds of ocean freighters entering the lakes each year—ships that inhaled ballast water at marine and freshwater ports around the globe and hauled it across the Atlantic Ocean before dumping it in the lakes—were inadvertently importing foreign species of fish, mussels, and plankton as well? If they were, Schormann wondered, could any of those species from overseas ports survive in the lakes? His intuition turned out to be tragically prophetic.

At the time Schormann posed his provocative question, no one even knew how much ballast water ocean ships were dumping in the Great Lakes. Subsequent studies estimated that ocean freighters dumped about 1.5 billion gallons of ballast water into the lakes annually. That estimate was based on ship traffic in the first decade of the twenty-first century, when fewer ocean freighters entered the lakes than in the 1960s, '70s, and early '80s.[8] Ocean freighters dumped between 30 billion and 50 billion gallons of untreated ballast water into the Great Lakes during the Seaway's first 30 years of operation. Most of that was deposited in Duluth Harbor.

Understanding why ocean ships discharged most of their ballast water in Duluth, instead of other Great Lakes ports, requires a basic understanding of shipping traffic. Most ocean freighters entering the St. Lawrence Seaway are destined for Duluth, where they take on grain and haul it back to Europe or other continents. Foreign ships that come to the lakes without cargo—like the *Aleksandrov*—haul up to three million of gallons of ballast water. As ocean freighters take on cargo in Duluth, the vessels unload an equal volume of ballast water before heading back to Europe, Asia, or South America.

Ships that enter the lakes fully loaded with cargo carry relatively small volumes of ballast water. Those ships are known as NoBOBs, because they carry No Ballast On Board. Still, there are generally about 150 tons of residual ballast water and muddy slop in the ballast tanks of NoBOB ships. The NoBOB ships typically unload their foreign cargo at ports around Lakes Ontario, Erie, and Michigan. As the cargo is unloaded, the ships take on ballast water in ports such as Toronto, Cleveland, or Chicago before heading to Duluth to pick up grain. En route to Duluth, the residual mud in a NoBOB's ballast tanks mixes with ballast water taken on in the Great Lakes, creating a biologically rich soup. Once in Duluth, the NoBOB ships dump the blended ballast water into the harbor as grain is loaded into cargo holds. The process is a recipe for ecological disaster.

Cholera in Ballast Water

In the early 1990s, a cholera outbreak that began in Peru and spread across South America killed 10,453 people and sickened more than one million others. The epidemic stunned health officials, who believed there were systems in place

National Oceanic and Atmospheric Administration's Great Lakes Environmental Research Laboratory

Researchers discovered that cholera and other potentially deadly pathogens flourished in the muddy slop that remained in the bottom of supposedly empty ballast tanks. The muddy water was often discharged, untreated, into the Great Lakes prior to 2006.

to prevent cholera outbreaks from spiraling out of control. Scientists had long known that *Vibrio cholerae*, the bacteria that caused cholera in humans, was one of the world's deadliest waterborne illnesses. Cholera pandemics killed millions of people over the past two centuries. Most of those incidents occurred before the advent of modern water-treatment technology, which killed the deadly bacteria. Occasionally, people died from cholera after eating contaminated shellfish. But bad shellfish did not trigger the Peruvian incident. Something else was at work. Researchers eventually identified contaminated ballast water discharged by an ocean freighter as the probable cause of Peru's cholera outbreak. The epidemic was a cautionary tale for the Great Lakes. The reason: Transoceanic freighters that carried *Vibrio cholerae* in ballast water tanks routinely discharged the contaminated water at ports in South America and the Great Lakes.

Could contaminated ballast water infect the Great Lakes and unleash a cholera epidemic among the 25 million people who relied on the lakes for drinking water? Researchers who tackled that question in the 1990s concluded that such a disaster was possible but unlikely. Still, their findings were, to put it mildly, unsettling.

Scientists at Virginia's Old Dominion University found the mother lode of *Vibrio cholerae* and other potentially deadly pathogens in the ballast water of oceangoing freighters entering the Great Lakes. What they encountered read like a "who's who" of freshwater pathogens: *Vibrio cholerae, E. coli, Cryptosporidium parvum, Giardia duodenalis*, and two strains of the toxic algae *Pfisteria.* Though *Vibrio cholerae* was the pathogen of greatest concern in terms of potential public-health impacts, the presence of the others also was cause for consternation. *E. coli* and *Giardia* caused stomach ailments; *Pfisteria* caused massive fish kills and threatened human health in North Carolina rivers; and *Cryptosporidium* killed 100 people and sickened 400,000 others when it infected Milwaukee's municipal water supply in 1993. The findings prompted Fred C. Dobbs, the professor who led the team of researchers from Old Dominion University, to issue a blunt recommendation to government agencies in 2003: "In formulating ballast water discharge standards," he said, "it would be prudent to regard all ships as potential carriers of pathogens."

Subsequent studies by Dobbs and scientists from the U.S. government found that *Vibrio cholerae* and other pathogens were common in ocean freighters' ballast tanks. One of the most unsettling discoveries came during a study that assessed whether flushing ballast tanks with seawater before ocean freighters entered the Great Lakes purged the tanks of harmful pathogens. *Vibrio cholerae* was found in 82 percent of the samples collected from the ballast tanks of two transoceanic freighters that entered the Great Lakes. Elevated concentrations of *Vibrio cholerae* were even found in ballast tanks that were flushed with seawater before the ships entered the St. Lawrence Seaway. Still, Dobbs and other scientists who worked on the study downplayed the risk of ballast water infecting a municipal water treatment system and causing a cholera epidemic in the Great Lakes region. They said a person would have to drink between one and ten liters of undiluted ballast water to become ill. In a report titled "Identifying, Verifying and Establishing Options for Best Management Practices for NoBOB Vessels," Dobbs and the other scientists offered the following risk assessment: "While the risk from *Vibrio cholerae* carried in ballast tank residuals is not zero, we can muster no evidence that the risk is particularly high. While we do not dismiss potential health concerns associated with this and other pathogens in ships arriving to the Great Lakes, it is relevant to consider that no outbreaks or epidemics of cholera, cryptosporidiosis or giardiasis have been associated with ship traffic or ballasting operations in the Great Lakes."

Just because there weren't bodies in the streets didn't mean pathogens in ballast water posed no threat to the 25 million people who relied on the Great

Lakes for drinking water. The absence of a public-health catastrophe didn't justify the resulting apathy on the part of government officials.

Unlike Peruvian officials, who knew nothing about pathogens in ballast water prior to 1991, the U.S. and Canadian governments were well aware of the risks associated with transoceanic ships disgorging disease-laden ballast water into the Great Lakes. Yet, government officials dithered for more than a decade before requiring all ocean freighters to flush ballast tanks with seawater before entering the lakes.

Veteran Great Lakes researcher David Jude said he feared government officials wouldn't require ships to disinfect ballast tanks until there was a public-health crisis. "Once we get a cholera epidemic in Chicago and it kills 200 people," he said, "then we'll deal with ballast water."

Ocean freighters, which are known as "salties" in shipping circles, aren't the only source of ballast water discharges in Great Lakes ports. The so-called "lakers," ships that only move cargo within the Great Lakes because they are too large to fit through the Seaway's locks in the St. Lawrence River, discharge 15 times more ballast water in the lakes than the salties. But there is a critical distinction between the two classes of vessels. Salties, prior to 2008, discharged ballast water and mud into the lakes from 460 different ports in Europe, Asia, South America, the Gulf of Mexico, Australia, and Africa. The lakers, by comparison, suck up ballast water exclusively from the Great Lakes and discharge that water back into the lakes. At worst, the lakers move water and aquatic species from one Great Lakes port to another in the process of transporting cargo.

This distinction was not lost on Schormann, a remarkably intuitive civil servant. Concerned about the prospect of ocean ships importing foreign aquatic organisms into the lakes, Schormann commissioned a $50,000 study. The investigation by Bio-Environmental Services, based in Georgetown, Ontario, was the first of its kind in the lakes. Scientists sampled the ballast tanks of 55 ships that sailed to the Great Lakes from 10 geographic regions of the world. They collected ballast water from the freighters in Montreal, before the ships headed up the St. Lawrence River and into the lakes. The results were stunning, unambiguous, and horrifying. Researchers discovered that ships' ballast tanks were like giant aquaria, bustling with foreign mussels, algae, fish, and zooplankton. They summarized their findings in a 1981 report. It said:

> Aquatic organisms were found in each of the ballast water samples examined. . . . Our overall assessment is that organisms are alive in ship ballast water when it is discharged. The results of the study clearly indicate that non-indigenous and non-endemic aquatic species are being imported into the Great Lakes system. Almost all ballast tanks sampled contained aquatic organisms in a viable state. It was apparent that organisms were capable of surviving in ballast tanks for extended periods of time up to at least 18 days. Ballast waters obviously can play an important role in the transport of aquatic organisms from one region of the world to another.[9]

Researchers found millions of viable organisms in ballast water tanks—larvae of foreign mussels, algae, zooplankton, and elevated concentrations of fecal coliform bacteria in seven ships. Their findings included 150 different species of phytoplankton, 56 distinct aquatic invertebrates, and millions of microscopic larvae of *Dreissena polymorpha*—zebra mussels. The scientists didn't find a few specimens of each foreign species. They discovered millions of specimens of nonnative phytoplankton, zooplankton, copepods, and mussels. In one ship, which picked up ballast water from the Mediterranean Sea, researchers found 2.6 million zebra mussel larvae, known as veligers. The tanks in a ship that took on ballast water from a port in France before heading to North America contained 380,000 zebra mussel veligers. The presence of zebra mussel veligers was disturbing. The mollusks were one of the world's most insidious invasive species. Liberated from the Black and Caspian seas by navigation canals dug across Europe in the eighteenth and nineteenth centuries, zebra mussels spread across much of that continent in the 1800s. The prolific mussels clogged water-intake pipes and disrupted water flow in several cities in Great Britain and Russia. The Bio-Environmental Services researchers summed up the threat:

"The occurrence of its veliger larvae in the plankton . . . greatly enhances its potential for introduction into the Great Lakes in ship ballast water. Some of these taxa contain species which are known problem organisms which would be capable of establishing themselves in the Great Lakes area, e.g. *Dreissena polymorpha*."[10]

Credible scientists were not prone to hyperbole. So when the authors of the Bio-Environmental Services report used words like "obviously" and "clearly" to explain that ocean ships were importing foreign species into the Great Lakes, it was like a military scout returning to base and shouting, "We're being invaded!" The threat of shipborne invaders sneaking into the lakes was that serious, the evidence clear beyond even a reasonable doubt.

Schormann shared the report with the U.S. and Canadian coast guards, both of which had the authority to regulate ballast water discharges in the Great Lakes. Both agencies shelved the report. "After it came out it was reviewed by a lot of people from both coast guards," Schormann said in a 1989 newspaper interview. "The opinion was 50–50 whether it was worthwhile to pursue it and do something or do nothing. Much to my regret, the do-nothing vote won the day and it was shelved."[11]

For the next few years, it seemed the U.S. and Canadian officials who chose to do nothing were vindicated. There was no evidence that foreign species were overwhelming the Great Lakes. Yet there were subtle, disturbing warning signs. Scientists had already discovered fourteen foreign species in the Great Lakes between 1959 and 1983, nine of which were attributed to ocean freighters.[12] The list of invaders included the humpback pea clam, Chinese mitten crab, flatworm, European flounder, Asiatic clam, and two species of copepods. Interesting finds, every one. Still, none of those imports rose to the level of public concern. In fact, those species seemed relatively harmless compared to the alewife, the last notable invader to show up in the lakes prior to 1981. Then, in 1984, came an unsettling development.

Lake Huron anglers reported thousands of specimens of a bizarre looking crustacean clinging to fishing lines. The creature wasn't large; it measured just a half-inch long. But its disproportionately large head, dominated by a single black eye, along with long legs and a barbed tail, gave it the appearance of a creature from the horror film *Alien*. And it had this odd habit of hooking onto its kin, much like the plastic monkeys in the old Barrel of Monkeys game. *Bythotrephes longimanus*, known as the spiny water flea, had invaded the Great Lakes. Somehow the species, which was native to eastern Europe, had managed to leapfrog the Atlantic Ocean and take up residence in Lake Huron.

Bythotrephes was an immediate headache for anglers. Its spiny tail hooked on fishing lines, fouling gear as the tiny beasts clung to monofilament line and each other by the thousands. It was impossible to reel in a line coated with

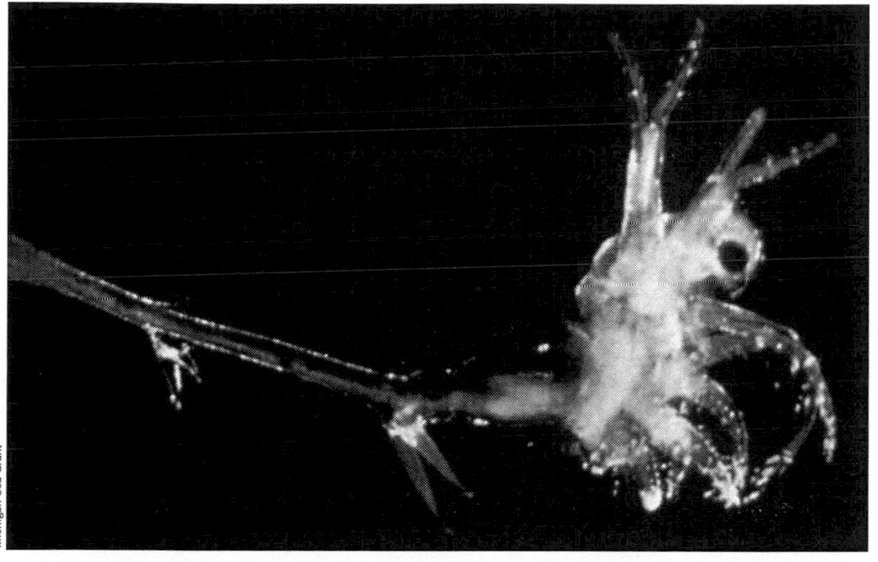

Michigan Sea Grant

Bythotrephes, known as the spiny water flea, was one of dozens of foreign species imported to the Great Lakes in the ballast water tanks of transoceanic freighters.

David Jude, University of Michigan, School of Natural Resources and Environment

The Asiatic clam, another product of ocean freighters' ballast water discharges, was discovered in western Lake Erie in 1981.

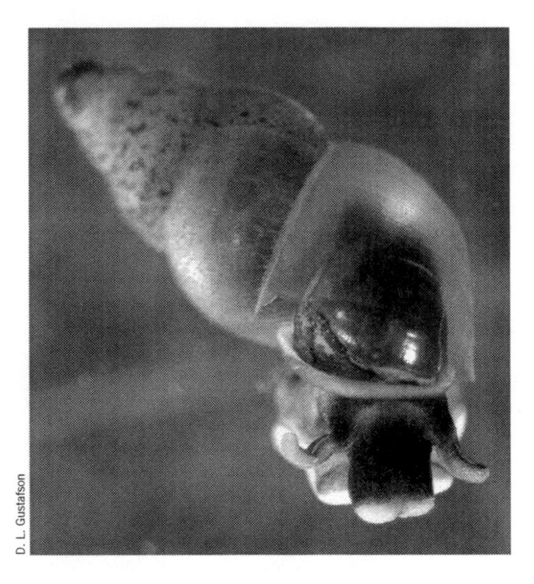

D. L. Gustafson

The New Zealand mud snail was one of 57 foreign species that ocean freighters imported to the Great Lakes during the St. Lawrence Seaway's first 50 years in operation.

spiny water fleas—the creatures formed what looked like prickly caterpillars on fishing gear. Some scientists feared that *Bythotrephes* could take out a link in the Great Lakes food chain. Spiny water fleas ate native zooplankton, and their barbed tails kept predators at bay. That phenomenon could prevent critical nutrients from being passed up the food chain—*Bythotrephes* had the potential to stop that process in its tracks. It was a potentially serious threat to the food chain that supported the multibillion-dollar Great Lakes fishery. Most alarming was the rate at which *Bythotrephes* spread across the lakes. The tiny beasts colonized parts of Lakes Michigan, Erie, Ontario, and Superior within three years of rearing their ghastly heads in Lake Huron. Frightening as it was, in appearance and potential implications, *Bythotrephes* was but a hint of a looming ecological disaster.

A second, more significant species was discovered in 1986. Scientists conducting routine fish monitoring found a species of perch native to Europe and Asia swimming in the steel blue waters of Duluth Harbor. The discovery of Eurasian ruffe—not to be confused with orange roughy—triggered mild panic among some government biologists. Their fears were warranted. The prickly fish, likely imported in the ballast water of a transcontinental ocean freighter, competed with native species wherever it lived. Oh, what a fierce competitor it was. The ruffe spawned up to six times per year, with each female pumping out as many as 200,000 eggs annually. The fish averaged just six inches long, but their spiny dorsal fin shielded them from being eaten by larger fish. Their bony composition kept them from becoming table fare for humans. All of those factors made ruffe a threat to several native fish species in all five Great Lakes. The invader had the potential to cause as much damage as the alewife had 20 years earlier.

The discovery of ruffe in a tributary of Lake Superior, spiny water fleas in Lake Huron, and at least a dozen other invaders in other parts of the Great Lakes fulfilled the predictions scientists had made years earlier. By 2003, the number of invasive species in the lakes had ballooned to 180, a third of which entered after the St. Lawrence Seaway opened, according to a study by scientists at the Smithsonian Environmental Research Center and the National Oceanic and Atmospheric Administration's Great Lakes Environmental Research Laboratory. Researchers Gregory Ruiz and David Reid summarized the rising tide of invasive species in the lakes:

> Over the past two centuries, the Great Lakes basin has been invaded by over 180 nonindigenous species, none of which have been extirpated. Analyses of life history, invasion history in other ecosystems and likely vectors, and opportunities for movement between ecosystems, lead to the conclusion that shipping, particularly ballast water release, is the vector responsible for most aquatic invasions in the Great Lakes since the opening of the St. Lawrence Seaway in 1959. Over 40 percent of all aquatic invaders have been discovered since 1960 and the average rate of discovery from 1960 through 2003 is one new invader every 30 weeks, higher than any other freshwater system for which long-term data exists and placing the Great Lakes among the most highly invaded ecosystems in the world.[13]

Every time a freighter emptied its ballast tanks in Duluth or any other Great Lakes port, it left behind a potential biological time bomb. The ballast water of an average ocean freighter entering the Great Lakes prior to 2006 contained billions of zooplankton, phytoplankton, and bacteria in every cubic meter of ballast water.[14] Collectively, those ships discharged about six million cubic meters of ballast water into the lakes each year. The fleet of ocean freighters that visited the lakes in 2002 deposited 79 million zooplankton, 49 trillion phytoplankton, and 14 quadrillion bacteria from ports overseas into the lakes. Only a fraction of the organisms that transoceanic ships disgorged into the lakes survived. But the bottom line was that ocean freighters dumped a phenomenal volume of biologically rich ballast water into the lakes every year. Logic dictated that as more species spilled into the lakes, the greater the likelihood that one would take hold and multiply in its new confines. The question was when one of those biological bombs would detonate with enough force to expose the myriad of ecological hazards lurking in the ballast tanks of ocean freighters. The awakening came in 1989. Three years after scientists discovered ruffe, a tiny mussel from eastern Europe brought an American city to its knees by cutting off access to the Great Lakes' most treasured asset: water.

The zebra mussel has had perhaps the most profound effect on the Great Lakes ecosystem, second only to human beings.

—Stephen Brandt, director of Great Lakes Environmental Research Laboratory, 2005

08 THE RECKONING

Lake Erie was the Rodney Dangerfield of the Great Lakes: It got no respect. Once derided by comedian Johnny Carson as "the place fish go to die," Erie struggled to escape the stigma of a lake that was so polluted in the 1960s, some people mistakenly declared it dead. A half century later, Erie supported the most vibrant fishery among the five lakes. It contained just 2 percent of the six quadrillion gallons of water in the Great Lakes, but was home to about half the fish (by weight) living in the freshwater seas in 2000. To those unfamiliar with the bounty of aquatic life in its pale green waters, Erie was like the short, homely kid in a family of beautiful children with sleek bodies, straight teeth, and perfect hair. The perception was entirely different among people who understood the enigmatic lake. To them, Erie was a phoenix that rose from the ashes of environmental ruin. Its dramatic recovery from the gross pollution of the 1950s and '60s lured droves of people back to the lake by the 1980s. But just as Erie was approaching the apex of its recovery, an army of invaders from distant waters launched a biological assault that would test the lake's resilience and, once again, bring the lake unwanted international notoriety.

Erie's transformation from industrial cesspool to recreational paradise was in full bloom in the mid-1980s, when Frank and Sandy Bihn built their dream on the shores of the lake's western basin. For Sandy, building within a stone's throw of the lake and Maumee Bay State Park was the ultimate homecoming. It was in Erie's waters that she had learned how to swim and catch perch. Its shoreline was where she had fallen for her future husband, Frank. The love-struck couple had often ended their dates on Erie's shoreline. Given a romance that was nourished by the lake, it was only fitting that they oriented their new house in a way that made Erie the center of their visual universe. The beach in front of the Bihns's new house was narrow but sandy, a lovely place for barefoot strolls. The water in Maumee Bay, which had an average depth of just five feet, was relatively clean but murky—ideal conditions for perch and walleye. Gone by 1980 was the era of rampant water pollution, when massive algae blooms fueled by phosphorus-laden detergents formed huge fluorescent green mats on the lake's surface. Lake Erie had been the poster child for water pollution in the 1960s. By the late '80s, Erie was gaining a reputation for supporting one of North America's best walleye fisheries. For the Bihns, their first year of living at the lake—1987—was about as good as life could get. But their honeymoon with the lake would be short-lived.

The next year, fingernail-sized mussels with jagged brown stripes on their shells began washing up on the shores of western Lake Erie. None of the locals had ever seen that kind of mussel in their treasured lake. Since its stripes resembled those on Bengal tigers, someone proclaimed the mysterious creatures "tiger mussels." The name stuck, for a while. The Bihns didn't have the slightest inclination that tiger mussels were the first hint of a biological plague sweeping across the bottom of Lake Erie. They were on the leading edge of an environmental catastrophe that would spoil thousands of miles of Great Lakes beaches, bring an American city to a standstill, and trigger global efforts to battle a new form of pollution that had nothing to do with chemical discharges or oil spills. Biological pollution, caused by foreign species invading new territories, would prove to be far more vexing and damaging to Great Lakes ecosystems than decades of sewage spills and toxic chemical discharges.

THE GREAT LAKES FISHERY COMMISSION CONVENED A WORKSHOP IN MAY 1988 to examine the possibility that transoceanic freighters were inadvertently dumping foreign species into the lakes when discharging ballast water. Members of the U.S.-Canadian commission, formed in 1955 to manage the war on sea lamprey, were concerned about the discoveries of the spiny water fleas in Lake Huron, and the Eurasian ruffe in Duluth Harbor. They wanted to know if it was possible, technically and biologically, for aquatic species to spread from one continent to another in the ballast tanks of transoceanic freighters. Were the ballast tanks of ocean ships acting as a conduit by which freshwater species from other parts of the globe could sneak into the Great Lakes? The answer was not the one commissioners had hoped to hear.

James Carlton, one of the first U.S. scientists to study whether ocean freighters' ballast water carried foreign organisms around the planet, was the featured speaker at the commission's workshop. His presentation was, to say the least, illuminating. "Jim gave about a twenty minute talk on his research to understand the diversity of live organisms found in ballast water," said David Reid, a Great Lakes researcher for the National Oceanic and Atmospheric Administration who was at the meeting. "His talk pretty much stunned the audience. This was something that had not been looked at."[1] That wasn't entirely accurate. The 1981 Bio-Environmental Services study found 150 species of phytoplankton and 56 species of aquatic invertebrates, including zebra mussel larvae, in the ballast tanks of transoceanic freighters entering the Great Lakes. But the U.S. and Canadian coast guards shelved that report, claiming it was inconclusive. Carlton's speech, coupled with the spiny-water-flea and ruffe invasions, was all the proof the Great Lakes Fishery Commission needed. The commission drafted a letter with the International Joint Commission asking the U.S. and Canadian governments to immediately require ocean freighters to flush ballast tanks with seawater before entering the Great Lakes. It said: "The Great Lakes Fishery Commission wishes to alert its parties [the United States and Canada] to a serious threat to the Great Lakes fishery. Scientists are convinced that harmful organisms are being introduced with discharge of ballast water from oceangoing vessels entering the Great Lakes." In the letter,

D. G. Chance, chairman of the Canadian section of the International Joint Commission, said: "We believe that sufficient studies have been conducted to confirm the threat posed to the Great Lakes and that action is required."[2]

Before the letter could be mailed, two college students made a shocking discovery: zebra mussels had invaded the Great Lakes. Their monumental find was the third example in five years of foreign species entering the lakes in ocean freighters' ballast water. Those invasions were the very problem the Bio-Environmental Services report had predicted seven years earlier. The chickens of biological pollution and government ineptitude were coming home to roost.

On June 1, 1988, two University of Windsor graduate students were studying native mussels on the bottom of Lake St. Clair, a heart-shaped lake a few miles north of Detroit. The lake was a thoroughfare for ships—all freighters crossed the 425-square-mile water body when traveling between lakes Erie and Huron. Lake St. Clair was situated near the middle of a 100-mile-long channel that connected the two Great Lakes. That channel was comprised of the St. Clair River, Lake St. Clair, and the Detroit River. Nearly all the water from Lakes Michigan, Superior, and Huron flowed into the St. Clair–Detroit River connecting channel before flowing into Lake Erie. Lake St. Clair's broad, shallow water was a playground for boaters and anglers. A dredged shipping channel allowed freighters to travel safely across the lake's shallow waters, which averaged just 10 feet deep. It was in one of Lake St. Clair's undisturbed shallow areas that the college students pulled up a rock to which clung a tiny, striped mussel. No one, not even their supervising professor—University of Windsor biologist Paul Hebert—could immediately identify the creature. They photographed the peanut-sized mollusk and carried on with their work. "We had no idea what is was, except that it was something new and very strange," Hebert said. "I figured it'd be the last and only one I'd ever see."[3]

Hebert identified the mussel after returning to his laboratory at the University of Windsor. It was *Dreissena polymorpha*, a zebra mussel. Theirs was an ominous discovery, the human equivalent of someone learning they had an aggressive, untreatable form of cancer. Zebra mussels, which were native to rivers in Russia that flowed into the Black Sea and Caspian Sea, were notorious for spreading rapidly in fresh water and saltwater environments, disrupting ecosystems and clogging water intakes. How did zebra mussels get to the Great Lakes from Europe? There was only one viable possibility: transoceanic freighters that shuttled cargo between Europe and North America.

Navigation canals built in the 1700s and 1800s allowed zebra mussels to spread from the Black and Caspian seas in eastern Europe to coastal ports in western Europe. Among the areas zebra mussels conquered were coastal ports where transoceanic freighters loaded cargo and ballast water before heading to the Great Lakes. Because *Dreissena polymorpha* was a euryhaline species—one capable of living in freshwater and saltwater—it had no trouble expanding its range across Europe. From there, it was just a matter of time before the mussels crossed the Atlantic in the bellies of transoceanic freighters that took on ballast water in Europe and spit it out in the Great Lakes.

The zebra mussel's arrival in North America was a crushing body blow to the world's largest freshwater ecosystem. The phenomenally prolific species was capable of reconfiguring entire ecosystems and causing economic distress by clogging water-intake pipes critical to industries, power plants, and municipal water systems. *Dreissena polymorpha* was the type of creature a mad scientist from the former Soviet Union might have bragged about, had he intentionally planted the mussels in the Great Lakes during the cold war. The truth was far less intriguing. Ocean freighters sucked larval zebra mussels into their ballast tanks while docked in Europe and expelled them while discharging that same ballast water in the Great Lakes. The importation of zebra mussels was a tragic accident—it was also predicted and entirely preventable.

The mollusks settled first in Lake St. Clair, using their super-clingy byssal threads to attach to rocks. There, the mussels proceeded to churn out millions of offspring. Each female mussel produced up to one million larval offspring, veligers, every year. The microscopic veligers drifted for days with water currents until they found a suitable place to settle; that process enabled the mollusks to quickly expand their range and numbers.

Hebert predicted in October 1989 that zebra mussels would quickly spread across the Great Lakes. Within a year of Hebert's landmark discovery, zebra mussel densities in Lake St. Clair exploded from 200 per square meter to 10,000 per square meter. It wasn't a question of if, but when zebra mussels would spread to the other Great Lakes. What really gnawed at Hebert was the knowledge that the zebra mussel invasion might have been prevented. Hebert told the *Toronto Star* that the U.S. and Canadian coast guards could have headed off the invasion if they had acted on the 1981 study by Bio-Environmental Services. That study revealed that ocean ships were carrying millions of foreign organisms—including zebra mussel veligers—into the Great Lakes. Hebert believed zebra mussels would not have invaded the lakes if the U.S. and Canadian governments had prohibited ocean ships from discharging untreated ballast water into the lakes in 1981. "I think it was a very, very expensive mistake," Hebert told a Toronto newspaper in 1989. "There's just no question this will be a multimillion dollar cost to the Canadian population forever. I think it's unacceptable that when this report was in, no action was taken upon it. It didn't take someone with a crystal ball to recognize this problem."[4]

A Canadian Coast Guard official told the *Toronto Star* that the agency hadn't acted on the Bio-Environmental Services study because the potential for foreign shipborne species to colonize the Great Lakes "wasn't clear." His comment was complete nonsense. Let's review the facts. The Bio-Environmental Services report said: "The results of the study clearly indicate that nonindigenous and nonendemic aquatic species are being imported into the Great Lakes system." The report could not have been any clearer or more direct.

Hebert's initial assessment that zebra mussels would cost Canadians millions of dollars as the invaders spread across the Great Lakes greatly underestimated the magnitude of the problem. Within a decade, zebra mussels had become a billion-dollar problem for power plants and municipal water systems that sucked water out of the Great Lakes. The potential scope of the problem came into focus on a frigid December morning in 1989, two months after Hebert warned that the mollusks would wreak havoc. Few people knew in 1989 that the mussels were spreading like a fierce rash across the bottom of Lake St. Clair and Lake Erie. How could such a massive infestation go undetected? Simple: The marauding army of mussels went about their business on lake bottoms that humans rarely visited. A catastrophe, a crisis involving one of the most basic human needs, was required for the U.S. and Canadian governments to take the foreign invaders seriously. By then, it was too late to prevent or slow the zebra mussel's conquest of the Great Lakes, and hundreds of other lakes and rivers across North America.

THE CITY OF MONROE WAS A GRITTY MICHIGAN COMMUNITY OF 50,000 people located south of Detroit, on the west end of Lake Erie. It was home to the Enrico Fermi nuclear power plant (one of the inspirations for David Bowie's song "Panic in Detroit"), the birthplace of General George Armstrong Custer (he of Custer's last stand), and a haven for anglers and boaters. In 1989, the city became the epicenter of a biological invasion that influenced the course of world events. No one saw it coming.

Employees at Monroe's water-filtration plant were conducting routine maintenance of the city's Lake Erie water intake in January 1989 when they discovered dozens of tiny mussels clinging to the inside of the pipe. They didn't attempt to identify the mollusks, the likes of which they had never seen before. Their job was to clean the inside of the water intake, which extended more than a mile into the lake. Crews removed the mussels and went about their business. They figured the problem was solved. Clearly, they didn't grasp the significance of the striped mussels. Wilfred LePage, who was superintendent of the Monroe water utility when zebra mussels showed up, recalled the events in a 1993 essay.

"Little significance was attached to the animal's presence since it was not uncommon for small gastropods and crustacea to pass through the intake screens and traverse the nearly 10 miles of pipe to the treatment plant," LePage said. He and his employees were about to get a crash course in zebra mussel ecology. The mussels laid siege to the city's water intake pipe six months after they were discovered. By July 1989, a growing layer of mussels inside the 30-inch diameter intake reduced water flow by 20 percent. The water plant's ability to pump adequate amounts of water rapidly deteriorated from there. The mussels made their presence felt in an unmistakable way on September 1, 1989. "Suction was suddenly and totally lost," LePage said.[5] "The plant had been pumping water at the rate of 10 million gallons per day . . . when suddenly, the raw water pumps totally lost suction. We lost everything; not a drop of water was to be had."[6] Crews worked for 16 hours to restore pumping capacity to 80 percent of normal. But the problem returned three months later, reaching a climax that cut off all water service in Monroe and made zebra mussels a household term.

The day of reckoning for Monroe's beleaguered water system arrived December 14, 1989. It was a bitterly cold day. Air temperatures hovered in the teens, and wind chill readings dipped to 10 below zero—a very bad day for a water crisis. The frigid air caused tiny shards of ice to form well below the frozen surface of Lake Erie. Those ice shards, called "frazzle ice," were a recipe for disaster when combined with the thick layer of zebra mussels that had formed in the three months since city workers had last cleaned Monroe's water intake. The mussels and ice formed a cork in the pipe that supplied water for every household, school, and business in the city. Deprived of all water, city officials promptly declared a water emergency. They went on radio and television stations that morning to alert residents to the dire situation. Officials also broadcast the news over loudspeakers that had been scattered throughout the community years earlier to warn residents in the event of an emergency at the Fermi nuclear power plant.

"Water crisis cripples area," read a large headline atop the *Monroe Evening News* the next day. City officials blamed the problem on ice in the water intake in Lake Erie and a pipeline break somewhere in the city. They surmised that zebra mussels were contributing to the problem. All water users in the city and surrounding areas were instructed to boil drinking water for three minutes to kill any potentially harmful bacteria. Students in all the local schools and the community college were sent home. All restaurants and bars were ordered closed. All elective surgeries were canceled at Mercy Memorial Hospital, and several patients were sent home.[7]

City workers scrambled to extend temporary water lines to the Raisin River, which would be used as a backup supply in the event that the mussels clogged the Lake Erie water intake for an extended period of time. That was far from an ideal solution. The river was polluted and certainly not fit for drinking without being boiled first. Faced with an environmental crisis that threatened public health and safety, the city tapped into the south Monroe County Water System, which obtained water from the city of Toledo's water system. Those emergency measures only provided Monroe with a fraction of the eight million gallons of water needed each day to keep the city humming.

The water crisis was a boon for local grocery stores, which quickly sold out of bottled water and other beverages. At one point during the first day of the crisis, the local grocery store sold 400 gallons of bottled water in a half hour. Grocers prospered while restaurants and bars that were ordered to shut their doors lost countless revenue. At Navarre's party store, owner Pete Navarre told the local newspaper that the water crisis cost him $3,000 in lost revenue in one day. "And I can't make it up," Navarre said. "It's a major loss, no doubt about it." Managers at the Pier House restaurant were forced to cancel several holiday Christmas parties. "Needless to say, it's a catastrophe,"

Michigan Sea Grant

Zebra mussels invaded the Great Lakes in the late 1980s. This photo shows a water pipe that was clogged by the foreign mussels.

Photo by the *Saginaw News*, courtesy of the *Muskegon Chronicle*

Zebra mussels cover a rock pulled out of Lake Huron's Saginaw Bay.

manager Terry Cryan told the local newspaper.[8] Angry, fearful residents overloaded phone circuits as they bombarded city officials with questions. Michigan Bell reported that Monroe residents placed 16,000 calls per hour during the height of the crisis, triple the normal volume.[9] *Dreissena polymorpha* taught Monroe residents and city officials that they could no longer take Lake Erie's seemingly endless bounty of water for granted. Suddenly, there was competition for the priceless natural resource.

The December 16 edition of the *Monroe Evening News* placed the blame for the water crisis squarely on zebra mussels. "Mussels worsen water crisis," the newspaper declared in a front-page headline. "That's the only condition that's changed," LePage told the newspaper.[10] By the time the crisis ended, zebra mussels and ice had choked Monroe's Lake Erie water intake for 56 hours. It was the first time since 1924 that Monroe's water system had been knocked out of service. Almost overnight, zebra mussels went from a largely unknown organism in North America to the poster child for invasive species in the Great Lakes. The mussel invasion also shone a bright light on a problem caused by ships moving cargo around the globe—biological pollution.

Monroe's water system was back to normal four days after the crisis erupted. But unbeknownst to city officials, the defeated mussels were preparing to strike again. In April and May of 1990, crews ran mechanical pipe cleaners through the water intake to remove a blanket of zebra mussels that was six inches thick in some areas. The machines did a fine job of cleaning out the pipe, but all the dead mussel shells flowed into the water treatment facility, clogging screens that kept debris out of the filtration system. Workers used a massive vacuum cleaner to extract several tons of zebra mussel shells. Crews then extended a small plastic pipe through the middle of the 10-mile-long water intake. The feeder line delivered 500 pounds of chlorine per day to the end of the water intake, bathing the inside of the pipe with a chemical that was toxic enough to kill existing zebra mussels and prevent veligers from clinging to it. But the chlorine treatment created another problem.

Mussels killed by the chemical clogged filter screens at the water treatment plant three days after the treatment began. The dead mussels congealed into a slimy mass that coursed through the water treatment facility. "The material was unlike anything previously encountered," LePage said. "It consisted of fibrous or threadlike material intermixed with a seemingly gelatinous binder and generously interspersed with whole and broken mussel shells. It formed a mat on the screens and severely impaired flow."[11] The slime dissipated within a few days without curtailing the flow of water. Still, the city was not out of the woods. An inspection in September 1990, three months after the city began chlorinating the water intake, revealed a huge accumulation of live zebra mussels on the bottom of Lake Erie. The blanket of mussels extended 50 feet in all directions from the underwater structure, called a crib, that held the intake pipe in place. Workers discovered that parts of the crib that had been completely cleaned of zebra mussels three months earlier were totally covered with young mussels. The chlorine treatment was keeping the water pipe free of mussels, but the mussels still swarmed the intake crib. The discovery was troubling, but minor to what city officials observed in January 1991.

Conventional wisdom at the time held that zebra mussels could not survive in water temperatures colder than 50 degrees Fahrenheit. Monroe city officials assumed it was safe to stop the chlorine treatment of the water intake on November 20, 1990, long after water temperatures dipped below 50 degrees. They planned to resume the chlorine treatment when lake water temperatures neared 50 degrees the following spring. But their bubble was burst in January 1991. A routine inspection of one of the suction wells at a pumping station turned up 600 live zebra mussels. Nervous city officials immediately reinstated the chlorine treatment to keep mussels from reinfesting the water intake. Treating the intake with chlorine on a continuous basis, and routinely scraping away mussels that survived the chemical warfare, kept Monroe's water system operating from then on. But zebra mussels were a costly problem. Dealing with the pests from 1989 through 1992 cost the city $321,795. Coupled with the cost of installing a new water intake, Monroe's zebra mussel bill totaled $787,000. All of that expense was borne by the city's water customers. Shippers who owned the ocean freighters that imported zebra mussels to the Great Lakes never reimbursed the city one cent for its costly war against the tiny but mighty invader. Biological pollution turned out to be a perfect environmental crime. No one could pinpoint which ship delivered zebra mussels or any other invasive species to the Great Lakes. In fact, no one bothered to try—there were too many suspects to make a case against any one ship.

Thirty years after dead alewives washed up on Great Lakes beaches, zebra mussels drove the issue of biological pollution back onto the front pages of newspapers across the region. Government agencies in the United States and Canada quickly jumped on the zebra mussel bandwagon, sounding the alarm and calling for prompt action and huge expenditures to stop the flow of foreign species into the Great Lakes. Bureaucrats determined to cover their own backsides never mentioned their failure to take action years earlier that could have prevented the zebra mussel invasion. Their inaction was a despicable blunder that allowed one of the world's most pernicious invaders to conquer the Great Lakes and, in the process, establish a North American beachhead. The mussels would eventually spread beyond the boundaries of the Great Lakes and infest lakes and rivers across the continent, from California to Quebec. The situation was thoroughly distressing, ecologically disruptive, and pathetic.

"If we had ballast exchange in place in the 1980s, my guess is that we would not have had the zebra mussel invasion," said James Carlton, a Williams College marine ecologist and invasive species expert.[12] Instead, the lakes were the recipients of what Carlton called "one of the outstanding invasions of North America in the past 200 years."

MONROE'S TRAUMATIC ENCOUNTER WITH ZEBRA MUSSELS WAS MERELY THE tip of the invasive-species iceberg. The mussels spread quickly across Lake Erie and increased their numbers exponentially in a matter of months. A few weeks after Hebert's mussel discovery, a group of limnology students sampling the bottom of Lake Erie near Ohio State University's Stone Laboratory had a "eureka" moment. They found a few zebra mussels clinging to a dock. Jeff Reutter, director of Stone Laboratory and Ohio's Sea Grant program, promptly launched a study of the mussels. Researchers scouring the bottom of Lake Erie the following year found huge clusters of zebra mussels; in some areas, there were one million mussels per square meter. Reutter feared that the lake's ecosystem was about to change at warp speed. He and his colleagues raced to document how the plague of mussels was altering Lake Erie and threatening its ecological bounty. They couldn't work fast enough to keep pace with the changes. Reutter explained: "We started a three year study of how zebra mussels were affecting native clams but had to stop the study after two years because all the native mussels were gone. Any time you knock out a native species there are going to be repercussions for the ecosystem—those species that were eliminated have impacts on other critters."[13]

Bewildered scientists watched helplessly as conditions in Lake Erie changed dramatically from one year to the next. Water clarity increased twelve-fold in the six years after zebra mussels arrived. The change was due to the mollusks' ability to filter huge quantities of plankton out of the water. A single zebra mussel filtered up to one liter of water each day. That was the human equivalent of a 150-pound person drinking 36,000 gallons of water each day, enough to fill a backyard swimming pool. Multiply the filtering capacity of one zebra mussel by, say, one trillion of their kin and you get the idea. "There were enough mussels in the western basin of Lake Erie to filter the entire volume of water in the basin every four to seven days," Reutter said. "That didn't mean they filtered all the water in the [2,500-square-mile] basin every week, but that they had the capacity to do that."

The mussels removed tremendous quantities of plankton, phosphorus, and chlorophyll from the water as they filtered it through their bodies. The mussels then deposited those nutrients on the lake bottom. Imagine a glass of chocolate milk, made with powdered cocoa, left untouched for an hour. What happens? The chocolate particles settle on the bottom of the glass. That's what zebra mussels did—they pulled all the nutrients in the water column down to the lake bottom. What the mussels didn't ingest they defecated on the lake bottom in tiny feces covered in slimy mucus. That didn't bode well for fish and other aquatic life: the algae and tiny invertebrates that zebra mussels hogged were the foundation of the Great Lakes food chain. The increased water clarity that resulted was aesthetically pleasing to swimmers and boaters, but posed major problems for fish that loathed clear water. Walleye were among the first victims. Clear water was the enemy of the crown prince of Lake Erie's sport fishery. The potential ecological and economic implications of a clearer Lake Erie were enormous and, at the time, beyond the understanding of even the most educated scientists.

At the same time zebra mussels were changing the chemical composition of Lake Erie, billions of their dead kin washed up on beaches along Lake Erie, Lake St. Clair, and Lake Huron's Saginaw Bay. The shells smothered once-sandy beaches with layers of sharp shells that were two feet deep in some areas. Shoreline property owners watched, aghast, as layers of dead zebra mussels blanketed their beaches. Elsewhere, utilities in Ohio and Detroit were waging fierce battles to keep the mollusks from clogging water intakes and shutting down coal-fired and nuclear power plants. Zebra mussel densities in the water intakes at Detroit Edison's Monroe power plant reached 800,000 mussels per square meter in 1989. The mussels coated nearly every hard surface in the utility's complex water intake system. Detroit Edison had a crisis on its hands. If a strong southwest wind triggered one of Lake Erie's legendary seiches—which could lower water levels in the western half of the lake by several feet in a matter of hours—officials would have to shut down the Monroe facility, causing widespread power outages. "Had a major seiche occurred in 1989, the plant probably would have lost water," plant engineer William Kovalak said in a 1993 review of the incident.[14]

Zebra mussel densities were lower at Detroit Edison's Fermi nuclear power plant, which was six miles northeast of the Monroe power plant. Still, the presence of the mussels put utility officials on edge. A sudden loss of water pressure at the nuclear power plant could have caused the nuclear reactor to overheat and, potentially, suffer a meltdown.

On the south side of Lake Erie, along the shores of Maumee Bay, Frank and Sandy Bihn watched in disbelief as the lake went from being murky and pale green in 1987 to looking more like a swimming pool by the early 1990s. "You could see your feet when you were waist deep in the water, that's how clear it was," Sandy said. But there was a catch. The mussels that made the lake clearer had razor-sharp shells. Their conquest of the lake made walking barefoot in the suddenly clear waters a hazardous proposition. "The shells were sharp as hell; they cut your feet," Sandy said. "After the mussels arrived, you had to wear shoes to go in the lake." For people who relished the feeling of warm sand and cool water on their bare feet, wearing shoes while swimming in the Great Lakes was like wearing a parka at the beach on a warm summer day.

Lake Erie's zebra mussel crisis sent ripples of fear across the eight U.S. states and two Canadian provinces that ringed the Great Lakes. Erie, after all, was the bellwether of change in Lakes Ontario, Michigan, Huron, and, to a lesser extent, Superior. What happened in Lake Erie's waters rarely stayed in Lake Erie—it usually spread to one or more of the other lakes. Scientists and policymakers knew something had to be done quickly to keep other invasive species from sneaking into the Great Lakes in the bowels of ocean freighters. Nothing less than the biological integrity of the freshwater seas, and the massive industrial and recreational economies they supported, was at stake.

The zebra mussel, which scientists later determined had invaded Lake Erie in 1986, was the first notable invader in a flood of aquatic invasive species that poured into the Great Lakes after the fall of the Soviet Union increased trade between Eastern Europe and the United States. Many more would follow. Twenty years after it was discovered, many scientists considered the zebra mussel and its heartier cousin—the quagga mussel—the most destructive foreign species ever to invade the lakes. The sea lamprey devastated Great Lakes fisheries, but zebra and quagga mussels transformed every level of the massive freshwater ecosystem. Researcher Michael L. Ludyansky compared the mussel invasion to legions of marauding Russian soldiers. "Just when the threat of Russian invasion of North America seemed to have disappeared with the end of the Cold War, an invasion has been found to be not only under way but proving to be successful. Rather than missiles, a naval force of hordes of zebra mussels has secured beachheads in many U.S. and Canadian lakes and rivers. One could call it biological warfare, but not directed by any human admiral."[15]

The United States won the cold war, but lost the battle to keep fish and mussels native to the former Soviet empire from colonizing the Great Lakes. Then again, when it came to battling invasive species, the United States and Canada didn't put up much of a fight in defense of the world's largest freshwater ecosystem—until the war was lost.

MONROE'S ZEBRA MUSSEL CRISIS STRUCK FEAR INTO THE HEARTS OF officials at hundreds of cities and industries that relied on Great Lakes water for their very survival. In the eighteen months after the mussels idled the city of Monroe, the issue of invasive species sneaking into the lakes via ocean freighters went from a fringe concern to an emerging economic and ecological disaster. Canada implemented voluntary guidelines that asked shippers to exchange ballast water at sea before entering the St. Lawrence River. It was an attempt to reduce the risk of more freshwater species from ports overseas sneaking into the Great Lakes. Sensing that Canada's voluntary ballast water exchange guidelines were inadequate, the International Joint Commission and Great Lakes Fishery Commission pressed their case again. In a joint 1990 report, the commissions urged the United States and Canada to immediately adopt mandatory ballast water exchange rules for every ocean freighter entering the lakes. The report said:

"The commissions conclude that immediate action is required by governments [*sic*] to reduce the risk of unwanted exotic species being introduced to the Great Lakes ecosystem through the discharge of ballast waters from oceangoing ships. Given the risks associated with the introduction of exotic species, the discharge of ballast water in the Great Lakes and connected fresh and brackish waters must become a privilege granted only to those ships that have taken reasonable and acceptable precautions to prevent ballast-borne introductions."[16] Two months later, the U.S. Congress approved the Nonindigenous Aquatic Nuisance Prevention and Control Act, known as NANPCA. It was the world's first federal law regulating ballast water. Its goals were as follows:

> To prevent unintentional introduction and dispersal of nonindigenous species into waters of the United States through ballast water management and other requirements; to coordinate federally conducted, funded or authorized research, prevention control, information dissemination and other activities regarding the zebra mussel and other aquatic nuisance species; to develop and carry out environmentally sound control methods to prevent, monitor and control unintentional introductions of nonindigenous species from pathways other than ballast water exchange; to understand and minimize economic and ecological impacts of nonindigenous aquatic nuisance species that become established, including the zebra mussel; and to establish a program of research and technology development and assistance to states in the management and removal of zebra mussels.[17]

The law was a huge, albeit belated, step into the uncharted waters of regulating ballast water. NANPCA came 17 years after the U.S. Environmental Protection Agency refused to regulate ships' ballast water discharges under the Clean Water Act. The EPA's ill-advised decision created a regulatory vacuum, with no rules governing how ships disposed of ballast water between 1973 and 1990. The 1972 Great Lakes Water Quality Agreement between the United States and Canada designated the coast guards of both nations as the lead agencies for regulating ballast water. Neither agency took the initiative. Absent any regulations, the NANPCA law was a good first step toward slowing the flow of shipborne invaders. Sadly, the law was enacted two decades too late.

Although it is unlikely that invasions by the ruffe would destroy the nation's entire freshwater sport fishery resources, even an overall loss of only 10 percent would cost over $7 billion, a significant economic impact.

—Great Lakes Fishery Commission's Ruffe Task Force, 1992

09
RUFFE SEAS

The calendar said it was the first day of autumn, but the air temperature felt more like summer when I joined an international trade delegation on a tour of the Port of Duluth. Beads of sweat rolled down our faces as we cruised slowly past massive freighters and towering grain elevators aboard the 92-foot-long *Vista Star*. With the temperature spiking at 81 degrees—19 degrees above normal—several of the visiting grain buyers took shelter in the tour boat's spacious lounge to avoid a sunburn. For the record, it was September 24, 2007, the fourth week in a month that has been known to produce blizzards in the northern Minnesota port city. Not that year. An unusually hot, dry summer that dropped Superior's water level to a record low was spilling into fall. The impossibly bright sun glinted off the dark blue water of the St. Louis River estuary as the tour guide educated the visitors from fifteen nations about the wondrous, mysterious, and dangerous Lake Superior. Their journey to the Great Lakes' busiest port was part of an annual effort to increase America's grain exports to Europe, Asia, the Middle East, and Africa.

The grain buyers traveled around the upper Midwest for nine days as part of a well-orchestrated sales pitch on the benefits of buying grain from American farmers and shipping it across the Atlantic via the Great Lakes and St. Lawrence Seaway. Freighters using the Seaway, they were told, could transport grain cheaper than what it would cost to ship it out of ports on the West Coast or the Mississippi River. As the foreigners stumbled in broken English to discuss grain quality and transportation costs with their American hosts, my eyes and mind were fixed on the harbor's opaque water. There were invaders from distant seas in the harbor. I couldn't see them, but they were most certainly there, lurking among the rocks on the harbor's undercarriage: Eurasian ruffe, millions of them. The small, spiny, aggressive fish—a species of perch native to the Black Sea in Europe—snuck into the harbor two decades earlier in the ballast tanks of an ocean freighter. A ruffe population that started with a few specimens ballooned to six million within four years; ruffe became the dominant species in the lower St. Louis River's 13,000-acre estuary, which formed Duluth Harbor. The ruffe's conquest of the harbor was the price western Lake Superior's ecosystem paid for Duluth being the Great Lakes' busiest port.[1]

The discovery of ruffe in Duluth Harbor wasn't big news initially. Even though ruffe had been discovered in the Great Lakes two years before zebra mussels, the mussels stole the spotlight in 1989 by clogging municipal and industrial water intakes. That didn't mean the ruffe was a benign invader. The foreign fish posed a potentially huge threat to one of the lakes' most beloved species: perch. Ruffe were also significant for another reason. Its spread across western Lake Superior was the first test of the Nonindigenous Aquatic Nuisance Prevention and Control Act. The 1990 U.S. law, a direct response to the zebra mussel and ruffe infestations, was supposed to prevent future invasions of the Great Lakes by foreign species. At the time of its passage, the law was the most powerful tool U.S. officials had in their regulatory arsenal for waging war against invasive species in the lakes.

A cornerstone of the 1990 law was ballast water exchange, which was voluntary for three years. After 1993, all transoceanic freighters entering the

Jeff Alexander

Ship lovers line the channel wall in Duluth to watch a transoceanic freighter from Europe enter the port of Duluth-Superior. More foreign ships visit the port of Duluth-Superior each year than any other Great Lakes port.

Seaway had to conduct a mid-ocean ballast water exchange at least 200 miles offshore of Newfoundland, and in seas at least 2,000 meters deep.

The rationale for exchanging ballast water at sea was simple. Most freshwater species couldn't tolerate salt water, and most saltwater species couldn't tolerate fresh water. Freighters destined for the Great Lakes from overseas often inhaled ballast water in freshwater, coastal ports. In the process of sucking in millions of gallons of ballast water, ships inhaled all that lived in the water: fish, mussels, plankton, and pathogens. Under the new law, ocean freighters had to replace freshwater ballast with sea water before reaching the Gulf of St. Lawrence. Mid-ocean ballast water exchange was supposed to flush freshwater species out of ballast tanks. Saline shock was expected to kill any freshwater organisms that survived the process. Any saltwater species inhaled during the ballast exchange would instantly die when discharged into the fresh waters of the Great Lakes. The process wasn't designed to keep ocean species out of the Great Lakes. Rather, the goal was to prevent freshwater species from other continents from reaching the lakes, where some could thrive and threaten native aquatic organisms. Biologists believed ballast water exchange was the easiest, cheapest, and fastest way to shield the lakes against future introductions by nonnative species, such as the Eurasian ruffe.

The authors of the 1990 law specifically referenced ruffe, saying: "Since their introduction in the early '80s in ballast water discharges, ruffe have caused severe declines in populations of other species of fish in Duluth Harbor . . . and are likely to spread quickly to most other waters in North America if action is not taken promptly to control their spread."[2] Let that last sentence marinate in your mind for a few seconds. The law made it clear that ruffe would likely spread across the continent "if action is not taken promptly." Authors of the legislation realized that surface waters in the United States and Canada faced a storm of ruffe invasions unless something was done to slow its spread. At issue was whether federal officials would back up their rhetoric on foreign species with aggressive action to combat the invaders.

The underwater battle for control of Duluth Harbor was over by the time the law took effect. Ruffe were already the dominant fish species in the lower St. Louis River in 1990. The river, which formed Duluth Harbor, was Lake Superior's largest tributary. Almost overnight, ruffe turned the river's mouth at the west end of Lake Superior into its North American beachhead. Left unchecked, the fish would likely spread beyond the harbor, migrate along the southern coast of Lake Superior, and eventually invade the other four Great Lakes. Once in Lakes Michigan, Huron, Erie, and Ontario, ruffe would have access to a vast assemblage of rivers and lakes across the eastern United States. Though it was too late to save Duluth Harbor from the wrath of ruffe, there was a narrow window of opportunity to contain the troublesome species. Biologists figured they had four years, until about 1994, to control the burgeoning ruffe population. Beyond then, scientists would have to surrender to the likelihood of ruffe spreading across the continent. Officials from the states of Wisconsin and Minnesota, along with the U.S. Fish and Wildlife Service, faced a difficult decision. Would they strike back aggressively at the ruffe, as their predecessors had done successfully 40 years earlier to prevent sea lamprey from eliminating lake trout in the lakes? Would they attempt to kill all ruffe in Duluth Harbor and the St. Louis River? Or would they sit on their

John Lyons, Wisconsin Department of Natural Resources

Eurasian ruffe (*shown above and below*) were discovered in Lake Superior's Duluth harbor in 1985. The foreign fish, a distant relative of the yellow perch that are native to the Great Lakes, became the most abundant fish species in Duluth harbor within a few years of its arrival.

Minnesota Sea Grant

bureaucratic hands and surrender the world's largest lake, indeed the fresh waters of North America, to a foreign fish species with no redeeming value? It was a seminal moment in the early days of the war against an invasive species. Would the government's response to ruffe take the battle to the invaders, or would it prove to be little more than hollow political rhetoric? Would the war on ruffe set the tone for all subsequent efforts to contain or eliminate new invaders? Answers to those questions emerged during a prolonged, divisive debate about how to deal with the ruffe's expanding population in Duluth Harbor and western Lake Superior.

No one knew precisely when ruffe snuck into Duluth Harbor, or which transoceanic ship dumped the little beasts into the St. Louis River while emptying its ballast tanks. The first sighting of the fish came in 1986, when a team of private-sector biologists found several larval specimens. They didn't know what to make of their discovery—they had never seen a fish like the ruffe. A year later, government biologist Dennis Pratt was conducting routine fish sampling in the harbor when he found several full-grown ruffe, each about six inches long. It was a discovery of the worst kind, a flashback to an earlier era when other foreign fish—sea lamprey and alewife—ruled the Great Lakes. The ruffe was a biological menace—a spiny, bony fish with no commercial or recreational value. Biologists feared that the prolific, domineering fish would compete with valuable native species, such as perch and walleye. Pratt summed up his reaction to discovering ruffe in a single, mournful sentence: "It was like finding out that your wife has AIDS," he said.[3]

In March 1988, biologists shared the news of their dismal find with a committee of scientists responsible for managing the Lake Superior fishery. Their message was clear: Something had to be done, quickly, if the U.S. and Canadian governments hoped to keep ruffe from spreading across Lake Superior and invading the four other Great Lakes. The committee faced a difficult choice: do nothing, or allow biologists to treat the lower St. Louis River and Duluth Harbor with rotenone, a compound that caused fish to suffocate and die. Rotenone would have killed all fish in the harbor, creating a ghastly scene. But it was the best hope for wiping out the ruffe population. Native species could be restocked after the invaders were eliminated. Time was of the essence. The longer government officials waited to strike back at the ruffe, the less likely they were to reduce its numbers.

Instead of moving quickly and aggressively, the committee opted to do nothing. The battle to control ruffe, in the committee's view, was already a lost cause.[4] U.S. Fish and Wildlife Service biologist Thomas R. Busiahn, who lobbied for the chemical treatment, explained the committee's decision-making process in a retrospective analysis of the case. He wrote:

"The [Lake Superior] committee seriously considered treating the St. Louis River with rotenone to eradicate ruffe. The committee decided not to treat to eradicate for two reasons. A treatment would be costly, probably over a million dollars for the chemical alone. The second and deciding factor was the judgment that ruffe could not be eradicated by treatment of the very complex estuary. Essentially, the managers decided that eradication, even at that early point, was not feasible, and they accepted as inevitable that ruffe would be a permanent part of the North American fauna."[5]

The Lake Superior committee surrendered without firing a single shot at the destructive invaders. The panel threw in the towel just as biologists were urging government agencies to raise the battle flag and launch an all-out attack on ruffe. Imagine what the Great Lakes fishery would be today if biologists in the 1950s had given up the fight against sea lamprey because it was difficult, expensive, and required a long-term commitment. Had those ingenious, determined biologists surrendered, there would be no lake trout, whitefish, or salmon in any of the lakes. Fortunately, state and federal agencies took up the call to arms against sea lamprey and launched a chemical war on the parasites that continues to this day. Gaining the upper hand on sea lamprey wasn't quick, easy, or inexpensive. Government and industry chemists spent four years testing 6,000 different chemicals before they found one that killed sea lamprey while causing minimal casualties among native species. The United States and Canada—working together under the auspices of the Great Lakes Fishery Commission—have spent more than $300 million since 1955 to control the lamprey population.

Sea lamprey may never be eradicated—there are still hundreds of thousands of the bloodsucking fish in the Great Lakes and their tributaries. But lamprey numbers were reduced to a level that permitted the restoration of lake trout in Lake Superior, allowed whitefish to thrive, and paved the way for the successful introduction of Pacific salmon in Lake Michigan. The payoff: the Great Lakes in 2008 supported a sport and commercial fishery valued at $7.5 billion. A new group of invaders that had stormed the lakes a half century after sea lampreys threatened the fishery in new and equally destructive ways.

NO. THAT WAS THE LAKE SUPERIOR COMMITTEE'S ANSWER TO THE REQUEST to use rotenone to kill ruffe in Duluth Harbor. Defeated but undeterred, a group of state and federal biologists formed an informal committee in 1988 and took matters into their own hands. Step one was persuading lawmakers to pass a law banning the possession of ruffe in Wisconsin and Minnesota. Done. The biologists then took steps to bolster native fish species in the harbor—they hoped the natives would regain control of the fishery in the harbor and the St. Louis River estuary. The states of Minnesota and Wisconsin then reduced the number of walleye, muskies, northern pike, and largemouth bass that anglers could take from ruffe-infested waters. The two states also increased stocking of native fish species in the river, from 4.5 fish per acre to 17 fish per acre. It was a clever effort. But like so many other efforts to eradicate invasive species, it proved to be futile.

Biologists in 1991 estimated that there were 1.8 million ruffe spawning in the St. Louis River estuary. In four years, the European cousin of native yellow perch had become the most abundant fish species in the mouth of Lake Superior's largest tributary. Worse, there was nothing preventing the ruffe from spreading across Lake Superior and invading the other Great Lakes. Still, none of the numerous government agencies working on invasive species in the early 1990s launched a counterattack that might have contained the invader to Duluth Harbor.

"No ruffe control initiatives were undertaken for three years after the initial response by the states and the [Fish and Wildlife] service," Busiahn said. "The [predator] stocking program and angling regulations would require time to take effect. The success of the ruffe colony was yet to be seen. The Lake Superior fishery agencies watched and waited. By 1991, it was apparent that ruffe were very successful. The population in the St. Louis River was reproducing rapidly."[6]

Ruffe accounted for more than half of all fish in the lower St. Louis River, by weight, by 1992. Its population mushroomed to six million in 1999.[7] Native fish species suffered in the shockwave of the ruffe's population explosion. The number of perch, shiners, trout, and bullheads in the harbor declined significantly as ruffe rose to dominance. Having conquered the estuary and harbor, the ruffe spread into western Lake Superior and other tributaries that fed the lake. Its march across the lakes had all the trappings of an emerging ecological disaster. Yet, calls for help fell on deaf ears in the upper echelons of the U.S. Fish and Wildlife Service, the federal agency charged with preserving and protecting the nation's flora and fauna.

The Great Lakes Fishery Commission asked the U.S. and Canadian governments in 1992 to provide $1.8 million annually to contain ruffe to Duluth Harbor. The funding request came with a report from the commission's Ruffe Task Force, which warned that the invasive fish were a threat to some of the Great Lakes' most valuable fish species: perch, walleye, and whitefish. The worst case scenario, biologists said, was that the ruffe would spread to all five Great Lakes and cause up to $2 billion damage to fisheries and related recreational activities. The task force also warned that the ruffe could cause $7 billion damage to the nation's sport fisheries. "The Ruffe Task Force urges that immediate action be taken to prevent further introductions or distributions by shipping or other human activities."[8] If no action was taken by 1993, the task force said, the ruffe would be free to expand its range across southern Lake Superior and move toward the other four Great Lakes.

The Fish and Wildlife Service rejected the $1.8 million funding request. Instead, the agency provided $11,000 for the war on ruffe. It was a pittance that had no measurable effect on ruffe control efforts. "Research administrators in the Fish and Wildlife Service did not place a high priority on ruffe research," Busiahn said.

The next two years were marked by political infighting among government officials, and heated debates over whether it made sense to use chemicals to kill ruffe. Just when the campaign to contain ruffe seemed lost, an accidental discovery in 1992 offered a glimmer of hope. A team of government biologists applying the chemical TFM to kill sea lamprey in Wisconsin's Brule River found that the lampricide also killed 97 percent of ruffe in the river. The Fish and Wildlife Service had used TFM since 1957 to kill sea lamprey in streams that flowed into the Great Lakes. The chemical killed lamprey while causing minimal harm to other fish and aquatic life. Suddenly, TFM became a possible weapon for the war on ruffe. The discovery was a huge break for scientists.

"There was actually some euphoria among [Ruffe Control] Committee members that chemical treatment of streams and ballast tanks might be successful in containing ruffe to western Lake Superior," Busiahn said.[9]

Fish and Wildlife Service officials met with representatives of the Wisconsin Department of Natural Resources in early 1993 to seek the state's approval to treat several Lake Superior tributaries with TFM. Busiahn, who was then director of the Fish and Wildlife Service's field station in Ashland, Wisconsin, said federal officials pressed for "urgent treatment" of three estuaries the ruffe had colonized along western Lake Superior. To their surprise and dismay, Wisconsin officials would only support treating the Sand River with TFM. But singling out the Sand River triggered opposition from the National Park Service and the Red Cliff Band of Lake Superior Chippewa Indians. Because the river flowed through land controlled by the tribe and the Apostle Islands National Lakeshore, which the park service controlled, both groups had to approve the chemical treatment. Both refused. Representatives of the tribe and the Park Service said they had no interest in becoming "guinea pigs" in the government's experimental use of TFM to kill ruffe. One of the region's most prominent environmental groups also opposed the use of TFM. "I would hate to see anybody run headlong into this," said Tim Eder, then manager of the National Wildlife Federation's Great Lakes office, in a 1993 newspaper interview. "There is a mild panic going on with exotic species. We don't want to create another problem by treating the ruffe problem."[10]

In the end, opponents of the chemical treatments prevailed. No streams were treated with TFM. Ruffe had the run of Lake Superior. Intransigent bureaucrats effectively gave the invasive fish the keys to the world's largest freshwater ecosystem. "The two year window of opportunity had passed," Busiahn said, "with no action to prevent ruffe from spreading along Lake Superior's south shore."

After ruffe were discovered in Lake Huron, the Great Lakes Fishery Commission's Council of Lake Committees announced in 1995 that it had taken action to keep the fish from spreading beyond the Great Lakes. The committee's "action" was a series of recommendations it forwarded to the National Aquatic Nuisance Species Task Force's Ruffe Control Strategy. The claim that more verbiage would keep the ruffe from spreading into inland lakes and streams across North America was an exaggeration that bordered on propaganda. The committee's recommendations included never using chemicals to control ruffe, because the fish had spread beyond the confines of Lake Superior; requiring freighters to exchange ballast water offshore of Lake Huron ports to keep ruffe from colonizing new areas; support for increased research on how ruffe affected native fish species; additional fish stocking that might control its population (a tactic that had failed in Duluth Harbor); and teaching anglers how to identify ruffe so they could kill the pests if they caught one. The commission then issued a self-congratulatory press release touting the importance of its recommendations. It read:

> The recommendations of the Council of Lake Committees will help fishery managers focus on the very real problems we face now that ruffe are in the lower Great Lakes," said Douglas Jester, a Michigan Department of Natural Resources biologist who cochaired the Council of Lake Committees in 1995. "The appearance of ruffe represents a permanent change in the Great Lakes fishery. The state, tribal and provincial authorities, through this meeting [and recommendations], have taken a big step in attacking this problem in a unified and constructive fashion.[11]

The Canadian cochair of the Council of Lakes Committee, Ron DesJardine of the Ontario Ministry of Natural Resources, made an even bolder claim. DesJardine said: "The Council of Lake Committees recognizes that strong, healthy fish communities can act as a bulwark against the strains ruffe will cause to the fishery. Our overall objectives go beyond dealing with ruffe and focuses on building a sustainable fishery that will not lose to foreign invaders."[12] DesJardine's comments were counter to scientific evidence that showed ruffe overwhelmed indigenous fish species in Duluth Harbor, even in the face of increased stocking of native species.

As government officials praised themselves for tracking the spread of ruffe across southern Lake Superior—while giving up the fight to control its conquest of the Great Lakes—anglers expressed outrage. Fishing groups railed at the Wisconsin DNR for opposing the TFM treatments. Several members of the Wisconsin Conservation Congress, a citizens' advisory board that monitored DNR policy decisions, accused the agency of malfeasance for blocking the chemical treatments. Bill Horns, a Wisconsin DNR biologist who opposed the TFM treatments, defended that course of inaction. "It was conjecture that it might work. And we knew that if we got into that business of attempting to control ruffe that way, it would be highly controversial, it would be potentially damaging and it would be expensive," Horns told a Cleveland newspaper in 1997.[13]

Busiahn never wavered from his stance that the government should have declared all-out war on the ruffe, even if it meant using chemicals to kill the fish. He believed the ruffe population could have been contained and controlled for about $15 million annually, the same amount spent on sea lamprey control. "Personally, my opinion was, 'Let's try it and see if it works,'" he said in a 1997 interview. "My feeling was that the collateral damage, so to speak, would be minimal and certainly acceptable. There were others that just didn't want to do anything."[14] Sandra Keppner, another U.S. Fish and Wildlife Service scientist who supported aggressive action to kill ruffe, said concerns about the environmental side effects of using TFM in a few Lake Superior tributaries would eventually be dwarfed by the ecological harm ruffe would inflict on the Great Lakes. "There is an irony there," Keppner said.[15]

Biologist Roger Bergstedt, who spent much of his career working on the U.S. and Canada's sea lamprey research and control program, said government agencies probably erred by not treating the Sand River with TFM. As a research biologist and supervisor at the federal government's Hammond Bay Biological Station in northern Michigan, Bergstedt knew TFM was the most effective weapon against sea lamprey. He said government agencies could have eliminated ruffe from Duluth Harbor with rotenone and used TFM to kill the fish in the Sand River. "A lot of us thought they should have taken immediate action," Bergstedt said.[16] He acknowledged, however, that the political realities of invasive-species management were far more complex in the 1990s than in the 1950s, when the United States and Canada declared war on the sea lamprey. Bergstedt said it was doubtful the United States or Canada could have responded as quickly and effectively to the ruffe invasion as they did to the sea lamprey infestation, even if the two nations had been willing to use TFM. There were more regulatory hurdles to clear in the 1990s, and far more government agencies involved. The era of quick decision making on invasive species apparently came to an end in the 1960s, when the state of Michigan unilaterally introduced Pacific salmon to rein in alewife and create a new sport fishery in Lake Michigan.

When ruffe and zebra mussels served notice in 1990 that the lakes were under siege from aquatic invaders, federal agencies in the United States were drowning in a sea of red tape, and Congress was paralyzed by political gridlock. The result: boat loads of rhetoric about the need to control invasive species, but no solutions or serious attempts to eradicate any invaders. "If today were the 1950s and sea lamprey just arrived, and we tried to control it with the current environmental protections in place, good luck," Bergstedt said. "There are far more protections in place. This is not a bad thing. But one reality of it is that unless there was specific legislation to encourage rapid control of invasive species, recognizing there would be some losses to other species, it would be a difficult thing to do."

As a soldier in the war on invasive species, Bergstedt said he was astounded at how much rhetoric, and how little action, surrounded the issue in the Great Lakes and globally. "I was at an invasive species conference once. I and a guy from Australia were the only presenters who talked about actually trying to eradicate or control invasive species," he said. "All the other presentations were about research into the problem and the distribution of invasive species."

Ironically, the domestic shipping industry took action on its own in 1993 to try to prevent ruffe from escaping Lake Superior and infesting the other Great Lakes. The Lake Carriers Association, a Cleveland-based group that represented the lake freighters that operated exclusively within the Great Lakes, enacted a ballast water exchange program to slow the ruffe's spread. The voluntary program asked operators of ships that took on ballast water in western Lake Superior ports to exchange that water in deep areas of the lake. The theory was that frigid water temperatures in the middle of Superior would kill any ruffe in the ballast water and prevent ships from spreading the fish to other Great Lakes ports.[17] It was a laudable effort on the part of the domestic shippers to limit damage from a problem caused by the international shipping industry. Critics derided the voluntary ballast water exchange as politically motivated window dressing, a short-term solution to a long-term problem. Whatever the motivation, the shipping industry's attempt to slow the spread of ruffe exceeded the government's lame effort.

Government biologists whose hands were tied by red tape could do little more than monitor the ruffe's spread along the western and southern coasts of Lake Superior. And spread it did, like a slow-moving wildfire in a parched forest. "In early 1994, the new information on ruffe was all bad," Busiahn said. "The population of the St. Louis River, which had leveled off in 1992, increased sharply again in 1993. Colonization of Lake Superior's south shore was surprisingly rapid. The Fish and Wildlife Service found ruffe in eight new locations in 1993." Ruffe were found up to 200 miles east of Duluth. This rapid spread further intensified the debate over whether chemical treatments could have, or would have, slowed the ruffe's spread. "This argument goes on to this day and can never be resolved," Busiahn said in his retrospective.

Ruffe continued their march across the Great Lakes as scientists bickered over what to do about the problem. In 1995, a ruffe was found in the lower Thunder Bay River, which flowed into Lake Huron near Alpena, Michigan. Researchers were looking for ruffe in that area because lake freighters regularly transported cement from Alpena to Duluth. Those ships took on ballast water after unloading in Duluth and transported that ballast water back to Alpena, where it was discharged. No one knew if the ruffe in Lake Huron had arrived before the domestic shipping industry began its self-imposed ballast water exchange program, or in spite of it. In the grand scheme of things, that didn't matter much. The important point was that ruffe were spreading around the lakes as government officials dithered. Ruffe made their way into northern Lake Michigan in 2002, to the surprise of no one. Busiahn and others predicted the fish would spread to all the Great Lakes—"when" was the only remaining question. The worst-case scenario, Busiahn said, was for ruffe to colonize lakes and river estuaries across the eastern United States, from the Mississippi River to the eastern seaboard.[18] "Ruffe have escaped from Lake Superior and prospects for controlling their spread in the Great Lakes are poor. Few tools exist to prevent it," he said. "Some say that a truly rapid, aggressive response to ruffe could have succeeded in containing the species. In the end, a rapid response was precluded because consensus was never achieved on the use of chemical control."[19]

A thin silver lining emerged in the black cloud of the ruffe invasion as the invaders spread across the Great Lakes. Researchers discovered that the ruffe had a modest effect on native fish populations. Granted, ruffe quickly became the most abundant fish species in the lower St. Louis River and Duluth Harbor. But its conquest did not come at the expense of native species. Ruffe simply filled a biological void in the lakes' ecosystem.

"It is remarkable that no immediate impacts on other fish were apparent when ruffe rapidly became the dominant fish," read a statement in an article in the *Journal of Great Lakes Research*.[20] "Intuitively, it would seem that as biomass of ruffe increased, the number or biomass of native fish would decrease correspondingly. Since that has not happened, it may be that ruffe have occupied an 'open niche.' Ruffe may not be as great a threat to yellow perch, walleye and coregonus species (such as whitefish) in the Great Lakes as was first thought. Even so, we must not become complacent and dismiss the potential for as yet unrealized impacts of ruffe on the Great Lakes ecosystem." Those researchers said that even if ruffe only reduced perch populations by 10 percent, and walleye and whitefish numbers by 1 percent, it would still cause significant economic damage. One of the scientists estimated that the government's refusal to spend $12 million on a decade-long ruffe control program could result in the fish causing between $24 million and $215 million damage annually to perch, whitefish, and walleye fisheries across the Great Lakes.[21]

Ruffe posed a serious threat to the ecosystem even if they didn't drive away native fish species. The fish were a potential bottleneck in the food chain. Few other fish species ate ruffe—it was like eating a porcupine. That meant nutrients, which moved up the food chain as big fish ate smaller fish, might hit a dead end in ruffe. Larger fish that usually ate perch could starve if ruffe displaced native species. Twenty years into its occupation of the Great Lakes, the ruffe had not delivered the ecological punch that many scientists feared. But it did cause a fair amount of damage and became major headaches for anglers who fished for perch and walleye in the St. Louis River estuary. Longtime Duluth resident and angler Dave Zentner said he could still catch walleye and smallmouth bass in the St. Louis River and Duluth Harbor two decades after ruffe showed up. But he said the invader soured many of his fishing experiences. "It's still wonderful fishing. It isn't all doom and gloom by any stretch of the imagination," Zentner said in a 2007 interview. "But the ruffe are always nibbling at your line, making you think you have a bite when you don't."[22]

Pratt, the biologist who discovered ruffe near Duluth, predicted in 2002 that it would spread far and wide: "It's fairly easy to predict that in the future they'll spread to the rest of North America (including Canada and Alaska). There are just too many connections between the Great Lakes and the rest of our freshwater ecosystems."[23] His prediction was worrisome for anglers on Lake Erie, home to the most productive fishery among the five Great Lakes. Erie's rocky bottom provided ideal habitat for the ruffe, as did the lake's warm, murky, and biologically rich waters. The Great Lakes Fishery Commission's

Ruffe Task Force delivered an ominous warning in its 1992 report to the U.S. and Canadian secretaries of state. "Should the ruffe become established in Lake Erie, the world's largest perch/walleye fishery, the potential economic impact could reach an annual economic loss of $90 million," the task force said.

On a larger scale, the government's response—or lack thereof—to the ruffe invasion was instructive and illuminating. The situation highlighted the difficulty of using chemicals to kill fish in a society that was increasingly chemophobic. The U.S. government's reaction also became the model for future efforts to prevent new invasions and control foreign species that were already in the lakes. That was unfortunate. In the two decades after ruffe were discovered in Duluth Harbor, government agencies in the United States and Canada repeatedly chose to battle invasive species with words instead of action. Bureaucrats on both sides of the border responded to the growing number of invaders by forming numerous committees, issuing strongly worded edicts, and spending millions of dollars on research. Instead of attacking the primary source of the problem—filthy ballast water discharges from ocean freighters—government officials did a political two-step that gave the impression they were systematically solving the problem. It was a grand illusion.

A remorseful Busiahn concluded in his retrospective that ruffe, if nothing else, offered government officials charged with protecting the Great Lakes a teachable moment. "Ultimately, preventing the spread of nonindigenous pests will pay greater dividends than attempts to [control] established populations," he said. That lesson wasn't learned quickly either. U.S. and Canadian officials took nearly two decades to enact the regulations necessary to disinfect 95 percent of the ballast water that ocean freighters hauled into the lakes. Their failure to deal with the plague of invasive species quickly and decisively allowed the economic and environmental damages caused by shipborne invaders to spiral wildly out of control.

From a regulatory point of view, a huge hole was still present.

—Chris Wiley, special projects manager, Canada's Department of Fisheries and Oceans, 2000

The Coast Guard's program has a loophole big enough to drive a cargo ship through.

—Jennifer Nalbone, Great Lakes United, 2005

10 SMOKE AND MIRRORS

It has often been said that regulations are only as effective as the people enforcing them. It is one thing to post a speed limit of, say, 70 mph on a highway. But if the police don't enforce the law, people drive as fast as they want. Similarly, it was one thing for the U.S. government to tell the shipping industry that every transoceanic freighter had to conduct an open-ocean ballast water exchange before entering the Great Lakes. Enforcing that mandate, a cornerstone of the Nonindigenous Aquatic Nuisance Prevention and Control Act of 1990, was another story. Ballast water exchange was voluntary for the first three years the law was on the books. The regulation became mandatory in 1993 and applied to all ocean freighters—every last one that entered the St. Lawrence Seaway en route to a port in the United States. Enforcing that mandate was crucial: Ballast water exchange was the primary tool for preventing future invasions by shipborne species. The law was tough, but its success in protecting the lakes depended largely on the U.S. Coast Guard. The military agency was charged with enforcing the ballast water exchange rule. But what if the Coast Guard didn't hold up its end of the bargain? Then what? That was precisely the situation in 1993, when the Coast Guard brought in a new sheriff to supervise the agency's ballast water inspection program.

Eric Reeves was in the twilight of a 20-year Coast Guard career when the military lawyer was promoted to what he called the most politically difficult and emotionally taxing job of his career: manager of the Great Lakes ballast water program. Stationed in Cleveland, Ohio, Reeves believed he was in position to help stop the flow of foreign species into the lakes via ballast water. He was a stern, fiercely inquisitive man who was as comfortable citing court cases as he was classical literature. Reeves approached his new task methodically. He read numerous reports about the problem of ocean freighters importing foreign species into the lakes he had come to love as a child growing up in western Michigan. Having digested the science, Reeves discussed the finer points of ballast water regulations with one of his assistants, Katherine Weathers. She raised the disturbing possibility that the existing ballast water regulations were, quite literally, missing the boat.

Weathers' suspicions were sparked by a 1991 Canadian study that concluded the muddy slop in ocean freighters' allegedly empty ballast tanks was a biological soup of foreign species. That concerned Weathers. She knew the vast majority of transoceanic freighters entering the lakes were loaded with cargo and, theoretically, had no ballast water on board. In Coast Guard circles, fully loaded freighters were known as "NoBOBs," because there was No Ballast On Board. The Coast Guard, Reeves and Weathers discovered, didn't require NoBOBs to flush ballast tanks with ocean water before heading up the St. Lawrence River and into the Great Lakes. That was the moment Reeves and his assistant shared an epiphany. "I have a clear memory of highlighting the relevant paragraphs in the 1991 report and discussing the issue with Katherine, at which point she said, 'I think we're missing ninety percent of the problem,'" Reeves said.[1]

The Coast Guard's decision to not regulate NoBOBs allowed at least 80 percent of transoceanic freighters to enter the Great Lakes without flushing their ballast tanks in the North Atlantic. That unwritten loophole in the law rendered the lakes' first line of defense against foreign species largely useless.

The Coast Guard's interpretation of the 1990 law turned it into regulatory Swiss cheese: It was full of holes. When Reeves grasped the magnitude of the regulatory gaps, he shifted his focus from being a compliant bureaucrat who championed a flawed law, to one trying to fix it. His would be a long, lonely crusade that put him at odds with titans of the shipping industry, narrow-minded environmentalists, and his superiors in the Coast Guard.

THE BASIC TENET OF THE NONINDIGENOUS AQUATIC NUISANCE PREVENTION and Control Act of 1990, known colloquially as NANPCA ("Nan-Packa"), was simple and powerful. It required the following: "All vessels equipped with ballast water tanks that enter a U.S. port on the Great Lakes after operating on the waters beyond the exclusive economic zone (200 miles offshore of the Atlantic Coast) . . . to carry out exchange of ballast water on the waters beyond the exclusive economic zone prior to entry into any port within the Great Lakes."[2] Ships unable to exchange ballast at least 200 miles offshore of Newfoundland had two options: complete the process in waters closer to land, provided it was approved by the Coast Guard, or retain all freshwater ballast from overseas ports in their tanks while in the Great Lakes. No ships destined for ports in the United States were exempt from NANPCA, until the Coast Guard quietly emasculated the law by creating what became known as the NoBOB loophole. Nowhere in NANPCA did the law say that NoBOB ships were exempt from the ballast water exchange requirement. Coast Guard officials unilaterally decided that only those ships carrying "pumpable" volumes of ballast water would be required to exchange it in the ocean before entering the lakes. That decision was a catastrophic regulatory failure. Worse, the NoBOB loophole remained in effect for the next 18 years. Reeves explained the situation in a 1999 report:

"NoBOBs never should have been exempt from the Great Lakes [regulatory] regime. Under the general definition in both NANPCA '90 and NISA '96 (the Nonindigenous Species Act of 1996) ballast water is any water and associated sediments used to manipulate the trim and stability of the vessel."[3]

Reeves's colleagues in the Coast Guard saw things differently. According to their interpretation of the law, only those ships that had "pumpable ballast water" on board were subject to the mandatory ballast exchange rules.[4] Agency officials allowed NoBOBs to skirt the law because those ships didn't

National Oceanic and Atmospheric Administration's Great Lakes Environmental Research Laboratory

A scientist samples muddy slop in the ballast tank of an ocean freighter that reported having no ballast on board. Scientists discovered that mud in the ballast tanks of so-called NoBOB (No Ballast On Board) ships harbored billions of foreign organisms and potentially harmful pathogens. The NoBOB ships weren't required to flush ballast tanks with sea water before entering the Great Lakes; they were unregulated until 2006.

have pumpable quantities of ballast water in their tanks. Technically, that was correct. The reversible pumps that inhaled and exhaled ballast water never completely drained the tanks. The reason: the pipes that pumped ballast water didn't reach to the bottom of the tanks. Rather, they were suspended a few inches above the bottom of ballast tanks, honeycomb-like structures that were nearly impossible to drain completely. As a result, the average NoBOB entering the Great Lakes carried 157 metric tons of muddy slop teeming with aquatic life in the bottom of its supposedly empty ballast tanks, according to the 1991 study by the Canadian Coast Guard.[5] The Canadian study was the one that alerted Reeves and Weathers to the gaping hole in the ballast water regulations. To put the significance of the NoBOB loophole in the proper context, consider this: The NoBOB ships that entered the Great Lakes in 1995 dumped 858 million gallons of untreated ballast water into the lakes. That was 84 percent of all ballast water discharged into the lakes that year.[6]

The fact that NoBOBs didn't flush their ballast tanks before entering the lakes was the product of two realities: it was difficult and dangerous to pump ballast water into ships fully loaded with cargo, and the U.S. Coast Guard showed little interest in addressing the issue, despite a mandate to do so.

Most ships were not designed to conduct a complete exchange of ballast water. There always remained a thin layer of residual sediment and water in ballast tanks. That coating of sediment and water, contained in dozens of ballast tanks, added up to a substantial quantity of biological soup. In some cases, NoBOBs carried as much as 200,000 gallons of muddy water in their "empty" ballast tanks.

The residual ballast water and sediment in the tanks of NoBOBs was critically important. When NoBOBs dropped off cargo at a Great Lakes port, the vessels sucked lake water into their ballast tanks to maintain stability during the next leg of their journey. That lake water mixed with the residual sediment and water in the ballast tanks, which was hauled into the lakes from ports overseas. The ship would then discharge some, if not all, of that mix of domestic and foreign ballast water—and all that lived in it—when taking on cargo at another Great Lakes port. Prior to Reeves's arrival, Coast Guard officials apparently did not take that issue into consideration when deciding to give NoBOBs a free pass into the lakes. Reeves believed that his predecessors in the agency didn't know how to deal with the NoBOB issue, so they didn't.

There were justifiable reasons for not requiring NoBOBs to conduct ballast water exchanges at sea. Understanding why required a brief lesson in the economics of shipping cargo. The most profitable way to operate a ship was to carry as much cargo, and as little ballast water, as possible. Cargo was revenue; hauling ballast water was an expense. As a result, the majority of ships entering the Great Lakes from other continents carried cargo. In some years, 90 percent of ocean ships entering the lakes carried cargo and, therefore, were declared NoBOBs. Ideally, ships were loaded to the brim with cargo, which made adding ballast water a potentially deadly proposition. Ships loaded with cargo ran the risk of capsizing or snapping in half and sinking if too much ballast water was added. The annals of maritime history included many incidents of ships that capsized or were ripped apart by improper ballasting.

Given the dangers associated with fully loaded ships taking on more weight in the form of ballast water, the Coast Guard erred on the side of safety. Thus was born the infamous NoBOB exemption. It was a fatal flaw in the ballast water regulations, which, intentionally or not, put the safety of transoceanic ships above the biological integrity of the Great Lakes. The guard's golden rule, after all, was safety first. Reeves said Coast Guard officials who drafted the ballast water exchange regulations never envisioned applying the law to NoBOBs. "As a matter of administrative practice and practical politics, they [NoBOBs] were simply ignored," Reeves said. For an agency that lived by its motto of "Semper Paratus," which meant "Always Prepared," the Coast Guard was woefully unprepared to protect the lakes from shipborne invasive species.

Establishment of the NoBOB loophole was a critical turning point in the battle against shipborne invaders. Roughly 80 percent of transoceanic freighters entering the Great Lakes were loaded with cargo, and thus free to haul ballast water from distant ports—and all that lived in it—into the lakes. Only 20 percent of ships flushed their ballast tanks with sea water before entering the lakes. The loophole was a huge and ecologically destructive gap in the law, one that remained in effect for 18 years. Why? Because the Coast Guard never got around to closing the loophole, despite Reeves's protestations and a 1996 law which affirmed that "all vessels equipped with ballast water tanks" were required to exchange ballast in the North Atlantic. Harmful as it was, the NoBOB exemption was, in Reeves's view, just one of three critical flaws in NANPCA and its successor, the National Invasive Species Act of 1996.

The other two flaws in the legislation were far less complicated. One was a safety exemption that allowed a transoceanic freighter entering the Great Lakes to forgo a ballast water exchange if the ship's master deemed the procedure too dangerous due to high waves or other conditions. Those ships were allowed to dump millions of gallons of filthy ballast water from foreign ports into the lakes. Why? Because the law, as written, said: "A vessel that does not exchange ballast water on the high seas . . . shall not be restricted from discharging ballast water in any harbor."[7]

Another glaring weakness in NANPCA was the reliance on a salinity standard to determine whether mid-ocean ballast exchanges were keeping new invaders from entering the Great Lakes in ships' ballast tanks. Beginning in 1993, ships required to exchange ballast water at sea were inspected at St. Lambert Lock, near Montreal, or in Massena, New York. Coast Guard crews boarded the ships, took water samples from a few of the ships' numerous

Jeff Alexander

A ship crew member drops a sampling device into a sounding tube that leads to a ballast water tank. The procedure allows inspectors to determine if ocean freighters have conducted a proper mid-ocean ballast water exchange before entering the St. Lawrence Seaway.

Inspectors with the St. Lawrence Seaway Development Corp. demonstrate for the media how they use a refractometer to measure salinity levels in the ballast water of transoceanic freighters entering the Seaway. The ships are sampled in Montreal.

Jeff Alexander

ballast tanks, and measured the salinity with a device called a refractometer. The salinity of the water had to be 30 parts per thousand to comply with the ballast water exchange regulations. The salinity of water in the North Atlantic, where ships destined for the St. Lawrence Seaway were supposed to exchange ballast water, had an average salinity of 35 parts per thousand. If the salinity of ballast water was at least 30 parts per thousand, Coast Guard officials assumed the ship had performed a proper ballast water exchange in the ocean.

Regulators assumed that ballast water with a salinity concentration of at least 30 parts per thousand meant that most freshwater organisms from overseas ports had been purged from the ship's bowels during the mid-ocean ballast water exchange. But that assumption was proven wrong in 1991, two years before the U.S. Coast Guard made mid-ocean ballast water exchange mandatory. The 1991 study by the Canadian Coast Guard found that exchanging ballast water at sea only eliminated 67 percent of freshwater species from ballast tanks.[8] Subsequent studies put the efficacy of mid-ocean ballast water exchange at 85 percent. Either way, a considerable number of viable foreign species survived the ballast exchange process and entered the Great Lakes. Ships rarely flushed away all the freshwater ballast and freshwater organisms when conducting a mid-ocean exchange. Accomplishing a complete exchange of ballast water was nearly impossible. Worse, it wasn't required. Reeves and Weathers pointed out the flaws in the U.S. ballast water regulations in 1996, in the journal *Marine Technology*. They wrote:

> With the active cooperation of the Canadian Coast Guard, Transport Canada Ship Safety and the Seaway Authorities, the Coast Guard has established a defense against the invasion of the Great Lakes in the form of ballast water controls on vessels entering the Seaway. However, this defense is less than complete. What we currently have in place is a strategy of limiting the probability of new introductions through what one might call the 85 percent solution. This does give the Great Lakes a substantial level of protection, and is a vast improvement in the situation as it stood only a few years ago, before the current controls went into effect. But it is not a winning strategy for the long-term war of attrition. . . . It should be kept in mind that a 100 percent exchange, not 85 percent, is the goal. The regulatory level of 30 parts per thousand, or an 85 percent exchange, was set as a practical accommodation

> for the shipping industry, because of the difficulty that many vessels have in accomplishing a 100 percent exchange.[9]

Their article noted the problems created by the safety exemption, the dangers associated with exchanging ballast water at sea, the limited effectiveness of ballast water exchanges in killing organisms in ballast tanks, and the questionable NoBOB exemption. "Technically, they [NoBOBs] are subject to the regulations," Weathers and Reeves said in the article. They offered possible solutions to the limits of ballast flushing. One suggestion: add a second pipe to pump water into the top of ships' ballast tanks, and then release the water through the bottom of the tanks. The seemingly simple remedy, which would have allowed a more complete flush of the ballast tanks—much like a household toilet—was ignored, Reeves said. His efforts to close the NoBOB loophole and restrict the safety exemption met with similar success. In other words, none.

The 1996 reauthorization of NANPCA afforded Congress the opportunity to strengthen the ballast water regulations. Reeves alerted officials in Senator John Glenn's office that the Coast Guard, by creating the NoBOB exemption, was not enforcing the 1990 ballast exchange regulation as written. Glenn, the former astronaut turned politician from Ohio, was a champion of the Great Lakes. Reeves figured Glenn would close the NoBOB loophole when NANPCA was reauthorized as the National Invasive Species Act of 1996. The strategy worked. Glenn's staff inserted language into the 1996 law that closed the NoBOB exemption. Specifically, the new law made the ballast water exchange requirements applicable to "all vessels equipped with ballast water tanks" that entered the Great Lakes from beyond an imaginary line 200 miles offshore of the East Coast of North America. The 1996 law also required the Coast Guard to establish national guidelines on "ballasting practices of vessels that enter the waters of the United States with no ballast on board." The law extended ballast exchange regulations to all U.S. waters and made it abundantly clear that Congress wanted all ships—every ship—entering American ports on the Great Lakes to exchange ballast water at sea. Still, the Coast Guard didn't develop ballast water exchange rules for NoBOBs. The 80 percent of ocean freighters entering the lakes without exchanging ballast water in 1996 continued to do so for another decade, despite clear evidence that the practice threatened the Great Lakes.

Reeves said he had tried to persuade the Coast Guard to require NoBOB ships to flush ballast tanks with a "swish and spit" of ocean water before entering the lakes. He also lobbied his superiors to force ships to install new pipes on ships that would allow a more complete exchange of ballast water and, thus, reduce the likelihood of new species getting into the lakes. Coast Guard officials in Washington overruled his proposals. "At that point, I felt very strongly that the Guard had sold out to industry pressure and I was not shy about expressing my opinion to [officials in Washington] D.C.L. But it was pissing in the wind," Reeves said. What proof was there that Reeves's allegations were more than sour grapes? For starters, none of the changes Reeves and Weathers recommended in their 1996 article in the journal *Marine Technology* were implemented, as of 2008. The Coast Guard didn't mandate ballast water exchange for NoBOBs or require ships to install ballast water treatment devices on transoceanic freighters entering the Great Lakes.[10]

Canadian authorities didn't fare any better than the Americans when it came to regulating NoBOBs. Canada's 1989 guidelines that recommended ships exchange ballast water at sea before entering the St. Lawrence River never took into account the possibility that ships loaded with cargo might have considerable amounts of ballast water and mud in their ballast tanks. A 1996 study commissioned by the Canadian Department of Fisheries and Oceans concluded that 90 percent of the ocean freighters entering the Great Lakes were NoBOBs and thus exempt from the 1989 ballast water exchange regulations.[11] Canada's efforts to further regulate ballast water were hampered by the 1995 reorganization of the Department of Transport. The restructuring moved the Canadian Coast Guard out of the Department of Transport and into the Department of Fisheries and Oceans. The division of Marine Safety, the section that regulated and inspected freighters on the Great Lakes, remained in the Department of Transport. "Responsibility for ballast water regulations got caught in the middle," said Chris Wiley, manager of special projects for the Department of Fisheries and Oceans. Tougher ballast water regulations essentially fell through the cracks of Canada's muddled federal bureaucracy. "The Department of Fisheries and Oceans Science Branch had the expertise with respect to the marine and freshwater environment, and the Canadian Coast Guard had the voluntary [ballast water exchange] guidelines, but no expertise or ships in the fleet with any appreciable ballast capacity," Wiley said in a 2000 report. "A memorandum of understanding was signed outlining the respective responsibilities. However, five years later, these political decisions have considerably complicated the ballast water regulatory picture in Canada."[12] Canada finally closed its NoBOB loophole in 2006, requiring every ship destined for a Canadian port on the Great Lakes to flush its ballast tanks before entering the Seaway.

The U.S. Coast Guard was even more sluggish than its Canadian counterparts. The Coast Guard in February 2005 published a notice in the *Federal Register* seeking ideas for how to regulate NoBOBs. "Our goal for these strategies is to prevent the introduction and spread of nonindigenous species via NoBOBs," the notice said. "The Great Lakes Ballast Water Management Program that became effective on May 10, 1993 has remained unchanged. . . . The Coast Guard will use information gathered from this notice to develop a comprehensive program to reduce the threat of introducing nonindigenous species into the Great Lakes via NoBOBs."[13]

The Coast Guard enacted a policy of "best management practices" for NoBOBs six months later. The agency touted its new regulations in a press release. "Coast Guard Establishes Policy to Stop Invasions into the Great Lakes," read the headline on the release. It was a misleading claim that bordered on propaganda. If you believed the Coast Guard, the NoBOB problem had been solved. No longer could invasive species get into the Great Lakes in the ballast water of transoceanic ships. But that wasn't the case. The new policy only encouraged NoBOB ships to conduct an open-ocean ballast water exchange or flush ballast tanks with sea water before entering the lakes. The key word in the policy was "encourage." The policy was not mandatory—it was like asking motorists to drive no faster than 70 mph and not penalizing those who ignored the request. The Coast Guard's policy was mere rhetoric, as evidenced by future pronouncements that the agency was finally cracking down on NoBOBs. Bivan Patnaik, manager of the Coast Guard's environmental division, said in a 2007 article that the best management policies for NoBOBs "strongly [encourage] NoBOBs to conduct saltwater flushing." The policy, which remained voluntary through 2008, was the latest in a long line of Coast Guard policies that appeared to deal with the problem of invasive species but were little more than regulatory smokescreens. Three years after the Coast Guard announced it would develop "strategies" to prevent NoBOBs from importing foreign species, the agency had yet to enact the mandatory regulation—flushing ballast tanks with ocean water—necessary to reduce the threat of more invasions.

Canada took the initiative in 2006 to close the NoBOB loophole for ships doing business on the Canadian side of the Great Lakes. The Canadian Coast Guard required all NoBOBs to flush ballast tanks with a small amount of ocean water before entering the St. Lawrence River. The rationale was that a "swish and spit" of salt water would flush out much of the residual water and sediment remaining from ballasting operations in previous ports. The St. Lawrence Seaway Development Corp., the agency that managed the U.S. portion of the Seaway, imposed similar regulations in 2008 for ships headed to American ports. Together, the U.S. and Canadian regulations ensured that every transoceanic ship entering the lakes—the NoBOBs loaded with cargo and ships carrying ballast water—flushed their ballast tanks with salt water before entering the lakes. The new rules were a major advance in ballast water regulations. Those rules should have been enforced sixteen years earlier in U.S. waters, as part of the 1990 federal law.

A study in 2007 found that flushing the tanks of NoBOBs with ocean water killed or purged 95 percent of organisms in ballast water. "One of the key findings here has been to confirm that saltwater can be quite effective at reducing the risk of invasions from ballast water," said David Reid, an invasive-species expert at the National Oceanic and Atmospheric Administration's Great Lakes Environmental Research Laboratory. "It's not 100 percent effective against all types of organisms, but it's far better than what's been going on, which has been basically no regulation at all for NoBOBs."[14]

The Canadian authorities and the St. Lawrence Seaway Development Corp. deserve credit for requiring NoBOBs to flush ballast tanks with ocean water before entering the Great Lakes. That the U.S. Coast Guard, the agency charged with regulating ballast water, didn't impose such rules speaks volumes about the agency's readiness, willingness, and ability to protect the lakes from shipborne invasive species. The fact that the St. Lawrence Seaway Development Corp. dealt with the problem before the regulators, the Coast Guard, is one of the most appalling examples of government neglect of the globally significant Great Lakes.

The irony is that the very structures that allowed ocean ships into the Great Lakes and opened the floodgates to foreign species—the locks and canals of the St. Lawrence Seaway—presented a golden opportunity to deal with the problem. Because all ocean freighters entering the Great Lakes had to pass through the St. Lambert Lock, there was a choke point where every ship could be stopped and inspected. It was the logical place to enforce ballast water treatment standards, if the U.S. or Canadian governments ever got around to enacting such rules. Neither government had done that as of 2008. The U.S. Coast Guard's numerous promises to develop ballast water regulations proved time and again to be empty rhetoric. Its failure to develop ballast treatment standards forced the issue to be settled in the courts. Great Lakes ecosystems suffered mightily in the interim.

HISTORY PROVED THE UNITED STATES AND CANADA'S FIRST BALLAST WATER regulations, and the agencies charged with enforcing them, to be little more than paper tigers. The regulations created the appearance that the two nations were clamping down on ballast water discharges from ocean freighters when nothing of the sort occurred. The proof: the number of invasive species discovered in the lakes increased markedly after the two nations began regulating ballast water.[15] It was illuminating to measure the outcome of NANPCA against results of the Oil Pollution Act of 1990. The U.S. Congress approved the Oil Pollution Act the same year as NANPCA. History has proven one of those laws to be a tremendous success, while the other was a failure, at least statistically.

Congress approved the Oil Pollution Act less than a year after the *Exxon Valdez* spilled 11 million gallons of crude oil into Alaska's Prince William Sound. The law required new oil tankers to have double hulls, increased corporate environmental liability, required oil companies and federal agencies to be better prepared for spills, and established a trust fund to pay for future cleanups. The result: far fewer oil spills from tankers, despite increased consumption and shipments of petroleum products via freighters. "During the past two decades, while U.S. oil imports and consumption have risen steadily, oil spill incidents and the volume of oil spilled have not followed a similar course," according to a 2007 Congressional Research Service report. The number of oil spills in U.S. waters that exceeded 100 gallons decreased nearly 50 percent after the law was passed—from 500 in 1995 to 275 in 2004, according to the report.[16]

The Nonindigenous Aquatic Nuisance Prevention and Control Act of 1990 and its successor, the National Invasive Species Act of 1996, could not claim such success, at least not in the Great Lakes. The number of invasive species in the lakes soared from 139 in 1990 to 186 in 2008, a 34 percent increase. "The rate at which new aquatic invasive species are colonizing the Great Lakes has not declined despite implementation of Canadian ballast water exchange guidelines in 1989, followed by mandatory ballast exchange requirements established in 1993 by the U.S. for ships entering the Great Lakes," according to a report by the International Association for Great Lakes Research.[17]

By 2007, scientists were discovering a new invasive species in the lakes every seven months. Not all of the invaders were imported by ocean freighters. But NANPCA was not aimed solely at ocean freighters. It was supposed to "prevent and control infestations" from all sources—freighters, recreational boats, the movement of bait fish, fish dumped from aquariums, and intentional releases. It was worth noting that between 1993 and 2008, scientists discovered 20 new species in the Great Lakes that they attributed to ballast water.[18]

For Reeves, who grew up along the shores of Lake Michigan and spent much of his career working on the Great Lakes, his agency's inability to deal effectively with the ballast water issue proved especially disheartening. He clashed repeatedly with his superiors for five years before retiring. Retirement didn't end his involvement in the ballast water issue. He authored a series of scathing reports critical of government and industry efforts to deal with a problem he considered solvable. A critical flaw in all the ballast water regulations prior to 2009, Reeves said, was the failure of the United States or Canada to establish discharge standards for ballast water. Put simply, Reeves and others believed that if the government set a standard for how many species a ship could discharge into the Great Lakes, treatment technologies would quickly evolve to help shippers meet that limit. Ballast water discharge standards were nonexistent as of 2008 because the U.S. EPA in 1973 declared ballast water discharges exempt from the federal Clean Water Act. Absent standards, shippers were reluctant to spend money on costly ballast water treatment systems, Reeves told the International Joint Commission in a 1999 report. His assessment, which was accurate then and remained accurate in 2008, said:

"There is a whole range of available technologies for treating ballast water, including filtering, heat, ultraviolet light, biocides and shore-side treatment of water, some of which may well be economically feasible for particular types of vessels and trades. None of them require new technology. However, in the absence of a legal regime requiring such changes, and thus creating a level playing field, there is little incentive for any shipping company to make the required investment."[19]

Who could blame shippers for not rushing to install million-dollar ballast treatment systems when there was no way of knowing if the devices would meet standards that had yet to be written? Certainly not Reeves, who spent five years in the trenches of verbal warfare over ballast water regulations. He concluded after retiring from the Coast Guard that the U.S. and Canadian governments—not the shipping industry per se—were responsible for the global shipping industry's biological siege of the Great Lakes.

"It's a classic case of regulatory failure," Reeves said in a 2008 interview. "If I had known better what I was doing I would have talked to the EPA right off the bat about revoking the [ballast water] exemption. The EPA did everyone a disservice when it put out that illegal ballast water exemption under the Clean Water Act. The Coast Guard was not equipped, technically or psychologically, to regulate the shipping industry. We were too close to the industry." Reeves said he deserved some of the blame for the Coast Guard's failure to effectively regulate ballast water. "I didn't understand the issue well enough," he said. "It took me a couple of years to understand that what we were doing was fallacious."

In 2008, a decade after Reeves retired, the Coast Guard still had not enacted ballast water treatment standards for freighters or closed the NoBOB loophole. Moreover, the agency wasn't enforcing the 1990 or 1996 laws that required all ocean freighters to exchange ballast water at sea before entering the Great Lakes. The wondrous lakes suffered mightily as a result.

Foreign food webs from the Black Sea that evolved together over millions of years are being reconstructed here like Lego bricks.

—**Anthony Ricciardi, Canadian research scientist, 2000**

11

MELTDOWN

The evening was young and there were few other boats on Lake Michigan when I joined research scientist David Jude and two of his colleagues for an unusual fishing expedition along a deserted stretch of beach in South Haven, Michigan. We ventured into the lake's placid waters in an 18-foot Boston Whaler cluttered with fishing nets, small buoys, and weights to hold the gear in place once it was deployed. The mission was simple: trap fish in a series of gill nets to get a snapshot of what lived in the nearshore waters of the lake. Jude had performed this drill annually since 1973. It was part of a long-term monitoring program that tracked the abundance of yellow perch and alewife—and whatever else turned up—in southern Lake Michigan. Jude invited me to join his annual excursion to see if the nets would produce any surprises. Sampling fish populations in the lake had become far less predictable in the early 1990s, after *Dreissena* mussels—zebra and quagga mussels—led an army of new foreign species into the Great Lakes. Sampling missions that once were routine in the 1970s and '80s had become fraught with uncertainty.

Monitoring aquatic life in the Great Lakes in the *Dreissena* era was a bit like the movie character Forrest Gump's theory of life. Gump's mother taught him that "Life is like a box of chocolates: You never know what you're gonna get." Scientists who studied the lakes after zebra and quagga mussels invaded faced a similar scenario. They never knew what might show up in their sampling gear. There was no telling what new species of fish, mussels, or zooplankton were lurking in these inland seas. "The lakes," Jude said on that temperate summer evening, "are in total chaos."[1]

As he steered the boat out of the Black River and into the gently rolling swells of a passing boat, Jude expressed deep concern for the future of native Great Lakes fish species and other aquatic life. Native species were wilting under the pressure of invasives, which thrived in their adopted homes and ruled like a gang of muscle-bound thugs in a neighborhood of scrawny computer nerds. Evidence of such change was found in the nets Jude's crew used to capture fish over the course of five hours that summer night. Theirs was a modest haul for such a considerable effort. The 100-foot-wide net that the sixty-something Jude and his assistants manually dragged through the water, perpendicular to the beach, caught few fish that night. Under normal conditions, they would have hauled in hundreds of young perch.

Frustrated by the meager catch, and with a half hour to kill before removing nets the crew had anchored in deeper water earlier in the evening, Jude lightened the mood with one of his trademark junk-food buffets. He broke out some Vienna mini-sausages, a bag of Doritos, and cans of Mountain Dew. The moment was classic Jude. The unconventional academic who had spent three decades studying the Great Lakes was known almost as much for his eccentricities as his expertise. To many in the scientific community, Jude was like the E. F. Hutton of Great Lakes research: When he spoke, people listened. His observations about the lakes—particularly during the biological roller coaster that followed the great mussel invasion in the late 1980s—were usually dead-on. Jude's assertion that zebra and quagga mussels were causing certain fish species to starve or decline in numbers was bolstered by his 2007 assessment of fish populations in a small area of Lake Michigan's 23,000 square miles. Trapped in the nets were a few perch and alewives, when there should have been many. Perch were once among the most abundant and coveted sport fish in Lake Michigan. Alewives, though an invader, fueled the lake's billion-dollar salmon fishery. Most disturbing, however, was the presence of several round gobies that had found their way into the nets. Gobies were bottom-feeding fish that, like *Dreissena* mussels, were native to the Black and Caspian seas in eastern Europe.

Jude had discovered gobies in the Great Lakes 17 years earlier, but the sight of them still made him wince. The goby was an ugly little fish, with bulging eyes atop its head. It was also a symbol of how profoundly zebra and quagga mussels had altered the lakes' ecosystem. Gobies might never have survived in the lakes had it not been for the foreign mussels arriving first and providing an abundant source of food for the invasive fish. Gobies feasted on the mussels in their native waters and, after 1990, in their new home: the Great Lakes. Zebra and quagga mussels were more than invaders—the mollusks were the leaders of an unprecedented ecological conquest that swept through these magnificent waters in the 1990s. For the lakes, the combination of *Dreissena* mussels and gobies was a synergistic match made in hell.

As zebra and quagga mussels spread across the lakes, they laid the foundation of a bastardized food chain that would allow other invaders from the Ponto-Caspian region of Europe to thrive in North American waters. Scientists had a term for the phenomenon of multiple foreign species thriving in the presence of one another while colonizing new territory: They called it "invasional meltdown." Fifty years after sea lamprey and alewife caused such a meltdown, a new suite of invaders from Europe was repeating the feat. The situation was analogous to a cancer patient learning he had contracted an incurable, contagious disease—such as AIDS—while undergoing chemotherapy. The uncontrollable infectious agent in the Great Lakes: tiny foreign mussels that packed a massive ecological punch.

ZEBRA MUSSELS NEEDED JUST TWO YEARS TO SPREAD TO ALL FIVE GREAT Lakes after settling in Lake St. Clair in 1988. Several of their European friends were hot on their trail. Quagga mussels were discovered in Lake Erie in 1989 and quickly spread to lakes Ontario, Huron, and Michigan; quaggas didn't show up in Lake Superior until 2005. One of the most alarming invasions in the early years of the *Dreissena* era involved the goby. Jude had stumbled across the fish while studying fish kills at a power plant on the St. Clair River, north of Detroit. Jude was plucking dead fish off the rotating screens that kept fish and debris from getting into Detroit Edison's Belle River power plant as it inhaled huge quantities of cooling water from the river. Power plants that used the Great Lakes and their connecting waters to keep coal-fired power plants and nuclear reactors from overheating killed huge quantities of fish every year. Jude was documenting the extent of those fish kills at the Belle River facility when his attention was diverted by a surprising find. He spotted an odd-looking fish with a prolonged nostril, which looked like a tiny straw extending from its snout. The three-inch-long, bug-eyed specimen with odd pectoral fins was something Jude had never seen in his three decades of fish research. "I thought maybe someone dumped it into the river and we would never see another one," he said.

A laboratory analysis identified the fish as a tubenose goby, another invader from the Caspian Sea. It was a troubling find, to be sure. But the worst was yet to come. Two days after a Detroit newspaper published an article about the discovery, a fisherman in western Ontario telephoned Jude. The man claimed to have a fish similar to Jude's tubenose goby living in an aquarium in his Sarnia, Ontario, home. He told Jude his daughter had caught the fish in a canal on the Canadian side of the St. Clair River. There was one other thing, the man told Jude: His fish was bigger.

Jude went to the guy's house the next day to see if his fish story held water. He was astonished by what he saw swimming in the aquarium: a larger, different version of the tubenose goby. Tests identified it as a round goby, the more aggressive cousin of the tubenose. It was a troubling discovery—round gobies had a bad case of Napoleon complex. The fish, which only grew to six inches, were pugnacious, aggressive, and known to chase larger fish off spawning beds and devour their eggs. Round gobies were like bad house guests: they showed up uninvited, ate too much food, were combative and domineering, and often evicted their hosts. The tubenose goby was not considered a threat to native fish species, but the round goby was another story.

First discovered in Lake Erie in 1990, round gobies colonized all five Great Lakes by 1995. The goby's rapid spread was aided by freighters that transported cargo within the lakes. Ships that took on ballast water at night, when gobies rose from the lake bottom to avoid predators, inadvertently sucked tiny gobies into their bellies. The goby-infested ballast water was then discharged in other Great Lakes ports. Freighters were to gobies what Johnny Appleseed was to fruit trees.[2] Though scientists were alarmed by the goby's colonization of harbors around the lakes, they were largely clueless in 1990 as to how much havoc the round gobies—working in concert with zebra and quagga mussels—could inflict on native fish and wildlife. The number of gobies in Lake Erie alone was enough to strike fear in the heart of fisheries biologists. There were an estimated *10 billion* gobies in western Lake Erie in 2002. That broke down to an average of 20 to 50 gobies per square meter in Erie's 2,500-square-mile western basin—a phenomenal number for a single fish species.

Round gobies also found Lakes Michigan, Ontario, and Huron to their liking. When researchers first looked for gobies in Lake Huron in 1996, they found none. They found thousands seven years later. A ten-minute trawl of the lake bottom offshore of Harbor Beach, Michigan, on western Lake Huron, brought up 425 gobies. A similar trawl at Point Lookout, on Saginaw Bay, produced 646 gobies in just ten minutes.[3] The round goby quickly became the worst foreign fish to invade the lakes since the alewife.

Gobies were a significant discovery for Jude, but one that filled him with more disgust than pride. He considered the fish one of many examples of how the U.S. and Canadian governments were failing to stem the tide of invasive species sneaking into the lakes in ocean freighters' ballast tanks. He explained his thoughts in a 1997 research paper. "The presence of gobies . . . is a symptom of our lack of dealing with the global transport of non-indigenous species. The recent examples of zebra mussels, ruffe and the gobies are testimony that we still have not curtailed this threat to our aquatic ecosystems. These organisms are wreaking more havoc with ecosystem health than toxic substances or nutrients, since at least we have done, and continue to do, something about controlling those problems," Jude said. "Once in, most successful exotic species establish permanent populations which are irrevocably detrimental to fish community stability and cause serious loss of irreplaceable genetic material and biodiversity. Their effects are difficult to deal with, complex and, like a broken heart on Valentine's Day, it is something we just can't fix."[4]

Ever the comedian, Jude sought to add levity to a serious ecological problem spreading unabated across the lakes. He deemed gobies the "Cyberfish of the Third Millennium. These fish are cyberfish because they came from a distant universe and have the unusual ability to attain high abundances . . . in the face of native fish communities and they also are able to disperse rapidly using Great Lakes freighters as transport vectors."[5]

Gobies and other Great Lakes invaders became somewhat of an obsession for the brilliant but unconventional Jude. He even called on his musical past, as a member of his college church band Plastic Jesus Pants, to pen a wistful song about gobies. Jude performed his tune, "The Galloping Goby Blues," at large research conferences and small-group seminars, prompting applause and laughter in both settings. His song was a much-needed dose of humor in the otherwise depressing saga of invasive species reconfiguring the lakes' ecosystems. It went like this:

We got evil fish coming into town,
Slippin' in our lakes without a sound,
And unless we stop them cold, they'll beat our natives down,
There's egg-suckin', fish bitin' exotics all around.

There's ruffe in our largest, greatest lake
Zebras, mitten crabs, it'll keep you awake
There's *Bythotrephes* in the summer, *Dreissena* in the fall,
If the quaggas don't get you, you've broken Darwin's Law.

Round gobies now rule among the rocks,
They can live out of water, they'll bite you in your socks,
And if we don't stop them soon, they'll only be bad news,
I've got those egg suckin,' fish chewin,' galloping goby blues . . .[6]

Though Jude made light of the goby's arrival, he and other scientists feared that the fish was another cog in the wheel of a biological meltdown steamrolling Great Lakes ecosystems. The lakes had suffered a similar meltdown in the mid-1900s. After sea lamprey infested the lakes and wiped out lake trout, alewives became the dominant fish species in all but Lake Superior. Alewives entered the lakes before sea lamprey and likely would have been kept in check if sea lampreys hadn't decimated the top native fish predator, lake trout. Absent a native predator, alewives thrived until Pacific salmon were stocked in the lakes to rein in the invader derided by biologists as a trash fish. Round goby threatened to duplicate the alewife's siege of the lakes.

Dave Brenner, Michigan Sea Grant

Round gobies and zebra mussels (*shown above the fish*) were part of a foreign food chain that took hold in the Great Lakes in the 1980s.

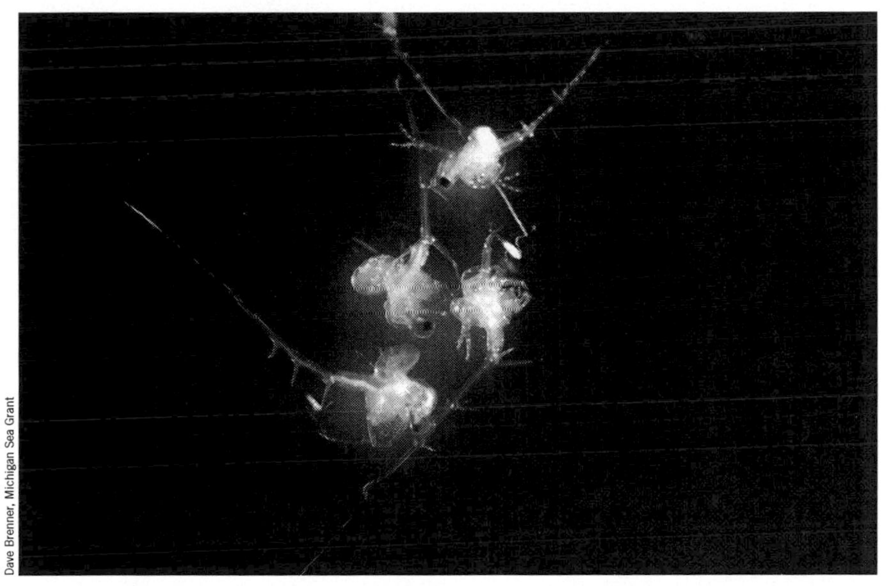

Dave Brenner, Michigan Sea Grant

Spiny water fleas (*shown here*) were part of the foreign food chain fostered by zebra mussels. The invasive zooplankton competed with native Great Lakes species and fouled fishing gear.

The invasional meltdown that zebra mussels triggered intensified in the mid-1990s. Two invasive species of zooplankton that competed fiercely with native species were found in the lakes and quickly spread. *Echinogammarus ischnus,* a shrimp-like creature, was discovered in the Detroit River in 1995. *Cercopagis pengoi,* the fishhook water flea, turned up in Lake Ontario in 1998. *Echinogammarus ischnus* worried scientists because it thrived in the presence of zebra and quagga mussels and was a voracious predator. It was more aggressive than native amphipods and was even known to eat its own kin. *Echinogammarus* and spiny water fleas weren't expected to cause nearly as many problems as zebra mussels or round gobies. What freaked researchers was the notion that entire communities of foreign species from the Black and Caspian seas were colonizing the Great Lakes.

Canadian research scientist Anthony Ricciardi concluded in 2000 that the goby's arrival on the heels of zebra and quagga mussels was clear evidence that the lakes were suffering an invasional meltdown. "I am convinced that the goby's rapid invasion of the Great Lakes was facilitated by this abundant food source [zebra and quagga mussels]," Ricciardi said in a newspaper interview.[7]

A group of distinguished scientists impaneled to assess the potential ramifications of foreign species creating self-sustaining food chains in the lakes echoed Ricciardi's assessment. Their conclusions, published in the *Canadian Journal of Fisheries and Aquatic Sciences,* were grim. "Round gobies, *Echinogammarus* and *Dreissena* (zebra and quagga mussels) formed an invasional meltdown community in the Great Lakes with the zebra mussels facilitating expansion of its Ponto-Caspian associates in the Great Lakes. The aggressive strategist round gobies in particular benefited because *Dreissena* is a favored prey that was not being utilized efficiently by native Great Lakes fishes. Based on the experience in Lake St. Clair, it seems likely that round gobies will become an important part of the Great Lakes food web."[8]

The authors of that study, one of whom was Jude, also warned that gobies feasting on zebra mussels could create a new mechanism for persistent toxic chemicals to move up the food chain and onto the dinner plates of people who ate Great Lakes fish. That was a frightening scenario, especially in light of the fact that billions of dollars spent on sewage-treatment improvements during the previous two decades had dramatically reduced toxic contaminants in fish. As zebra and quagga mussels filtered water through their tiny bodies, they concentrated minute amounts of persistent toxic chemicals, such as PCBs, in their flesh. Gobies that ate the tainted mussels absorbed the contaminants, as did the bass and other fish that ate gobies.

"What we've done is create a new mechanism to get PCBs into fish and people who eat fish," said Jeff Reutter, director of the Ohio Sea Grant program at Ohio State University. "We have less PCBs coming into the lake, but by changing the food web, zebra mussels have doubled the concentrations of PCBs in fish that people are going to eat. That is really significant."[9] It was a frightening, discouraging scenario. And yet, history proved that the ramifications of the invasional meltdown facilitated by the zebra mussel/round goby/*Echinogammarus* connection would be far worse than the initial, dire predictions. Just those three species managed to reconfigure, in less than a decade, massive freshwater ecosystems that had been 10,000 years in the making.

What Lies Beneath

Frank Bihn had an unusual hobby that led him to a remarkable encounter with one of the worst invasive species in the Great Lakes. In the early 1990s, Bihn liked to walk in Lake Erie's Maumee Bay when a seiche drained water from much of the bay. A seiche is one of the most fascinating and dangerous natural phenomena in the Great Lakes. It is the freshwater equivalent of a tsunami, only far less intense than the tidal waves that rage across oceans. Pronounced *saysh*, the surges are caused by powerful winds or fast developing squalls that force air pressure down on the surface of a Great Lake, much like a platter plunged into a tub of water. In either case, the result is the same: the lake's surface tilts, with water levels dropping on one side and rising by an equivalent amount on the opposite side. The events, which are fairly common on the lakes, are an act of nature to behold and fear.

Most seiches are small, subtle events. Occasionally, they spawn huge, deadly waves. That was the case on July 4, 1929, when a seiche produced waves 20 feet high at Grand Haven State Park in western Michigan. Ten people drowned at the park that day, one of whom was swept off the city's pier by the powerful seiche. A similar event in Chicago, in 1954, caused Lake Michigan's water level to rise four feet in just 30 minutes. That was followed by a massive wave estimated at 20 feet high and 25 miles wide. The wave swept several people off the city's pier, killing eight.

David Jude, Michigan Sea Grant

The round goby was discovered in the Great Lakes in 1990 and spread to all five of these lakes. Scientists estimated there were 10 billion round gobies in western Lake Erie in 2002.

Lake Erie is the most common site of seiches on the Great Lakes. The lake's relatively small surface area (10,000 square miles), shallow western basin, and east-west orientation make it vulnerable to dramatic water-level fluctuations. When strong winds blow out of the southwest, water levels in the western basin can drop several feet near Toledo and rise by the same amount in Buffalo, New York. The effects of a strong seiche can be so pronounced as to leave boats docked in Erie's western basin sitting on the lake's muddy bottom. This is not an urban legend—photographers have documented such events. In 2001, the National Oceanic and Atmospheric Administration documented a seiche that dropped water levels in western Lake Erie by six feet in just nine hours, sending a massive surge of water toward Buffalo.

A strong seiche could cause the water in front of Frank and Sandy Bihn's house to retreat more than 1,000 feet into the deeper areas of Lake Erie. On those days, Frank Bihn would walk into the parched bay to fetch lures and other tackle anglers had lost while fishing for walleye or perch. He was walking across the muddy bottom of Maumee Bay one day in the mid-1990s, collecting lures, when he noticed things squirming in puddles of water that remained in the drained bay. They were odd-looking fish, four to six inches long, with a squared off snout and bulging eyes near the top of their heads. The fish looked like something from the horror film *Creature from the Black Lagoon.* By the late 1990s, Bihn gave up looking for fishing lures when seiches sucked the water out of Maumee Bay. "All the standing water left in the holes on the lake bottom were filled with gobies," he said. Sandy Bihn said the wriggling lake bottom was creepy. "You'd walk out there and there were gobies all over the place," she said. "It was like, 'Holy shit.'" She was not exaggerating. Scientists in 2002 estimated that there were 10 billion gobies in Lake Erie's 2,500-square-mile western basin. That worked out to nearly 50 gobies per square meter. Imagine having 50 gobies, each ranging from three to six inches long, occupying an area roughly the size of a residential bathtub. That was the situation in Lake Erie after 1990, when gobies stormed the lake. It was the stuff of nightmares—for humans and native fish forced to cope with the invaders.

THE GOBY INVASION MARKED THE END OF TREASURED FISHING EXPERIENCES for many Great Lakes anglers. Among those hardest hit by the invasion were children who learned to fish on piers that jutted into the lakes from hundreds of coastal communities. The fish of choice for pier anglers often was perch, a delicious species that was abundant in Lake Michigan before zebra mussels invaded. The goby displaced perch in many areas of the lakes, forcing countless anglers to fish elsewhere. There was no joy to be had catching goby: the fish were too small to eat, and too ugly to mount on a wall. University of Notre Dame biology professor David Lodge explained the changes gobies caused in stark terms when he testified before a congressional committee in 2007. He said:

> The addition of just one species [the goby] to North America matters for several related reasons. In southern Lake Michigan, where recreational and commercial anglers used to harvest vast numbers of prized native yellow perch, now only invasive round gobies are caught. In the lakeside economically depressed areas in northern Indiana and southwestern Michigan, where poor boys and girls used to be able to catch their dinner off the breakwaters, fishing is now futile, unless they want round gobies on the menu. I've had this experience myself: pulling in small and useless goby after goby, with not a single native or valuable

> fish species in hours of fishing. . . . Why? Because round gobies eat the eggs and fry of smallmouth bass and other highly valued fishes, out-compete valuable fishes for food and out-compete native bottom dwelling fishes for shelter.[10]

David Vogt, a longtime Lake Michigan angler, recalled a frustrating goby encounter in 1998. Vogt volunteered to help a group called Perch America take a group of 50 Chicago boys fishing in Lake Michigan. The outing was a part of the Big Brothers program, which paired boys whose fathers weren't involved in their lives with successful men who served as role models and friends. That year, Big Brothers joined with Perch America to take a group of boys fishing from a pier on 95th Street. Perch America supplied the bamboo fishing poles; the Big Brothers provided instruction and encouragement. The boys caught an astounding number of fish—1,100 in just three hours. All were gobies, every last one. Vogt recalled the experience when he testified before Coast Guard officials at a hearing in Cleveland on the issue of ballast water treatment standards. "Three hours fishing in one place and a nonnative fish that came from somewhere else has just taken over, completely," he said. "It is just unbelievable to see that many gobies caught. Nothing else. There were no perch, sunfish, anything. It was just unbelievable. I had to tell you."[11]

Migrating armies of gobies swarmed into the mouths of Great Lakes tributaries like angry honeybees swirling around a hive. Bewildered anglers expressed disbelief and disgust as popular perch fishing holes were overrun by gobies. Many simply gave up and moved to other fishing spots where gobies had yet to invade. But one man decided to take matters into his own hands. Tom Matych, a factory worker in the gritty blue-collar town of Muskegon, Michigan, decided in 2005 to wage war on gobies. He feared the invaders would drive away the native fish he loved to pursue in Muskegon Lake: yellow perch and walleye. Muskegon Lake was a 4,150-acre embayment on the east coast of Lake Michigan that supported one of the region's most diverse and abundant fisheries. Determined to save perch and walleye, Matych organized a fishing tournament on Muskegon Lake. The target: round gobies.

Matych's goby tournament didn't promote the type of fishing experience where anglers soaked up nature's beauty and tranquility while testing their skills against fish. Not even close. The goal at Matych's tournament was to catch and kill as many gobies as possible. The team that caught the most gobies, and the individual who caught the largest goby—"Goby Dick," in

Matych's words—would win hundreds of dollars in cash. Four hundred anglers signed up for Matych's "Goby Assault Party." Collectively, they hauled in 5,000 gobies in just five hours. It was a tremendous haul, and a drop in the proverbial bucket. Carl Reutz, a biology professor at Grand Valley State University in West Michigan, had two of his students fish for gobies off the Muskegon pier a few weeks before the goby tournament. "It was amazing how many they could catch," Reutz said. "At times they were approaching 60 gobies per hour."[12]

Matych wanted the state of Michigan to stock millions of yellow perch and walleye in Muskegon Lake and Lake Michigan to drive away gobies and an earlier invader, alewives. State officials rejected his request, saying there was no evidence that planting more native fish would drive out the foreign species. "While it is true that goby and alewife are exotic species that we wish had never invaded the Great Lakes, there is no human management activity that will ever be able to make them disappear," said Jim Dexter, Lake Michigan basin coordinator for the Michigan Department of Natural Resources, in a 2004 letter to Matych.[13] What Dexter didn't mention was that the state didn't want to eliminate alewife from Lake Michigan. The reason: Lake Michigan's billion-dollar salmon fishery would vanish, and scores of port communities around the lake would suffer the economic consequences. Biologists were desperately trying to keep the lake's salmon fishery healthy in the face of a second invasional meltdown that affected all levels of the food chain. Dealing with sea lamprey and alewife was a relative cakewalk compared to coping with the ecological mayhem that zebra and quagga mussels inflicted on the lakes. Chemical applications controlled the lamprey population, and the artificial salmon fishery reined in alewife. There was no known way of reducing the number of zebra or quagga mussels in the lakes. Nor was there any known method for reducing a massive goby population that dined on zebra and quagga mussels. The foreign mussels were the nucleus of a biological New World Order in the Great Lakes. How the new order of things would play out was anybody's guess.

There are seven species of *Dreissena* mussels in the world. Two species, *Dreissena polymorpha* (zebra mussels) and *Dreissena bugensis* (quagga mussels) invaded the Great Lakes. The results were disastrous.

In terms of the whole food web, I don't think there's any question that zebra and quagga mussels have had the largest impact on the biological communities of the Great Lakes.

—Tom Nalepa, research biologist, National Oceanic and Atmospheric Administration's Great Lakes Environmental Research Laboratory, 2007

03

THE *DREISSENA* EFFECT

We have created an environment that is ideal for algae to grow.

—Joseph Makarewicz, State University of New York biologist, 2008

12 SOMETHING AMUCK

For a few days each summer, when still winds allowed the surface of southern Lake Ontario to lie flat, the lake's steel blue water took on a clarity approaching that of the Caribbean Ocean. The phenomenon became prevalent in the mid-1990s along Grand View Beach, a narrow spit of land on the outskirts of Greece, New York. Grand View Beach was a popular summer retreat for tourists and dozens of families who, through good economic fortune or inheritance, owned cottages within a stone's throw of the lake. For them, sweltering summer days were countered by dips in the lake's cool water. At night, the rhythmic sound of the inland sea's small waves lapping at the shoreline serenaded the cottage dwellers. Life on that beach was grand, indeed. Alas, there was a price to be paid for the lake becoming a mirror image of a tropical ocean. Longtime residents knew that the unusually clear water was the calm before a sickening storm. A west wind arriving after a prolonged period of calm water often delivered a load of algae that stained the scenic beach with mats of putrid, dark green vegetation. Most residents stayed inside on those days, windows closed, to avoid the fetid odors of decaying algae. It was a cruel hoax, the pattern of breathtakingly beautiful water spitting out nauseating mats of algae that had been dislodged from the lake bottom and blown ashore by wind and waves. "It wasn't like this when I was a kid," said Joe Martin, a retired dentist who had spent most summers since 1927 at his Grand View Beach cottage.

The cycle of clear water followed by massive algae blooms was the handiwork of *Dreissena* mussels—zebra and quagga mussels—imported from Europe a decade earlier. The prolific filter-feeders dramatically increased water clarity in numerous areas of the Great Lakes, a change that many residents of the region interpreted as a sign that water quality was improving. Clearer water was aesthetically pleasing to boaters and swimmers. Who wouldn't want to swim in a lake that looked like a swimming pool? The problem was that zebra and quagga mussels made the lakes too clear. It was an underwater version of the Butterfly Effect. The foreign mussels consumed huge quantities of plankton as they filtered water through their tiny bodies—triggering a cascading series of profound ecological changes.

One of the early signs of trouble surfaced in 1995, seven years after zebra mussels were discovered in Lake St. Clair. Large mats of native algae, *Cladophora glomerata*, returned to shallow Great Lakes bays with the vengeance of a spurned lover. Algae was an essential part of the aquatic ecosystem—too much of it suggested the system was out of balance. The resurgent algae blooms initially puzzled scientists. The phenomenon was counterintuitive. How could the lakes produce so much algae years after tough environmental laws had dramatically reduced the amount of phosphorus and other nutrients flowing into the lakes? With less phosphorus in the water, one would expect to see less algae, not more. Researchers knew that zebra and quagga mussels increased water clarity, which fueled the growth of submerged aquatic plants. But that didn't explain the tremendously abundant resurgence of *Cladophora*, which thrived in lakes with elevated phosphorus concentrations. Something else, something new, was at work. Was it possible that the imported mussels were fueling the algae blooms? It was an intriguing and entirely unpleasant possibility. If that was the case, the mussels had the potential to unravel dramatic water-quality improvements achieved in the 1970s and '80s, after government agencies spent $20 billion to reduce phosphorus inputs to the lakes.[1] That disturbing scenario had uncertain ramifications for Great Lakes beaches and the humans, fish, and wildlife that congregated along the shoreline.

Researchers from the University of Waterloo in Ontario launched a detailed study of the situation in 1995, after observing huge algae blooms on the north shore of Lake Erie. They spent seven years tracking the return of *Cladophora*. Their findings could not have been more disheartening: Zebra and quagga mussels were fueling the algae's phenomenal growth. Adding to the misery, a separate study discovered that decaying *Cladophora* on beaches was a reservoir for the potentially deadly *E. coli* bacteria. Water quality in parts of the Great Lakes was suddenly backsliding after two decades of steady improvement. Scott Higgins, a research scientist at the University of Waterloo, laid out the situation in a 2005 study:

> Despite the large reductions in phosphorus loadings and concentrations, it is troubling to note that large stands of *Cladophora* are currently widespread in the lower Great Lakes, especially in oligotrophic systems such as the western basin of Lake Erie. The invasion and proliferation of *Dreissena* [zebra and quagga mussels] within the Great Lakes has resulted in environmental conditions that are conducive to the growth of *Cladophora*. . . . In eastern Lake Erie, for example, we estimate that 13,000 tons of this algae grows along the shorelines during the peak summer period.[2]

That same year, a team of scientists from the United States and Canada reached a more disturbing conclusion: "Lake Erie increasingly resembles the polluted lake known during the late-1970s and early-1980s rather than the cleaner lake of the early 1990s."[3]

Compelled by public outrage and health concerns, researchers scrambled to figure out how foreign mussels triggered algae blooms when there was less phosphorus—fertilizer, essentially—entering the lakes from sources on land. That answer was eventually provided by Higgins's mentor at the University of Waterloo, biology professor Robert E. Hecky. He theorized that the ability of zebra and quagga mussels to filter huge quantities of water was short-circuiting the distribution of phosphorus and other nutrients in the lakes. The mollusks created what Hecky termed a "near-shore shunt." Bands of water-filtering mussels sucked phosphorus out of the water and concentrated it along the shoreline where the mollusks lived. When mussels excreted bodily waste, they effectively fertilized the nearshore area with a concentrated dose of soluble reactive phosphorus, the type that fueled algae growth. In Lake Michigan's Green Bay, phosphorus concentrations increased 22 percent following the zebra mussel invasion.[4]

By concentrating phosphorus near the shoreline, the mussels starved deep areas of the lakes of the nutrients that fed the entire food chain. Rocky coastal areas where mussel densities were greatest became underwater algae forests, while some deeper areas of some Great Lakes became biological deserts. The concentration of phosphorus along the shoreline was a subtle but incredibly significant change in the way nutrients were distributed in the lakes. The resulting effects would be felt at all levels of the food chain, from microscopic plankton to massive sturgeon. The storm of algae generated by zebra and quagga mussels fouled vast areas of shoreline from Wisconsin to New York. Elevated bacteria concentrations in the algae prompted government agencies to close popular beaches. Owners of million-dollar lakefront homes were driven inside by gut-wrenching odors emanating from the very water that had lured them to these freshwater seas.

THE BEACH COMMUNITIES WEST OF ROCHESTER, NEW YORK, ON THE south side of Lake Ontario, were among those hit earliest and hardest by the resurgence of *Cladophora*. Recurrent algae blooms in the mid-1990s soiled treasured beaches in the town of Greece, several miles west of Rochester. Residents in the small community of vacation homes, situated atop a narrow strip of land between Lake Ontario and Long Pond, encountered putrid odors and discolored water every time the algae blooms returned to their beloved beach. Homeowners who had witnessed the lake's recovery over the previous two decades were outraged. Dr. Martin was one of the first residents to express concern about the problem. He had lived through waves of biological pollution that plagued Lake Ontario during the twentieth century: piles of dead alewives that littered the beach from the 1940s through the late 1960s, after sea lamprey wiped out lake trout and plunged the fishery into chaos; and huge blooms of blue-green algae, fed by phosphorus discharges from sewage treatment plants, that turned the lake the color of pea soup in the

1950s and '60s. He said those problems paled in comparison to the conditions zebra mussels caused.

"When we have a summer with average precipitation, the algae is worse now than it was in the 1960s," Martin said. "I live next to a beach that is probably 20-feet wide and when the algae comes in it becomes stagnant on the beach. It smells like dead fish. It's pretty bad—it's a pain. We have to close our windows sometimes."[5]

Officials in Monroe County, where the town of Greece was located, waged an all-out war on the noxious algae. The county's algae eradication program focused on the popular Ontario Beach, which was five miles east of Greece and bore the brunt of algae blooms in the 1990s. The cleanup effort was not a new campaign for the county. Crews often stormed the beach in the 1970s and '80s with tractors to remove tons of algae. Theirs was an annual struggle to make the beach fit for human use. The volume of algae washing ashore declined dramatically in the late 1980s, only to return with a vengeance in 1993. By 1995, after zebra mussels had colonized the nearshore areas of Lake Ontario, massive slugs of *Cladophora* routinely washed ashore. The decaying algae occasionally turned black, creating the appearance of an oil slick in the molasses waves. The county bought a custom aquatic weed harvester that year to fight the algae. That, too, proved futile. The tractor was incapable of ridding the water of thin strands of *Cladophora,* which had the consistency of human hair. The county sold that tractor and opted to go back to using front-end loaders to scoop algae off the beach. That didn't work much better. On July 2, 2001, winds blew a huge quantity of algae onto the beach, transforming the water near the shoreline into a thick soup that harbored dangerously high levels of bacteria. The bacterial pollution forced officials to close the beach for the next 19 days.[6] The recurring algae blooms, which made the beach aesthetically unpleasant and nurtured potentially dangerous pathogens, remained a problem into 2007.

Greg Pien, who took up the algae fight after Martin stepped down as president of the Grand View Beach Association, said local efforts to deal with the mess were well-intentioned but of dubious value. He questioned whether small-scale actions could solve a problem that affected much of Lake Ontario's 712 miles of shoreline. "Some of the efforts were sort of quixotic," Pien said. "It's such a huge lake, I'm not sure you could do anything that would have a measurable effect in your lifetime."[7]

Shoreline property owners watched in disgust as the menacing algae from years past blanketed beaches and discolored the water in parts of four of the five Great Lakes—only Superior was spared. With no feasible way to remove zebra and quagga mussels from the lakes, the algae plague was free to run its course. University of Wisconsin researcher Wendy S. Stankovich explained the situation at an algae workshop: "As of 2004, no known control method has been developed to solve the *Cladophora* problem along the Milwaukee shoreline and, at this point, it is difficult to suggest any suitable mechanism to control *Cladophora* in a large water body such as Lake Michigan. Unless a control method is developed, the *Cladophora* problem is not expected to wane at any time in the near future."[8]

Rotting mats of *Cladophora* routinely soiled hundreds of miles of beaches on Lakes Erie, Ontario, Huron, and Michigan in 2007. Aside from being a nuisance and a potential health threat, the algae blooms gave the Great Lakes another black eye and hurt the region's tourism economy. How bad was the situation? One summer, crews removed 25 tons of *Cladophora* from Lake Michigan beaches near Milwaukee. They could have removed far more; their noxious harvest was limited only by the number of people willing to help with the gut-churning task of scooping rotting *Cladophora* off beaches and hauling it to a landfill.

THE VAST EXPANSE OF SANDY BEACHES THAT LINE THE NORTHWEST COAST of Michigan's Lower Peninsula are among the region's most beautiful. Miles of off-white sand that has the consistency of sugar give way to water of stunning clarity. Towering dunes in some areas offer breathtaking, panoramic views of sparkling blue waters that extend beyond the imagination. One of the most charming spots along the 60-mile stretch of heavenly beaches is tiny Cross Village—a burg with one restaurant, one gift shop, and one intersection on the one street that runs through town. The remote village was the place Dr. Henry Singer chose to spend his retirement. A successful ear, nose, and throat specialist in nearby Petoskey, Singer had the financial resources to buy a retirement home just about anywhere. He chose a log cabin on the sparsely occupied shoreline south of Cross Village.

Singer said he felt a calling to retire along the same lakes where he had spent much of his life. He grew up in Toledo, where algae blooms in the 1950s

Jeff Alexander

Zebra and quagga mussels that have washed up on beaches, like this one along eastern Lake Ontario, have also increased water clarity in parts of the Great Lakes. The clearer water has fueled the growth of *Cladophora*, a native algae species that washes up on beaches when wind-driven waves rip it off lake bottoms. The algae has harbored harmful bacteria and has fouled the water, often forcing health officials to close popular beaches.

and '60s turned Maumee Bay the color of pea soup and rendered the lake unfit for swimming. Singer was working at the Cleveland Clinic in 1969 when the Cuyahoga River caught fire. After a stint in Chicago, Singer ended up being a medical specialist in a small tourist town in northern Michigan. He loved the area's relaxed lifestyle and the people, who were more inclined to judge people by their character and actions than by material possessions.

Singer planned to spend his retirement years 20 miles north of Petoskey, soaking up the wondrous, changing moods of Lake Michigan from his waterfront cabin. His first few years of lakefront living were glorious—long morning walks on the sugar-sand beach as loons wailed in the distance. He spent afternoons watching his grandchildren swim in the lake and skip stones across the translucent water. Then, in 2001, the foundation of Singer's personal paradise began to crumble. Windrows of tiny shells, the armor of zebra and quagga mussels, began washing up on his beach. Wave after wave of shells piled up on the shoreline, smothering some parts of the beach with a layer of shells two feet thick. Algae soon followed, transforming the idyllic beach into a swamp of *Cladophora* that looked like a mix of regurgitated pea soup and cooked spinach. "To see our wonderful Lake Michigan spoiled like this is just horrendous," Singer said in a 2007 interview. "Since 1981, we've had people stop here and say they can't believe how clean the water is. Now the lake is turbid. The algae makes the water look like pudding."[9]

Outraged by the nauseating mess unfolding on his beloved beach, Singer in 2002 aired his disgust with fellow members of the local lake association. He did so with a clever bit of show and tell. Singer took a glass from the restaurant where the group was meeting and filled it with tap water. The water was clear, clean, and odor-free. He then unveiled a glass of water he had collected from his beach earlier in the day. The water from his beach was lime green, opaque, and smelled like rotten food. His fellow lake lovers were aghast. They, too, had seen the piles of mussel shells and mats of algae on the beach. But no one knew what to do about the problem.

A local environmental group and scientists from the National Oceanic and Atmospheric Administration's Great Lakes laboratory were called in to study the algae that blanketed Singer's beach and the expansive Sturgeon Bay, home to the wildly popular Petoskey State Park. Researchers found four different types of algae on the beach. Determining the species of algae was easy. Figuring out how to keep it from washing ashore, that was a far more complicated challenge. In reality, there was no solution, short of eliminating nearly a half-billion pounds of zebra and quagga mussels living on the bottom of Lake Michigan.

Desperate to reclaim their sullied beach, Singer's wife, Naomi, and some neighbors tried repeatedly to rake the mussel shells off the tarnished shoreline. They were spinning their wheels—the ailing lake vomited tons more mussels onto the beach every time a storm stoked angry waves that pounded the shoreline. Reeking mats of algae that washed ashore in the summer of 2007 drove Singer from his cottage for three days. "It was a terrible, putrid smell. It's a pungent odor—very, very offensive. And I have a strong nose," Singer said. "This is very scary for us. We love our lake but this ecosystem has been degraded and this algae is a symptom of a serious problem."

Having witnessed the Great Lakes recover from the wretched pollution of the 1950s and '60s only to see invasive species foul the lakes anew turned the soft-spoken Singer into a fire-and-brimstone environmental activist. The doctor who had spent a career helping sick humans wondered if people who loved the Great Lakes realized the extent to which invasive species had sickened the majestic waterways. "People drive over the Mackinaw Bridge and see the beautiful water and they don't think about the catastrophe taking place in these lakes," Singer said. "We take these lakes for granted."[10]

LIKE MANY OF THE 40 MILLION PEOPLE WHO LIVED IN THE GREAT LAKES basin, I was occasionally guilty of taking these amazing glacial relics for granted. Living in West Michigan, where the beaches were sandy, I was not subjected to the *Cladophora* blooms that plagued beaches further up the coast and along the shores of Lakes Ontario, Erie, and Huron. West Michigan beaches were thought to be immune from algae blooms because the southeast coast of the lake had few rocks to which zebra mussels and *Cladophora* could cling. Ours was a false sense of security.

West Michigan residents learned in the spring of 2007 that quagga mussels could unleash algae blooms in areas where the lake bottom was dominated by sand, not rocks. The first hint of trouble came in early June, when several wind-free days allowed the lake to settle into a state of astonishing clarity. So transparent was the water, it was possible to see the lake bottom that lay 30 feet beneath the end of the Grand Haven pier. I had never seen the lake so clear in my 15 years of living in West Michigan. I was walking along the pier on a warm summer night, marveling at the hundreds of round goby huddled around boulders along the base of the pier, when a friend snuck up from behind and startled me.

"Can you believe the lake?" he said. "It's so clear."

I knew the words about to come out of my mouth would make no sense to him.

"This is not good. It's too clear."

"Are you kidding?" he said. The doubt in his voice was reinforced by his body language, which included arms folded tightly across his chest. "How could this be bad? What's causing it?"

"Quagga mussels," I said.

"What's that?"

"The bigger, meaner cousin of zebra mussels," I said.

I struggled to explain the complex situation in simple terms. My intent was not to be condescending—this truly was a complex problem. Explaining it in a way that nonscientists could understand was difficult.

Here's what I told him: The mussels filtered particles out of the water column, concentrating the phytoplankton and zooplankton that fish and other critters needed to survive on the lake bottom. That made the water clearer.

"But clear water is a good thing, right?" he said.

To a certain extent, I said. But the clear water allowed sunlight to penetrate deeper, which made more algae grow. Wind-driven waves ripped the algae off the lake bottom and pushed it up to the beach.

"The next time the wind blows out of the west, we're going to get algae on the beach. It won't be pretty," I said.

My friend wasn't convinced. That wasn't surprising. There had not been an algae bloom in this part of the lake for more than two decades.

The next week, my wife and I walked down to Lake Michigan to spend Father's Day lounging on the beach. As we reached the crest of our hilly road, we saw the lovely, familiar, and always inspiring lake. But there was something different about the lake that day. We noticed it almost as soon as we caught our first glimpse of the alluring water.

"What's that in the water?" my wife said.

My reaction was more impulsive. The words shot out of my mouth: "Oh, crap."

My prediction proved accurate. For one of the few times in my life, I hated being right.

A band of algae roughly 100 feet wide cast a pale brown stain on the turquoise water; it ran parallel to the beach and extended as far as the eye could see to the north and south.

A stiff breeze the day before had slimed a 30-mile stretch of the shoreline in western Michigan, from Holland to Muskegon. Our beach was in the heart of the worst algae bloom to hit this area in more than 20 years. It was a sickening, unsettling sight. The algae instilled in us a sense of vulnerability, of innocence lost. From that day on, every trip to a West Michigan beach was tempered by the fear that I would encounter algae in the lake. Little did I know that the worst of the Father's Day algae storm had yet to rear its ugly head.

The day after the algae slick ruined a gorgeous Sunday afternoon for thousands of beach-goers, health officials discovered high levels of *E. coli* bacteria in the lake at Pere Marquette Beach in Muskegon.[11] The finding prompted officials to urge people to stay out of the water. It was a shock for Muskegon residents—Pere Marquette consistently ranked among the nation's cleanest beaches, in terms of water quality. The algae bloom, if nothing else, taught all who loved these beaches that nothing in the Great Lakes was sacred in the era of zebra and quagga mussels. Everything was at risk—fisheries, the health of humans and wildlife, and the peace of mind that came with knowing it was safe to swim in the water or walk the sands of treasured beaches.

What we're seeing is a system out of balance.

—Gregory Boyer, chemistry professor, State University of New York, 2008

13

BLUE, GREEN, AND DEADLY

Family vacations spent at lakes are supposed to create pleasant memories, lasting mental images of giddy children leaping off docks and stressed-out parents frolicking with their kids in a sort of liquid heaven. The best trips can turn even the most unfortunate mishaps into treasured, comedic adventures. Broken-down vehicles and bouts of stomach flu that made everyone miserable become the stuff of family legend and laughter. Yet, there are times when tragedy prevails, trumping the highlights of the most revered vacations. Such was the case in 1999, when a Vermont family headed to a cabin on the shores of Lake Champlain. The Couture family of Burlington, Vermont, went to the scenic lake with their beloved Labrador, named Bear, for a few days of relaxation. There, they became the unwitting victims of an environmental tragedy that was spreading across the Great Lakes and many of the region's inland lakes. Like most Labrador retrievers, Bear was drawn to the water like a moth pulled toward light. The dog bolted for the lake and lapped up some of Champlain's liquid bounty. Four hours later, Bear was dead.[1] The killer wasn't a boat, a car, or old age. Toxic algae killed the Couture's family pet. Several more dogs died that year after drinking water from Lake Champlain; one died within an hour of slaking its thirst in the lake. All of the animals were casualties of the same deadly menace: cyanobacteria, or blue-green algae—a naturally occurring bacteria capable of releasing compounds more toxic than some types of poison nerve gas.

Gregory Boyer, a chemist at the State University of New York who determined that the dogs that died after drinking Lake Champlain water had ingested blue-green algae, said the incidents were heartbreaking. "These pets were parts of people's families; it was fairly traumatic," Boyer told researchers at a Great Lakes conference in 2007. "Kids were writing letters to the editors of newspapers saying, 'Please save my dog.' It was very sad."[2] The dog deaths on Lake Champlain were cause for concern in the Great Lakes, which were afflicted by the same type of toxic algae. In fact, the toxin in Lake Champlain that killed the dogs surfaced the same year in mats of blue-green algae that formed on the surfaces of shallow bays around Lakes Erie, Huron, Michigan, and Ontario. No longer were the toxic remnants of industrial discharges the only threat to humans, fish, or wildlife in the Great Lakes basin. Potentially deadly algae blooms posed equally acute health risks in places like Saginaw Bay, Maumee Bay, Green Bay, and the Bay of Quinte.

A GROUP OF SCIENTISTS WERE ANALYZING COLOR SATELLITE IMAGES OF THE Great Lakes in 1995 when they noticed bright green patches on the surface of Lake Erie's Maumee Bay. The water wasn't the deep blue characteristic of the rest of the lakes—it was fluorescent green. The color contrast between water in the bay and deeper areas of the lakes was striking, provocative, and unsettling. The lime green water suggested that a pernicious problem from the 1960s had returned: blue-green algae blooms. Its sudden return that year left boaters and shoreline-property owners wondering why the resurgent lake was covered with a blanket of scum that looked like green paint and smelled like sewage.

Scientists quickly figured out that the bright green cloud scum on the surface of western Lake Erie was a bloom of cyanobacteria—a harmful algae bloom. Cyanobacteria are a class of bacteria that have the ability to release potent toxins when they float to the surface, form a mat of blue-green algae, and die. One of the most common strains of blue-green algae is *Microcystis aeruginosa*. It often surfaces in lakes and ponds with high concentrations of phosphorus. *Microcystis* blooms were common in Great Lakes bays prior to 1972, before new pollution-control regulations dramatically reduced phosphorus discharges into the lakes. The controls improved water quality and greatly reduced the incidence of blue-green algae blooms in western Lake Erie, Lake Huron's Saginaw Bay, and Green Bay, in northern Lake Michigan. For some odd reason, the problem returned in 1995, even though phosphorus concentrations in the lakes were on the decline. Scientists were baffled. There was a sense of urgency surrounding the problem—about half of all strains of *Microcystis* release potentially harmful compounds. Cyanobacteria are among the planet's oldest and most widely distributed species. Some species have been found in fossils that date back 3.5 billion years. The bacteria are an important food source for some plants. Others produce some of Earth's most toxic compounds.

Some cyanobacteria produce saxitoxin, a potentially lethal poison. So potent is saxitoxin, the U.S. government listed the compound as a biological warfare agent. Saxitoxin is 1,000 times more potent than sarin, the nerve gas a Japanese cult used to attack people on a Tokyo subway in 1995. That attack killed 12 people and sickened 5,000 others. As of 2008, there were no known cases of saxitoxin from blue-green algae harming humans, fish, or wildlife in the Great Lakes; the toxin that killed the dogs that drank from Lake Champlain was anabena. Though anabena and saxitoxin were found in some Great Lakes algae blooms, there had never been a documented case of humans becoming ill after coming in contact with blue-green algae in the lakes. But the fact that some cyanobacteria blooms released toxins capable of causing organ damage, cancer, and death was cause for concern. Worse yet, zebra and quagga mussels created ideal growing conditions for cyanobacteria. There seemed to be no limit to the havoc the tiny invaders could cause in the Great Lakes.[3] Boyer, an expert on cyanobacteria, summed up the potential dangers in a 2006 publication:

> Cyanobacteria, or blue-green algae, are ubiquitous in nature and found in nearly all environments. Many have selective advantages [over other algae species], such as the ability to use atmospheric nitrogen for growth, or the production of gas vacuoles to control their exposure to light that allows them to dominate other phytoplankton during the later months of the season. . . . These cyanobacteria blooms can lead to taste and odor problems in drinking waters and the formation of surface scums. Cyanobacteria can also produce extremely potent toxins and if this occurs, the blooms can be hazardous to animals and humans alike. One need only look at their historic names: slow death factor, fast death factor and very fast death factor, to appreciate their effects.[4]

The secret to controlling cyanobacteria was keeping phosphorus concentrations in lakes and streams at healthy levels. Phosphorus fertilized cyanobacteria, just as it fueled the growth of desirable algae. The importance of keeping cyanobacteria in check could not be overstated. Harmful algal blooms, or HABs, had killed farm animals in Australia, dogs in northern New York, and humans in Brazil. In 1988, a toxic algae bloom in a reservoir that supplied water to Bahia sickened more than 2,000 people and killed 88. And in 1996, 75 dialysis patients in Brazil died after drinking water laced with toxic algae.[5]

There were no documented cases, as of 2008, of blue-green algae killing humans in the Great Lakes region. One reason: Municipal water-treatment systems easily destroyed cyanobacteria. Also, experts said a human would have to ingest a measuring cup of toxic algae to endanger their lives. Humans could avoid the dangers associated with cyanobacteria by staying out of water where surface scums formed. The only documented cases of cyanobacteria killing humans or animals were those in which the victims ingested the toxic algae, usually by drinking contaminated water. Still, given the potent toxicity of some blue-green algae, it was understandable that scientists were anxious to unravel what had triggered its resurgence in the Great Lakes. Six years of intense research on western Lake Erie and Saginaw Bay revealed the root cause of the algae blooms: zebra and quagga mussels. The mollusks gave blue-green algae a competitive advantage by eating tremendous quantities of desirable green algae. Their role was analogous to spraying an agricultural herbicide that killed corn and soybeans but allowed weeds to prosper.

Scientists were somewhat relieved to learn in 2001 what was causing the *Microcystis* blooms on the lakes. But that sense of relief was tempered by the notion that invasive *Dreissena* mussels were responsible for yet another major, troubling change in the lakes. The news soon went from bad to worse. Microcystis blooms followed the plague of zebra mussels as it spread beyond the confines of the Great Lakes and infected inland lakes.

Shortly after the 1995 *Microcystis* bloom on Lake Erie, mats of *Microcystis* surfaced on the Bay of Quinte, on the north side of Lake Ontario. Over the ensuing decade, blue-green algae blooms became a nearly annual occurrence in Saginaw Bay and western Lake Erie. The blooms also surfaced in Green Bay, in Lake Michigan; at six sites around Lake Erie; and at seven sites in Lake Ontario, one of which was the Bay of Quinte. Most alarming was the appearance of *Microcystis* blooms on some of the region's cleanest, most popular inland lakes. *Microcystis* blooms sprouted like dandelions on inland lakes from Michigan to New York and Quebec, prompting health officials to issue advisories urging people to stay out of the water. Health officials had good reason to be concerned: One of the toxins that had killed the dogs in New York, anatoxin-a, was found in algae samples collected at sites across Lake Erie, in Lake Huron's Saginaw Bay, and in parts of Lake Ontario. "Significant levels of anatoxin-a have been reported in the Great Lakes, particularly in the western basin of Lake Erie," Boyer said in a 2006 research paper. The concentrations of anatoxin-a found in algae blooms in western Lake Erie were five times higher than the levels that killed dogs that consumed toxic algae in Lake Champlain.[6] With no hope of immediately controlling the mussels that caused the toxic algae blooms, scientists focused on understanding precisely how zebra mussels and other factors—such as phosphorus runoff from farms, residential lawns, and golf courses—might fuel the growth of *Microcystis* and other cyanobacteria. Zebra mussels, it turned out, were picky eaters.

Dreissena mussels survived by eating tiny particles of plant matter and microscopic organisms—phytoplankton and zooplankton—in the water they filtered through their pea-sized bodies. Henry Vanderploeg, a biologist at the National Oceanic and Atmospheric Administration's Great Lakes Environmental Research Laboratory in Michigan, discovered that zebra mussels ate green algae, but didn't like the blue-green strains. The mussels inhaled *Microcystis* into their bodies, but quickly spit it out. It wasn't clear whether *Microcystis* was too large for the mussels to digest, was poisonous to the mollusks, or just didn't suit with their tiny palates. Whatever the reason, the mussels refused to eat blue-green algae—Vanderploeg videotaped zebra mussels spitting out tiny balls of *Microcystis* to prove his point. His discovery had potentially dire consequences for the Great Lakes. It meant zebra mussels would only eat desirable algae, thereby allowing *Microcystis* and other blue-green algae to flourish wherever the mussels were present. Vanderploeg and his team of scientists outlined their concerns in a landmark 2001 research paper:

> *Microcystis* and other cyanobacteria blooms may have serious consequences to aquatic ecosystem function and health, to aesthetics, and to wildlife and human health. *Microcystis* and some other cyanobacteria produce a potent class of hepatotoxins called microcystins that can poison aquatic organisms as well as wildlife, domestic animals and humans that drink or ingest algae in the water.[7]

Subsequent research revealed that zebra mussels excreted nutrients in their bodily waste—phosphorus and ammonia—that fertilized blue-green algae. The phenomenon created ideal growing conditions for toxic algae: The mussels ate the good algae that naturally competed with *Microcystis* and then fertilized the undesirable algae with their feces. With zebra and quagga mussels conquering western Lake Erie and Saginaw Bay in the 1990s, it was only a matter of time before huge *Microcystis* blooms became the norm. The mussels didn't disappoint: Massive blooms of blue-green algae covered parts of western Lake Erie every year from 2003 to 2007.

TOXIC ALGAE BLOOMS WERE NOT UNIQUE TO THE GREAT LAKES. THE phenomenon was documented in 49 of the 50 United States, and in lakes in Canada and numerous other countries. What made the situation unique in the Great Lakes, and inland lakes in the region, was that zebra and quagga mussels, not phosphorus, fueled the resurgence of blue-green algae. That was terribly bad news. Within 20 years of being discovered in the Great Lakes, the mussels had colonized more than 240 inland lakes in Michigan, and numerous other lakes, rivers, and reservoirs in 27 states and two Canadian provinces. Scientists feared that zebra mussels could trigger toxic algae blooms wherever they lived, regardless of water quality. That concern became reality in 2004, when scientists from Michigan State University documented that zebra mussels caused *Microcystis* blooms in clean, low-phosphorus lakes. Inland

Jeff Alexander

Zebra and quagga mussels have contributed to the proliferation of toxic blue-green algae blooms (shown here on an inland lake in West Michigan). Toxins in the algae killed several dogs that drank the contaminated water in Lake Champlain; the compounds are capable of causing liver damage and death in humans.

Annoesjka Steinman

This photo shows blue-green algae, *Microcystis*, in a water sample taken from an inland lake in western Michigan. Zebra and quagga mussels ate beneficial algae, which allowed *Microcystis* and other types of toxic algae to proliferate in some of the region's cleanest lakes.

lakes infested with zebra mussels had higher concentrations of blue-green algae capable of releasing toxins, according to a study by Orlando Sarnelle, an associate professor in MSU's Department of Fisheries and Wildlife. Sarnelle's take on the situation was sobering.

"If these blooms of blue-green algae [cyanobacteria] are a common side effect of zebra mussel invasion, then hard-fought gains in the restoration of water quality may be undone," Sarnelle said. "It appears that the numbers of blooms in Michigan have been increasing and appear to be correlated with the spread of zebra mussels."[8]

One of the focal points of Sarnelle's research was Gull Lake, a clean and picturesque lake in southwestern Michigan. Before zebra mussels infested the lake, there was almost no *Microcystis*; after the invaders arrived, the toxin-producing algae became the dominant species. That discovery, and similar findings in other lakes, led Sarnelle to conclude that every lake with zebra mussels could experience blooms of toxic blue-green algae. "We used to worry about *Microcystis* in lakes with high levels of phosphorus," he said. "Now we've got to worry about lakes that are really high quality . . . if they have zebra mussels."[9]

Unlike industrial pollution, which was more common in low-income communities, toxic algae blooms knew no social or economic boundaries. In fact, the algae blooms often were most severe in clean lakes that were the centerpieces of affluent communities. One of the most striking examples was Lake Leelanau, a gorgeous lake in one of Michigan's toniest counties. Long a magnet for the wealthy and members of the wine-and-cheese crowd, Lake Leelanau filled a 13-mile-long gash in the rolling hills of northwest Michigan. Leelanau County was home to the scenic Sleeping Bear Dunes National Lakeshore and the tourist towns of Leland, Glen Arbor, and Suttons Bay. There was little mystery as to why the wealthy built second homes around Lake Leelanau. The area was scenic, sparsely populated, and close to one of the most breathtakingly beautiful stretches of Lake Michigan coast. Its rolling hills, turquoise lakes, and crystalline streams offered a sumptuous respite from urban madness and suburban monotony. So lovely was the region, Leelanau County had some of the highest waterfront-property values in the world in the 1990s, according to the Leelanau Conservancy. It wasn't unusual for vacant land on Lake Leelanau to sell for $10,000 a linear foot. Imagine spending six or seven figures to build a dream home on the lake, only to see the water despoiled by huge mats of blue-green algae. That gloomy scenario unfolded in Lake Leelanau less than a decade after zebra mussels colonized the lake bottom.

Zebra mussels were discovered in Lake Leelanau in 1996; their numbers swelled to 38 billion by 2005. Each of those 38 billion mussels was capable of filtering one liter of water per day and, in the process, consuming large quantities of desirable green algae. That allowed *Microcystis* and other cyanobacteria to take over the lake's bottom. In one decade, zebra mussels transformed the scenic lake from a hot spot for wealthy vacationers to a hot zone for potentially toxic algae. A 2005 study found eight different species of blue-green algae in the lake, many of which were known to produce potent liver- and neurotoxins.[10]

Similar problems surfaced on inland lakes across the Great Lakes region. Massive blooms of blue-green algae blanketed clean lakes, leaving property owners and local officials wondering what the heck was happening. An experiment by a curious scientist in Muskegon, Michigan, shone a bright light on the issue. Ecologist Gary Fahnenstiel was driving his boat to work one morning in 2004, crossing Bear Lake and Muskegon Lake en route to his office at the Great Lakes Environmental Research Laboratory's Lake Michigan Field Station, when he spotted the telltale signs of a blue-green algae bloom. A fluorescent green mass that looked like a cloud of pollen just beneath the water's surface stood out in the murky brown water of Bear Lake. Fahnenstiel, a distinguished algae expert for the National Oceanic and Atmospheric Administration, knew a blue-green algae bloom was forming in the lake. His suspicion was confirmed when he arrived at his office on Muskegon Lake. There, he found the boat basin covered with a bright green mat of algae. A closer look revealed that the mat was actually a collection of millions of tiny green balls of blue-green algae clinging together. The next day, he collected a sample from a layer of algae scum in the middle of Muskegon Lake, which was wildly popular among sailors and anglers.

Fahnenstiel suspected that the algae was *Microcystis aeruginosa*, the same species he had studied in Lake Erie and in Lake Huron's Saginaw Bay. What he really wanted to know was whether the lakes where he often went boating with his wife and two sons were spawning algae blooms that produced highly toxic compounds called microcystins. He sent a sample of the algae to a laboratory in North Carolina for analysis. The results were shocking:

The algae scum on Muskegon Lake contained some of the highest levels of microcystins ever documented anywhere in the world. The microcystin concentration of 238 parts per billion was roughly 12 times higher than the level the World Health Organization considered potentially harmful to human health. The microcystin levels were lower in Bear Lake, but still high enough to warrant concern.

Fahnenstiel, a distinguished researcher whose passion for protecting the Great Lakes occasionally landed him in hot water with his politically correct colleagues, went public with the results. "I don't want to scare people but the levels of microcystins we found are significant," he told the local newspaper. "These are very high concentrations and are on the same order of magnitude as the highest concentrations of microcystins ever reported."[11] He urged people to avoid swimming or boating in the algae scums and to keep dogs out of the algae. Ingesting microcystins in the algae, he said, could cause vomiting, diarrhea, fever, rashes, throat irritation, and in extreme cases, liver damage or death.

The next summer, Muskegon County health officials tested a large algae scum along the Bear Lake shoreline and found even more alarming results. The microcystin concentration in the algae scum was 1,024 parts per billion—51 times higher than the level the World Health Organization considered potentially dangerous. Local officials posted warning signs urging people who were brave enough to venture out into the lake to avoid the algae scum. One of the warning signs was posted near the newly built dream home of Jim and Diane Allen. Jim Allen was melancholy about the suite of conditions in the Great Lakes that contributed to the algae bloom a few feet from his back door. "It seems like the lakes are getting attacked by all kinds of things—the goby attacking perch and now this algae," he said. "It's discouraging; it seems like we're fighting a battle with no weapons."[12]

As word of toxic algae on the lakes spread across the region, boaters and anglers began to ask lots of questions. Among the most common inquiries: Could toxins in the algae contaminate municipal drinking-water supplies? Could the poisons move up the food chain and harm people who ate fish from lakes where blue-green algae was present? Scientists in 2003 found microcystin concentrations as high as 1.1 parts per billion near municipal drinking-water intakes along the southeast shore of Lake Ontario, in New York. There were no standards for microcystins or other cyanobacteria in drinking water, but the World Health Organization had an advisory limit of 1 part per billion. The WHO considered microcystin concentrations above 1 part per billion in treated drinking water a potential human health risk. Fortunately, municipal water-treatment facilities killed most types of cyanobacteria.

A more immediate concern for water-treatment facilities were the taste and odor problems that blue-green algae blooms caused. There were 45 different strains of blue-green algae that caused taste and odor problems in water. Some could persist in the water for weeks after visible algae scums dissipated from the surface of a lake. Blue-green algae blooms sparked complaints about taste and odor problems in drinking water from at least ten Great Lakes communities. Residents in Grand Rapids, Michigan; Cornwall and Kingston, Ontario; and several New York communities along Lake Ontario complained about the water having a musty odor and tasting a bit earthy. The odor was caused by decaying blue-green algae, which produced geosmin. Geosmin is a naturally occurring compound that is responsible for the smell that emanates from freshly turned soil.

On the question of microcystins from blue-green algae contaminating fish, initial studies found low levels of the toxin in Lake Erie perch and walleye. Researchers called that good news. But NOAA's Great Lakes laboratory in Michigan continued monitoring fish to check for signs of a more serious problem. If microcystins accumulated in fish to levels that warranted concern, the problem could cast another cloud over a Great Lakes fishery long plagued by the stigma of industrial contaminants.

Whether cyanobacteria and the harmful algal blooms they formed might someday pose acute health threats to humans in the Great Lakes region remained an open question in 2008. There was disturbing evidence, however, that people who drove boats or personal watercraft in lakes where harmful algal blooms were present could absorb some of the toxins in their blood. The U.S. Centers for Disease Control sent a team of researchers to Muskegon's Bear Lake in 2006 to study whether boaters might be absorbing microcystins from blue-green algae blooms into their bodies. As it turned out, weather conditions were not right for an algae bloom when the scientists arrived from Atlanta, Georgia. There were no blue-green algae on the surface of Bear Lake. The scientists decided to conduct the study anyway.

CDC officials paid several volunteers $25 each to drive boats and personal watercraft back and forth across Bear Lake.[13] Researchers then took blood samples from all of those individuals. The results were unprecedented. No microcystins were found in the blood of the study participants. But the tests showed that boat traffic transferred microcystins from the water into the air, creating an aerosol form of toxic algae that could be inhaled. CDC officials said the study confirmed, for the first time, that microcystins could be transferred from water to air. That such a phenomenon occurred in a lake with no visible algae bloom further heightened concerns that lakes where harmful algal blooms were common could pose potential health risks to human health and animals. The CDC was conducting more research to figure out how significant a health risk harmful algal blooms posed to people, pets, and wildlife. There were many questions to be answered. One thing was certain: With zebra and quagga mussels firmly entrenched in the Great Lakes and spreading to inland lakes across North America, toxic algae blooms would likely impair water quality in even the cleanest lakes for the foreseeable future.

"The occurrence of toxic cyanobacteria in our natural waters is here to stay," Boyer said. "The recent resurgence of *Microcystis* blooms in Saginaw Bay and western Lake Erie is of concern due to the widespread use of these waters for recreation, fishing and drinking water."[14] With no apparent solution on the horizon, algae experts and health officials could only offer this advice: If your favorite swimming hole looked like pea soup, or the surface was covered by a layer of green scum that resembled paint, keep out of the water. That was a bitter pill to swallow for millions of people who lived in the Great Lakes region for the simple fact that it was the center of the freshwater universe.

If dead birds lined up on beaches don't get us passionate about addressing these problems, it's hard to see what will.

—Mark Breederland, Michigan Sea Grant educator, 2008

14 A CRUEL HOAX

The narrow peninsula of Long Point stretches into the eastern basin of Lake Erie like one of the long, boney legs of great blue herons that frequent its coastal marshes. The 24-mile-long sand spit is a haven for migratory birds and the birders who stalk them with binoculars and spotting scopes. Its tree-lined dunes, miles of undeveloped shoreline, and sprawling wetlands provide refuge and food for migratory birds that cross Lake Erie every spring and fall en route to summer nesting areas or winter retreats. Campers flock to Long Point Provincial Park in southwestern Ontario in the summer to soak up the beauty of a remote beach that extends more than a mile along the peninsula's western side. It is a spectacular place to observe birds, walk barefoot on the sandy shoreline, and watch the sun set over the glistening waters of Lake Erie. Long Point is also one of the few places where a person can see across any of the Great Lakes using nothing more than the naked eye. From the peninsula, it is possible to see the lights of Erie, Pennsylvania, at night without using binoculars.

The tip of Long Point is just 35 miles from the northern edge of Presque Isle State Park in Erie, Pennsylvania. The opposing sand spits that reach toward each other create one of the narrowest points in all of the Great Lakes. For birds crossing Lake Erie, it is a short hop from the tip of Long Point to Sunset Point at Presque Isle State Park. That close proximity explains why the two peninsulas are part of the Atlantic Flyway, an invisible corridor that millions of birds follow every spring and fall as they cross North America along a north-south trajectory. Dozens of bird species—common loons and black terns, sandhill cranes and tundra swans, yellow warblers and red-breasted mergansers—cross Lake Erie each year, placing Long Point and Presque Isle among North America's most important bird sanctuaries.

Given the global ecological importance of the parks, it was especially disturbing when dead birds started washing up on the beaches of Presque Isle in 1998. At first, it was a few dead gulls. But the mysterious problem quickly mushroomed. By the end of the year, thousands of dead birds—gulls, loons, and mergansers—littered miles of beaches of Long Point and other national parks in Ontario. Spectacular bird refuges that lined the north and south shores of Lake Erie were transformed into avian killing fields. No one knew why.

Over the next decade, the epidemic of bird kills that had surfaced first at Presque Isle spread to three other Great Lakes—Ontario, Huron, and Michigan—and claimed more than 70,000 fish-eating birds. Among the casualties were thousands of ring-billed gulls and Caspian terns, long-tailed ducks, black-backed gulls, red-breasted mergansers, bald eagles, and endangered piping plovers. Common loons were hit especially hard. The mysterious ailment killed more than 8,000 loons. It was the latest threat to a species imperiled by shoreline development in the northern boreal forests. "This is just another hit; it's pretty scary," said Dr. Grace McLaughlin, a wildlife disease specialist at the National Wildlife Health Center in Madison, Wisconsin, in a 2002 article in the *New York Times*.[1] The carnage was bewildering and horrifying to all who appreciated the beauty, grace, and ecological significance of birds. It was also an alarming sign that the Great Lakes were critically ill.

BOB WELLINGTON WAS FIGHTING A NASTY COLD WHEN HE MET ME AT Presque Isle State Park on a raw November day. A third-generation native of Erie, Pennsylvania, Wellington was a skilled outdoorsman who had worked for 30 years as a biologist for the Erie County Health Department. He agreed to give me a guided tour of the park. Wellington was something of a local legend, a salt-of-the-earth type known for a sharp tongue and an encyclopedic knowledge of Lake Erie. He also knew the bird-kill story as well as anyone—Wellington was one of the first public officials to sound the alarm about the mysterious problem. He was working at the health department in July 1998 when local residents began reporting dead channel catfish washing up on Presque Isle's beaches. Fish kills in the Great Lakes were not uncommon in the spring, when rapid changes in water temperatures could unleash a lethal dose of thermal shock. But the death of numerous catfish in July, 10 to 15 of which washed up on beaches every day for three weeks, was something Wellington had never witnessed in his six decades of fishing and swimming in the lake. Still, he considered the dead catfish little more than an oddity. That was until gulls started dropping dead on Presque Isle's expansive beach. It was around that time that Wendy Campbell, a local wildlife rehabilitator, expressed concern about gulls staggering across the beach like drunken sailors.

"We knew something wasn't right but we didn't get too excited because there were thousands and thousands of gulls," Wellington said.[2] A longtime volunteer at Presque Isle, Wellington had seen much in his 60 years of living along Lake Erie—horrific pollution in the 1950s and '60s, the lake's recovery in the 1980s, followed by the zebra mussel infestation in the 1990s that made the parts of the lake as clear as a swimming pool. In all those years, he had never encountered dozens of birds that died for no apparent reason. Wellington reported the problem to the Pennsylvania Department of Environmental Protection, where his first contact gave him a polite brush-off. "The first guy I talked to told me the birds probably ate some bad fish. He thanked me for calling and hung up," Wellington said. No one in Pennsylvania knew it at the time, but dozens of dead loons had begun washing ashore at Ontario's Pinery Provincial Park, on Lake Huron, in October 1998. A mysterious killer was claiming birds in Lake Erie and Lake Huron. The situation exploded into an ecological crisis the next year.

A massive fish kill in April 1999 left a 40-mile swath of dead alewives and gizzard shad along the Pennsylvania coast of Lake Erie. That was followed in June and July by a huge algae bloom that coated miles of shoreline at Presque Isle State Park and neighboring counties with thick mats of filamentous *Cladophora*. Within a couple of weeks, gulls started dying along miles of Pennsylvania beach—150 gulls died at Presque Isle alone. Several hundred more gulls dropped dead at an abandoned industrial parking lot a half mile from the coast. Local officials initially suspected that those birds had been poisoned. Wellington feared that something more complex was at work. He telephoned Pennsylvania game warden Larry Smith, who went to Presque Isle to see the carnage with his own eyes. After observing hundreds of dead gulls on the beach, Smith related his concern to the Pennsylvania Game Commission, which launched the first investigation of the mysterious bird kills.

Wellington began his own quest for information, scouring the Internet for clues as to what could kill large numbers of seemingly healthy birds. His search led to an article about a 1963 disease outbreak that had killed 12,000 birds along the Lake Michigan coast. The cause of death in that case: *Clostridium botulinum,* the bacterium that caused Type E botulism, better known as food poisoning. Type E botulism caused sporadic bird kills in Lake Michigan and western Lake Huron in 1963 to '64, disappeared for more than a decade, and then resurfaced periodically in Lake Michigan birds from 1976 through 1983. Collectively, all of those incidents killed about 16,300 birds. The deadly bacterium then went into remission for 15 years until it was confirmed as the cause of death for thousands of birds in Lake Huron.[3]

One of seven types of botulism, Type E lives in spores that are ubiquitous in lake and marine environments. The spores reside in bottom sediments and are harmless unless they germinate and enter the food chain. Botulism spores enter a vegetative state when exposed to oxygen-deprived environments, such as the carcasses of dead animals and mats of decaying algae. The vegetative form of Type E spores produces the bacterium *Clostridium botulinum,* which causes muscle paralysis in vertebrates—fish, birds, and mammals. Type C botulism also kills birds and is common throughout North America, particularly in the muck of dried-up lakes. The type of food poisoning most common among humans is Type A or Type B, but Type E can kill people who eat fish or animals infected with the bacterium. If Type E botulism was to blame for the Lake Erie bird kills, Wellington knew it could escalate into an environmental nightmare. Lake Erie, after all, supported more fish than all the other Great Lakes combined. It was also a magnet for migratory fish-eating birds.

The problem escalated in the spring of 1999 as Wellington and others scrambled for information on Type E botulism. Hundreds more dead birds washed up along a 40-mile stretch of shoreline between Erie and Dunkirk, New York. On the Canadian side of the lake, near Long Point, dead gulls and carp were washing up on beaches. Pennsylvania wildlife officials sent several dead gulls to the National Wildlife Health Center in Madison, Wisconsin, for testing. The laboratory, which documented botulism as the agent that killed hundreds of birds in Lake Michigan in 1963, confirmed the worst: Type E botulism was back. The difference was that the bacterium was stalking birds in Lake Erie for the first time in recorded history. Within weeks of Pennsylvania officials receiving the laboratory results they feared, Canadian officials reported a massive die-off that claimed 6,000 birds between Rondeau Provincial Park and Point Pelee. More than 5,000 of those birds were mergansers—large diving ducks with black crowns and tufts of hair on the back of their heads that look like scraggly mohawk haircuts. Biologists knew by then what was killing birds on Lake Erie and Lake Huron. But they didn't know why Type E botulism had resurfaced in the Great Lakes after lying dormant for nearly two decades. The uncertainty left some scientists wary of the future.[4]

The body count from the botulism epidemic escalated rapidly over the next three years, with increasing numbers of birds dying in 2000, 2001, and 2002. Tens of thousands of fish from 18 different species, 20 species of mud puppies (aquatic salamanders), and several giant sturgeon washed up dead on Lake Erie beaches in 2000 and 2001. Some people who observed the bird kills were beginning to panic. "I was getting calls at the health department from people saying there were dead alligators on the beach," Wellington said. "They were mud puppies. They had four legs and kind of looked like alligators but they weren't. I was getting a lot of interesting calls back then."

More than 25,000 birds died in Lake Erie between 1999 and 2001. The worst was yet to come. In 2002 alone, the epidemic claimed another 25,000 birds. The victims included loons, gulls, long-tailed ducks, mergansers, and cormorants.[5] Canadian officials that year counted more than 1,000 dead loons along the north shore of Lake Erie; 700 of those birds were found along an 18-mile stretch of Long Point.[6] The sight of hundreds of common loons, the icon of the North, washed up on the resplendent beach of Long Point was gut-wrenching.

Health officials and biologists became increasingly alarmed by the rising death toll and the potential threats that fish and birds infected with Type E botulism posed to other animals and humans who might unknowingly consume tainted fish or fowl. "We were concerned about people smoking fish that had Type E botulism not cooking it to the proper temperature. We didn't want people dying," Wellington said. Health officials in Pennsylvania and New York issued advisories warning anglers and hunters not to eat fish or birds that appeared lethargic before they died. As the situation grew increasingly grave, researchers at several universities and government agencies scrambled to figure out what had triggered the resurgence of Type E botulism.

Dr. Ward Stone, a wildlife pathologist for the New York Department of Environmental Conservation, found a major clue in the process of making the first diagnosis of Type E botulism in fish and birds in New York. He was performing a necropsy on a long-tailed duck when he found quagga mussels in its stomach. He also found mud puppies and round gobies—which he determined were vulnerable to the botulism toxin—in the guts of gulls, mergansers, and loons that had died from Type E poisoning. Stone and other researchers began to connect the dots of a new Great Lakes horror story, one born of foreign species.

Researchers theorized that zebra and quagga mussels were ingesting Type E botulism from sediments on the lake bottom and concentrating the bacterium in their tiny bodies. Botulism didn't harm the mussels, because the toxin only affected vertebrates. Round gobies that ate the infected mussels became paralyzed as the toxin shut down the neural response that controlled muscle movement. Infected gobies swam in circular patterns near the water's surface as their muscles froze, which made them easy pickings for fish-eating birds. Within a day or two of ingesting contaminated fish, the toxin paralyzed birds' muscles. Unable to hold their heads up, many of the birds suffered what biologists call "limber neck": the birds' neck muscles went limp, their heads fell in the water, and they drowned.

Wind-driven waves delivered dead birds to beaches, where gulls and other scavengers that picked at the carcasses also could get a lethal dose of the botulism toxin. Stone believed that the foreign mussels and gobies created a new link in the food chain that allowed the deadly toxin to spread from the

Jeff Alexander

Michigan wildlife biologist Tom Cooley conducts a necropsy on a common loon found dead on a Lake Michigan beach in 2007. Outbreaks of Type E botulism killed thousands of fish-eating birds on Lakes Erie, Ontario, Michigan, and Huron. Scientists linked the problem to invasive zebra mussels and gobies, which delivered the ubiquitous but potent toxin to fish and fish-eating birds.

foreign mussels to gobies and, ultimately, kill native fish and birds. "I don't think the Type E bacteria is new, I think the changing ecological conditions in the lakes have allowed it to become far more important than it was in the past," Stone said. "The mussels have really changed the lakes. The toxin moves readily from mussels into gobies."[7]

Loons and mergansers, which dined exclusively on fish, were especially vulnerable to the toxin that Type E botulism produced. The evidence was found on beaches around the lakes and in the laboratories where pathologists collected blood samples from the dead birds. A 2002 die-off, which claimed 343 loons in New York alone, was one of the worst bird kills in the state's history. Stone called the death of loons, given their ecological and emotional significance, tragic and symbolic. "We love loons because they represent wild areas; they have a wonderful call that represents wilderness and those brilliant red eyes and distinctive black and white markings," he said.

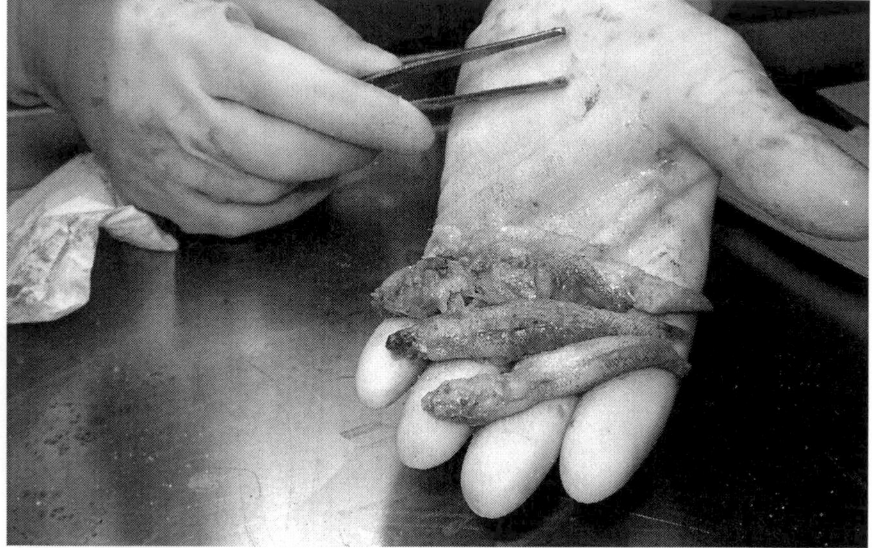

Jeff Alexander

Scientists often found round gobies in the stomachs of Great Lakes birds that died from Type E botulism. The toxin paralyzed birds' muscles, causing them to drown when their heads fell in the water.

Jeff Alexander

Dead loons found along popular Lake Michigan beaches at Sleeping Bear Dunes National Lakeshore in October 2007 await necropsies at the state of Michigan's wildlife laboratory in East Lansing. Biologists dissected the birds to check for signs of Type E botulism.

Most of the dead loons recovered in the New York waters of Lakes Erie and Ontario were from Canada. Few of the birds were banded—they were wild birds that spent summers breeding on remote lakes in Canada and northern Michigan, Minnesota, and Wisconsin. No one knew whether the die-offs posed an immediate threat to the North American loon population, but it seemed clear their numbers would drop. Canada in 2000 was home to about 545,000 loons; the United States supported another 32,000 loons.[8]

Loons are long-lived birds that breed annually for about 20 years. A healthy pair of loons produces an average of one chick per year. Losing more than 8,000 loons, assuming they all bred and produced one offspring per year, could reduce the loon population by 80,000 birds over a 20-year period. Such a loss wasn't enough to imperil the species, but the botulism epidemic certainly would leave some northern lakes with fewer loons. Or none. "Some area is losing one heck of a lot of loons," Stone said. "When you look at a beach and it's strewn with birds it's telling you the ecology is changing tremendously and not for the better, especially if you like fish and wildlife. These dead loons are telling us the world is changing rapidly and we'd better pay attention. There have been some major ecological changes in the Great Lakes and this Type E botulism is one of the worst."

Biologists gained an understanding of how Type E botulism was reactivated in the lakes, but were powerless to do anything about the problem. There was no way to keep gobies and sturgeon from eating zebra and quagga mussels. Likewise, there was no way to keep loons, mergansers, and other water birds from dining on fish infected with a potentially lethal dose of the bacterium. "All we could do was pick up dead birds on the beach and take them to an incinerator," said Eric Obert, extension director of the Pennsylvania Sea Grant program and a longtime biologist. Sometimes there were so many dead birds that the incinerators used to burn the carcasses couldn't keep up. Some of the birds had to be sent to crematoriums, which were usually reserved for incinerating the remains of dead humans.

The number of Lake Erie birds killed by Type E botulism declined sharply in 2003 and remained lower through 2007. Researchers in 2007 still didn't know precisely why the bird kills had spiked in some areas in 1999 through 2002. One possibility was that zebra and quagga mussels, by increasing water clarity, increased the growth of *Cladophora*. Mats of decaying *Cladophora* could provide ideal growing conditions for Type E botulism, which thrived in anaerobic (oxygen-deprived) environments. Another possibility: Blankets of mussels that carpeted the lake bottoms created anaerobic micro-environments, triggering the growth of Type E botulism, which the mussels then ingested and passed up the food chain. Those theories were little more than educated guesses. A third theory was that warmer water temperatures and lower lake levels helped fuel the growth of the deadly vegetative form of the botulism bacterium. The truth was that no one had figured out, by 2008, precisely what caused botulism-related bird kills to spike.

The toxin eased its death grip on Lake Erie birds in 2003, but it wasn't gone. That was clear during a park tour Wellington afforded me during a November 2007 trip to Presque Isle State Park. As wind-driven waves pounded the beach, Lake Erie coughed up several dead loons, long-tailed ducks, and gulls. One of the gulls was frozen upright on the beach, looking as if it was protecting a nest. The bird was dead, its tiny white body a monument to a lethal bacterium turned loose by invasive species. A few feet away, a loon lay chest up on the sand, its black and white checkered wings drawn close to its snow-white breast, the brilliant red eyes replaced by a lifeless shade of gray. Sad as it was, the presence of a few dead loons on that particular beach was, by Lake Erie standards, a good year. The body count there was declining. Along the shores of Lakes Michigan and Ontario, the number of dead birds was headed in the other direction.

LITTLE GALLOO ISLAND IS A ROCKY OUTPOST AT THE EAST END OF LAKE Ontario, a bird haven 15 miles southwest of where water from all five Great Lakes funnels into the St. Lawrence River at Cape Vincent, New York. Among islands, Little Galloo is sort of an ugly duckling, at least aesthetically. Its terrain and brushy vegetation don't make the 40-acre island a destination for tourists. For birds, it is an entirely different story. Caspian terns, gulls, loons, and cormorants love the place. It is uninhabited by humans and close to several bays that teem with fish. Home to one of New York's two nesting colonies of Caspian terns, the island also is a barometer of bird health in the eastern Great Lakes. It foretold a major bird kill in eastern Lake Erie in the summer of 2006, when hundreds of gulls, Caspian terns, Canada geese, and sandpipers on the island died. State wildlife officials collected 800 dead

Caspian terns from Little Galloo that year, nearly one-third of all terns on the island. The killer: Type E botulism.

Biologists nursed a few of the sick birds back to health, pumping Gatorade into their stomachs to flush out the botulism bacteria. Stone took a few Caspian terns back to his laboratory near Albany, where he nursed the birds back to health before releasing them near Lake Champlain. Saving those birds was a noble but largely symbolic effort; there were far too many sick birds and not enough wildlife rehabilitators to cope with the enormous problem. "The numbers are overwhelming," laboratory technician Dave Galinski told the *Rochester Times Union* in 2006. His comments turned out to be prescient. Three months later, dead loons started washing up on beaches near Rochester. Crews picked up 343 dead loons and 30 other birds along a two-mile stretch of beach.[9]

That pattern was repeated the following year: hundreds of gulls and 300 Caspian terns died on Little Galloo in the summer of 2007. But the bird kills were sporadic and hopscotched across Lakes Ontario, Erie, Huron, and Michigan. Some researchers theorized that the die-offs followed the spread of round gobies across the Great Lakes. That explained why Lake Superior was immune to the problem through 2007—there were only a few quagga mussels and gobies in the largest and coldest of the Great Lakes. Still, it was a puzzling situation. How could so many dead birds wash up on a beach one year and so few the next, when there were no obvious changes in the lake or weather conditions? That was a question researchers investigating the epidemic could not answer with any degree of certainty. A decade after it surfaced in Lake Erie, the conditions that triggered the botulism epidemic and caused it to spread so rapidly remained cloaked in scientific uncertainty. "The big question is how long is this botulism going to last and is there anything we can really do about it?" said Irene Mazzochi, a wildlife biologist for the New York Department of Environmental Conservation.[10]

THE TRAIL OF BOTULISM-RELATED BIRD KILLS LED ME IN LATE 2007 TO Michigan's Wildlife Disease Laboratory in East Lansing, roughly 500 miles west of Lake Ontario's Little Galloo Island. There, I observed biologist Thomas Cooley perform necropsies—animal autopsies—on several loons suspected of dying from Type E botulism. The process of dissecting the birds to search for the telltale signs of food poisoning was at once sickening and fascinating. After observing Cooley for a while, and listening to him tell the tale of Michigan's botulism epidemic, I popped the $64,000 question. The time had to be right, for my query would challenge all that Cooley and many other scientists believed about how invasive species—zebra and quagga mussels and the round gobies that ate them—formed a toxic food chain that delivered deadly botulism to thousands of fish-eating birds. I nervously asked the question: "Are you sure these birds are getting botulism from zebra and quagga mussels that pass the toxin on to gobies? Couldn't something else be the cause?"

Cooley froze, lifted his hands slightly from the bloody chest cavity of a loon, and looked at me resolutely. "The evidence is too strong to be circumstantial at this point," he said. "The mussels and gobies concentrate the toxin and pass it up the food chain to birds."[11] The only uncertainty about the epidemic, Cooley said, was how the mussels obtained the botulism bacterium. It was possible the mollusks inhaled botulism spores off the lake bottom, or ingested the toxin after storms churned up sediments and released the spores into the water column, where the mussels could readily ingest the poison.

Previous Type E botulism outbreaks that killed birds in Lake Michigan and Lake Huron in the 1960s and 1980s were likely caused by storms that released botulism spores from the lake bottom, Cooley said. Those spores then morphed into the toxic vegetative state that contained the deadly *Clostridium botulinum* bacteria. Bottom-feeding fish unknowingly ingested the bacteria, were paralyzed, and floated toward the surface, where birds nabbed them. Birds that ate infected fish died an agonizing death that usually culminated in drowning. It was an inglorious death for such majestic fliers.

Given that there was no way to rid the lakes of zebra or quagga mussels, or gobies, biologists were resigned to the fact that the botulism epidemic would likely haunt Great Lakes birds for years to come, if not decades. "This botulism problem is probably something we're going to be faced with for a long time," Cooley said. "There's no way to get rid of the exotic species that are causing this."

With the biology lesson wrapped up, Cooley returned to the task at hand—completing a necropsy on one of the 16 dead loons lined up on surgical tables in his lab. "Look at this," he said. "See that water coming out of the lungs? That means the bird drowned." It was one of the hallmarks of botulism poisoning in water birds. Paralyzed by the toxin, the birds' heads often fell into the lake, where their last gasps for air instead produced a lethal dose of water.

To confirm the loon had died from Type E, Cooley extracted a sample of blood from its heart. That blood was then injected into three mice as part of an analytical procedure known euphemistically as the Mouse Protection Test. Before being injected with the bird blood, one mouse was inoculated with the antidote for Type C botulism; one mouse received the antitoxin for Type E botulism; the third mouse received no antidotes. The mice were injected at 9 A.M. and usually provided answers by the end of the workday. If Type E botulism had killed the bird, only one mouse—the one injected with the antitoxin for Type E—survived. The other two mice suffocated as the botulism toxin paralyzed their respiratory systems. "It's as humane as we can be," Cooley said, almost apologizing. "It's the best test we have."

Type E botulism could kill a mouse in under an hour. It could take several days for the toxin to kill loons, which had a four-foot wingspan and could weigh up to 12 pounds. Infected birds that reached land often staggered as the toxin slowly paralyzed their muscles, much like a drunken person whose motor skills were impaired by alcohol. To Cooley, who made a living dissecting dead animals and diagnosing disease, the notion that foreign mussels and fish were indirectly killing thousands of native birds was, in his words, bothersome. "An awful lot of birds have been affected. There are thousands of birds in Lake Michigan dying from this. It's disturbing," he said. "We don't normally see diseases that kill thousands of animals."[12]

Type E botulism killed 2,900 birds over the course of four months when it surfaced in 2006 at the Sleeping Bear Dunes National Lakeshore, in Michigan. The die-off caught park managers and area residents off-guard. The epidemic returned in 2007, but killed birds at a slower pace than the previous year. But the problem also spread across a wider swath of northern Lake Michigan that year, much as it had done in Lake Erie in 2000. The widening epidemic was like a recurring nightmare. In October 2007, dozens of dead birds washed up on beaches near Cross Village, about 60 miles northeast of Sleeping Bear Dunes as the crow flies. The 82 dead birds recorded that month included red-headed grebes, loons, gulls, and cormorants. Retired physician Henry Singer, who owned a lakefront cottage south of Cross Village and spent much of his free time keeping tabs on environmental issues affecting northern Lake Michigan, delivered the grim news in a profoundly sad e-mail message. He wrote:

> We see dead gulls, cormorants . . . but most disheartening of all—loons! Our precious loons, whose eerie calls echoed off the lake and turn early summer dawns into a meditation on the northern wilderness experience. My wife took a long walk on the beach last Sunday and saw many dead birds. She came back very upset. Our neighbors have noticed it as well. Tragic. Another example, if we needed one, of the devastating effects of the ecological imbalance evinced as a consequence of the exotics. What a shame . . . and shame on us that we have been unable to do what is necessary to prevent further havoc from the unregulated maritime industry.[13]

The shipping industry was an obvious target for people frustrated by the mounting toll invasive species were inflicting on the lakes. But to claim that the shipping industry was solely to blame for the problem was, in my opinion, overly simplistic. There was plenty of blame to go around when it came to sorting out how and why shipborne invaders had inflicted such a heavy toll in the Great Lakes. Many who professed a love of the lakes—government bureaucrats and politicians, shippers and environmental advocates—had contributed to a tragic series of events that allowed the largely preventable problem to snowball into an environmental catastrophe without equal.

We may not be able to dodge this bullet.

—Dan Thomas, Great Lakes Sport Fishing Council

15 CASPIAN SEA DIET

Every September, after most Great Lakes anglers call it a season to avoid foul autumn weather, teams of biologists head out on the lakes for a series of unusual fishing trips. The month-long excursions take them, year after year, to the same fishing spots—scattered across the vast expanse of the lakes. Their targets are not the coveted chinook salmon, elusive steelhead, or whitefish. The biologists go fishing for bottom feeders, those inglorious species of prey fish that support the more glamorous predators on the upper rungs of the food chain. The crew from the U.S. Geological Survey's Great Lakes Science Center targets homely fish with odd names: bloater, slimy sculpin, smelt, alewife. Trophy fish they are not. But without billions of prey fish in the lakes, there would be no whitefish, perch, walleye, trout, or salmon.

I joined a crew from the USGS for a day in September 2007 as they dragged the bottom of Lake Michigan in search of prey fish. Chuck Madenjian, the lead USGS biologist on the fishing trip, welcomed me aboard the 78-foot-long research vessel *Grayling* as it prepared to depart from Grand Haven, Michigan, before the crack of dawn. The *Grayling* headed out of the Grand River channel as most of the town was still asleep. The boat passed a few hearty anglers pier fishing by the light of camping lanterns and flashlights. Having cleared Grand Haven's lighthouse, the ship veered left and headed south for a two-hour journey to the first of several trawl sites offshore of the resort community of Saugatuck. It is one of seven sites around the lake that Madenjian and his crew visit every autumn. Other crews carry out similar surveys on the other Great Lakes. The trawl surveys, conducted annually in the Great Lakes since 1973, give biologists a statistical tool by which to gauge the health of the lakes' $7 billion sport and commercial fisheries.

As the *Grayling* approached Saugatuck, the five-man crew prepared the gear used to capture bottom-dwelling fish. They would drag a large net on the lake bottom at eight sites, ranging in depth from 30 feet to 360 feet. Once we reached the proper GPS coordinates, Captain Keith Peterson put the *Grayling* in neutral and waited as ship engineer Jim Page prepared the trawl net for deployment. The trawling device was a funky looking contraption that consisted of a rectangular net 39 feet wide, four feet high, and 70 feet long. Heavy wood and steel sleds attached to the sides of the net took it to the bottom and kept it there as the boat dragged Lake Michigan's sandy undercarriage.

All of the fish species the biologists hoped to land on that unusually warm autumn day were small, save one: burbot. Burbot is one of the larger bottom-feeding fish, reaching lengths of 22 inches. Its long, slender body limits its maximum weight to about three pounds. Some people call the fish a "poor man's lobster" because it has a sweet flavor when broiled and buttered. Most anglers call burbot "lawyers," Madenjian said, "because they're slimy bottom feeders."[1] It was an old joke among Great Lakes biologists, one that always drew a laugh when told to a newby. I was the victim that day. But it was a well-delivered joke by the affable Madenjian, a living textbook on fisheries whose bedraggled hair, blue jeans, and faded T-shirt gave him the appearance of an aging rock star.

For a layperson like me, who was fascinated by the unseen biological drama that played out daily in the depths of the Great Lakes, the bottom trawl was an opportunity to catch a glimpse of rarely seen fish. It was also a chance to observe whether foreign zebra and quagga mussels were colonizing Lake Michigan's bottom to the extent scientists claimed. Researchers reported finding quagga mussels nearly everywhere they looked on the lake bottom, even in the black, frigid depths 300 feet below the surface. I had to see that with my own eyes. Madenjian and his crew would not disappoint. As Page used a crane to lower the trawling net over the stern, Petersen put the *Grayling* in gear and proceeded to drag the net across the lake bottom at a speed of two miles per hour. The net was pulled across a half-mile section of the lake bottom for precisely 10 minutes before it was hoisted back onboard the boat. The trawl captured everything living on the lake bottom and swimming within four feet of the bottom. "We'll probably see lots of quagga mussels today," Madenjian said.

The first few trawls brought up modest amounts of alewife, deepwater sculpin, and quagga mussels. As the *Grayling* moved into deeper water, the volume of quagga mussels mined from the lake bottom steadily increased. On the second-to-last trawl of the day, at a depth of 250 feet, the crew hit the mother lode of bad news: nearly 1,000 pounds of quagga mussels and only a handful of fish. The worst was yet to come. The last trawl, conducted 13 miles offshore of Saugatuck in 280 feet of water, produced nearly a ton of quaggas. A crane on the rear deck of the boat, capable of hoisting 1,800 pounds, was no match for the load of mussels. "We're going to have to cut the net," Page yelled to the crew. Biologist Jeff Holuszko sprinted into the ship's cabin and returned with a butcher knife duct-taped to a long wooden pole. He leaned far over the railing on the ship's stern and sliced open the side of the net. Quagga mussels and silt from the lake bottom gushed out of the net and sank—a cloud of tan, muddy water the only evidence of their visit to the surface. It took four men a half hour to empty all the mussels from the net. They had to cut open the trawl net and release its bounty in order to save the crane onboard the *Grayling*. The net could be repaired with a fairly simple sewing job; fixing the crane would be a more expensive and time-intensive proposition. The load of mussels was an impressive show for a skeptical journalist who wanted to see if the stories of quagga mussels taking over the bottoms of Lakes Michigan, Huron, Ontario, and Erie equaled the hype. A trawl net loaded with quagga mussels, and just a few fish, seemed to support the claim that quagga mussels dominated the lake bottom. It was an ominous indication of what was happening in the depths of Lake Michigan.

Madenjian and his crew caught phenomenal quantities of quagga mussels during their month-long journey. After analyzing the data, he concluded that there were 500 million pounds of quagga mussels on the bottom of Lake Michigan in 2007. The volume of foreign mussels was eight times the volume of all species of prey fish combined. The 60 million pounds of prey fish in Lake Michigan in 2007 was the lowest recorded since biologists began dragging the lake bottom in 1973. In 1989, the year before zebra and quagga mussels were discovered in Lake Michigan, the volume of all prey fish in the lake totaled 900 million pounds. Biologists were nervous about the wildly divergent trends in prey-fish and quagga-mussel abundance. Quagga mussels had displaced nearly all the zebra mussels in Lake Michigan, something biologists had thought impossible. Experts believed zebra mussels would reside in Lake Michigan for at least a century. Quagga mussels had evicted their smaller cousins in less than two decades. By 2007, quagga mussels accounted for 98 percent of all mussels in the lake. The story was the same in Lakes Huron, Ontario, and Erie. "Most of the stuff we bring up in our bottom trawl now is quagga mussels," Madenjian said. "Their population has just exploded in the lake in the last five years."[2]

Cures for Mussel Pain

Scientists began researching ways to kill zebra mussels shortly after the invaders were discovered in the Great Lakes in 1988. There had never been a documented case of humans eradicating the mollusks from a lake or river anywhere in the world. Eliminating the little monsters once they colonized a body of water seemed a hopeless proposition: The mussels arrived undetected, multiplied rapidly, quickly expanded their range, and reconfigured ecosystems. Fierce though they were, zebra mussels were not indestructible. They had an Achilles heel, a weak spot that could be exploited. That an American scientist found it so quickly after the mussels invaded the Great Lakes was serendipitous.

Ohio State University toxicologist Susan Fisher led a group of researchers in the late 1980s that went to great lengths to understand what might kill zebra mussels. Her team subjected the creatures to a variety of toxins and bacteria. They even conducted electrocardiograms—the same tests used to detect heart disease in humans—in their quest to find a chink in the mussels' biological armor. "We actually attached tiny electrodes to their hearts to see what would kill them," said Jeff Reutter, director of Ohio State University's Stone Laboratory. Those tests, Reutter said, produced a startling discovery: Pure water killed the mussels.

Zebra mussels croaked when put in what was called EPA Reference Water—water the U.S. Environmental Protection Agency considered pure, uncontaminated. Fisher determined that potassium compounds in clean water destroyed the mussels' gill tissue, causing asphyxiation and heart failure. The finding was a major breakthrough. Because the potassium that killed the mussels was found in water the EPA considered clean enough to drink, Fisher believed she had discovered the magic bullet for the war on zebra mussels. The initial euphoria quickly faded when scientists realized the implications of their discovery. There were billions, if not trillions, of zebra mussels carpeting the bottom of Lake Erie in 1990. And Erie, the world's 11th largest lake, contained 116 cubic miles of water. The amount of potassium needed to rid the lake of zebra mussels was incomprehensible. Said Reutter: "The potassium was harmless and we could have killed a lot of mussels with it, but you can't put enough potassium into Lake Erie to kill all the invasive mussels."

Killing adult zebra mussels required relatively high doses of potassium chloride (the same compound used to kill death-row criminals via lethal injection). Fisher discovered that the mussels were so resilient, they could come back to life after receiving a dose of potassium chloride that stopped their tiny hearts for an hour. The dose needed to kill mature zebra mussels was so large, it could have killed native mussels in Lake Erie. Given that risk, and the practical impossibility of treating a lake that spanned 9,910 square miles of surface area, potassium chloride was ruled out as a cure for zebra mussels in the Great Lakes. However, the chemical was put to good use in a small Virginia reservoir.

The state of Virginia in 2006 used potassium chloride to attack zebra mussels in a 12-acre pond known as Millbrook Quarry. State officials said eradicating the mussels from the quarry was essential to preventing their spread to nearby lakes that provided drinking water for the cities of Fairfax and Manassas. Dealing with zebra mussel infestations would have cost each of those cities millions of dollars. To prevent that from happening, contractors pumped 174,000 gallons of potassium chloride into Millbrook Quarry over the course of three weeks. To the delight of all involved, the $365,000 treatment worked—all the mussels in the lake died. Traces of potassium chloride were expected to remain in the water for 30 years after the treatment, but state officials said the concentration was so low it wouldn't harm fish, wildlife, or nearby drinking-water wells. Government biologists said the project was the world's first successful, albeit small-scale, eradication of zebra mussels from a lake or stream.

There was no such remedy for the Great Lakes. Some species of diving ducks and fish in the lakes—particularly sturgeon and round gobies—feasted on zebra mussels and their cousins, quagga mussels. But fish, birds, and humans were no match for the prolific invaders from Eastern Europe. Scientists at New York's State Museum Field Research Laboratory discovered a naturally occurring soil bacterium that gave the mussels fatal bleeding ulcers. The bacterium was patented as a natural alternative to using chlorine to prevent mussels from clogging water intakes. Like potassium chloride, the bacterium could not be used on a large enough scale to kill zebra and quagga mussels in the Great Lakes. The fate of the lakes' ecosystems would, for the foreseeable future, be determined largely by the marauding army of tiny, conquering invaders from Eastern Europe.

The trend toward more foreign mussels and fewer prey fish, which began in Lake Michigan in the early 1990s, threatened to wipe out the lake's prized whitefish and salmon fisheries. Whitefish and salmon started shrinking shortly after zebra and quagga mussels began colonizing the lake in 1989 and 1990, respectively. The average size of seven-year-old whitefish in the lake in 1988, the year before zebra mussels invaded, was five pounds.[3] By 2005, the average adult whitefish weighed barely one pound, though there was evidence the fish were recovering some of their body weight by 2007. Whitefish, which were starving as zebra and quagga mussels hogged plankton the fish consumed, were forced to eat the foreign mussels to survive. Salmon didn't have that option. Salmon didn't feed off the bottom. They cruised the lake's pelagic zone like sharks, dining almost exclusively on alewife.

The size of salmon also began to shrink in the late 1990s as the alewife population plummeted in the lakes. Anglers in the early 1990s routinely hauled in 30-pound chinook salmon in Lakes Michigan, Huron, and Ontario. In 2008, anglers were happy to land a 15-pound king. The average size of a chinook salmon caught in Lake Ontario dropped from 24 pounds in 1991 to 18 pounds in 2007; in Lake Michigan, the average size decreased from 18 to 12 pounds; and in Lake Huron, the figure dropped from 17 to 10 pounds during that same period.[4]

Scientists debated whether the prey-fish abundance was decreasing because there were more quagga mussels in the lake, or too many salmon and other predatory fish. Whatever the cause, the result was the same: fewer prey fish and smaller salmon and whitefish. The Great Lakes fishery was being subjected to a Caspian Sea diet, one served up by zebra and quagga mussels that were native to those distant waters. The burning question was the degree to which such a diet would affect a food chain that supported the lakes' $7 billion sport and commercial fisheries.

WHAT IMAGES FLOOD YOUR MIND WHEN YOU STAND ON A GREAT LAKES beach? Do you ponder the underwater mysteries hidden by these massive inland seas? Does your inner cinema conjure visions of giant sturgeon, powerful salmon, or succulent whitefish cruising the depths of the seemingly boundless waters? Perhaps your imagination is drawn to the many busted ships swallowed by these salt-free oceans, or the fierce invisible currents that have claimed thousand of sailors and swimmers alike. Now envision this: a vast biological stage on the lake bottom, a place where the actors—shrimp-like creatures, microscopic plankton, tiny fish, crustaceans, and invertebrates—work together to form an intricate food chain that supports the Great Lakes fishery. The star of the biological drama is a tiny shrimp-like creature called *Diporeia*. The half-inch-long amphipod has been the main food for whitefish and other important fish species since melting glaciers formed these lakes roughly 10,000 years ago. That began to change in the early 1990s, after transoceanic freighters imported zebra and quagga mussels from the Caspian Sea. In the blink of an eye on the geologic clock, the foreign actors took over the stage, gave *Diporeia* its walking papers, and rewrote the biological script for these freshwater seas.

The first act of the ecological tragedy opened in 1992. Fish populations in all five Great Lakes were relatively healthy at the time. The Lake Superior

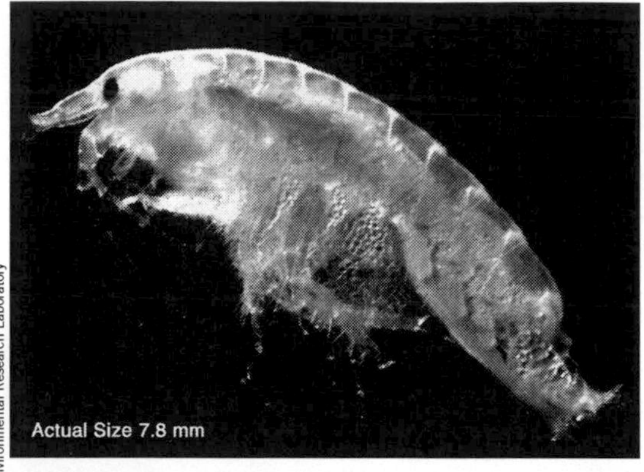

Diporeia (*shown here*) once was the most abundant source of fish food in the Great Lakes. The tiny shrimp-like creatures, which live on lake bottoms, were nearly eliminated from Lakes Michigan, Huron, and Ontario after zebra and quagga mussels colonized those lakes.

National Oceanic and Atmospheric Administration's Great Lakes Environmental Research Laboratory

This image shows the subtle differences between quagga mussels (*left*) and zebra mussels. The mussels are related, but quaggas have rounder shells and can live in deeper, colder waters. Zebra mussels invaded the Great Lakes first, but quaggas proved to be heartier and more ecologically disruptive.

lake-trout fishery was self-sustaining again, following a long recovery from the ravages of sea lamprey. The imported salmon fishery was thriving in Lakes Huron and Ontario. Lake Erie was recovering from the gross pollution of the past and supported phenomenal walleye and yellow-perch fisheries. And Lake Michigan's prized salmon fishery was showing signs of healing from a mysterious kidney ailment that killed scores of fish from the late 1980s into the early 1990s. On the surface, the lakes' ecosystems seemed reasonably healthy. Granted, the lakes were still plagued by persistent toxic chemicals in the bottom sediments of several dozens of Great Lakes harbors; some fish were still too contaminated with toxins to be eaten on a regular basis. But all things considered, the lakes were probably as healthy in 1990 as at any time

David Jude, University of Michigan, School of Natural Resources and Environment

Quagga mussels are shown on a sediment sample collected from the bottom of Lake Michigan, more than 200 feet below the lake's surface.

since the onslaught of European immigrants in the 1800s delivered a myriad of environmental insults. It was during that period of ecological recovery that a biologist from the National Oceanic and Atmospheric Administration noticed a curious change on the bottom of southern Lake Michigan, 130 feet below the surface. The population of *Diporeia* in that area of the lake had fallen into a sudden, precipitous decline.

The disappearance of *Diporeia* was terribly significant: The six-legged amphipods with an orange hue on their shrimp-like bodies accounted for up to 70 percent of the living biomass on the bottoms of healthy lakes. Because of their abundance, and the fact that they converted plankton into high-fat body tissue, *Diporeia* were the single most important food source for prey

Jeff Alexander

The abundance of quagga mussels is evident in this trawl net that was dragged along the bottom of Lake Michigan. The trawl pulled up more than 1,000 pounds of mussels but few fish. Scientists in 2008 estimated there were 500 million pounds of quagga mussels living on the bottom of Lake Michigan.

fish in the Great Lakes. Whitefish, bloater, alewife, and sculpin—all gorged on *Diporeia*. Its high-fat content was a gold mine of calories for fish, which burned calories the way Americans and their SUVs consumed gasoline. *Diporeia* were the bread and butter in a complex food chain that supported lake trout, whitefish, and salmon. A one-time population decline in an isolated pocket of one lake did not constitute a crisis. But sampling revealed signs of an alarming trend: *Diporeia* were vanishing from entire regions of Lake Michigan.[5]

Tom Nalepa, the NOAA biologist who tracks the health of bottom-dwelling species in the Great Lakes, said researchers routinely found up to 20,000 *Diporeia* per square meter on the bottoms of the Great Lakes in the 1980s. The *Diporeia* population crashed after zebra and quagga mussels infested the lakes. By 2005, *Diporeia* had been nearly eliminated from Lakes Michigan and Huron; the species was struggling in the other lakes. The most dramatic loss occurred in Lake Michigan, where the *Diporeia* population dropped by 94 percent between 1995 and 2005. The density of quagga mussels on the lake bottom exploded during the same period, reaching 16,000 mussels per square meter. "There's a tremendous shift in the living biomass taking place on the bottom of Lake Michigan," Nalepa said.[6] The most important source of fish food in the lake was suddenly, inexplicably, gone. The emerging famine hit Lake Huron in 2000. By 2003, the *Diporeia* population there had dropped 57 percent.[7]

A similar decline was documented in Lake Ontario. Lakes Erie and Superior were spared. *Diporeia* were naturally less abundant in Erie, which is much shallower and warmer than the other Great Lakes. Lake Superior's frigid water, which lacks the high levels of calcium needed to form mussel shells, was initially spared the widespread plague of zebra and quagga mussels that took over much of Lakes Michigan, Huron, and Ontario.

The sudden decline of *Diporeia* began shortly after zebra mussels started colonizing the nearshore areas of Lake Michigan in 1990. The arrival of the quagga mussel two years later accelerated the demise of *Diporeia*. Unlike zebra mussels, which prefer warmer water and only cling to hard surfaces, quagga mussels can live in the frigid depths of the Great Lakes and attach to almost any surface—hard or soft. Quagga mussels have longer byssal threads than zebra mussels, which allow them to attach to softer surfaces; the quaggas also are more efficient filter-feeders than their *Dreissena* cousins.

The trillions of mollusks blanketing the lake bottoms removed tremendous quantities of phytoplankton and zooplankton—microscopic plant material and tiny animals—from the water. Their feeding habits effectively short-circuited the natural cycle of food and energy in the lakes. *Diporeia* were the first victims of the mussels' gluttony. The amphipods were no match for the invaders. *Diporeia,* which burrowed into the lake bottom during the day, emerged at night and swam up into the water to capture tiny particles of aquatic vegetation. The giant sucking effect that zebra and quagga mussels brought to the lakes intercepted the diatoms *Diporeia* needed to survive. Suddenly, there was little food left in the water: the foreign mussels were inhaling most of it and depositing the leftovers on the lake bottom in the form of feces coated in slime. Scientists have not figured out precisely why *Diporeia* crashed after zebra mussels arrived. But given the circumstantial evidence, and the fact that the mussels consume the plankton that *Diporeia* need to survive, most researchers have attributed their near disappearance to zebra and quagga mussels.

The virtual elimination of *Diporeia,* the human equivalent of subtracting corn syrup from the diets of sugar-addicted Americans, sent shock waves up the Great Lakes food chain. Commercial fishing operations and sport anglers felt the repercussions across much of the Great Lakes. From Green Bay, Wisconsin, to Rochester, New York, anglers and researchers began to notice in the early years of the new millennium that whitefish and salmon were shrinking. It seemed the Great Lakes food chain was crumbling under the weight of zebra and quagga mussels. John Gannon, a senior scientist at the International Joint Commission, expressed the dismay many scientists felt when *Diporeia* began to vanish. "It's one of those issues that is just so scary because we have seen such a recovery [of fish species] in the 30 years since the Clean Water Act was passed," Gannon said in a newspaper interview. "We had this wonderful success story running and then one of the main food sources [*Diporeia*] starts to disappear."[8]

It was bad enough that ocean ships imported foreign mussels that depleted one of the most important sources of fish food in the Great Lakes. Even more alarming, and outrageous, was the fact that there wasn't a thing anyone could do to fix the problem. There wasn't a fish, bird, or chemical known to man that could rid the lakes of zebra and quagga mussels and save *Diporeia*. Several species of fish and birds ate zebra and quagga mussels, but not nearly enough to keep their numbers in check. "Unfortunately, all we can do is keep an eye on it, monitor it, note the decline and hope it reverses," said Marc Tuchman, an environmental scientist at the U.S. Environmental Protection Agency's Great Lakes office in Chicago. "There is nothing we can do."[9] That was the same stance EPA officials took in 1973, when the agency made freighters' ballast water discharges exempt from the stringent regulations of the Clean Water Act. Had the EPA done its job then, quagga mussels and a suite of other shipborne invaders might still be where they belong: in Europe.

Whitefish are the bread and butter of the tribes, the only really valuable commercial fishery left.

—Mark Ebener, tribal fisheries biologist, 2007

16 WHITEFISH AND GREEN SLIME

The fishing tug *Thomas A.* slipped out of the northern Michigan village of Naubinway long before sunrise, its steel hull slicing through the glass-flat waters of northern Lake Michigan. A canopy of stars illuminated an ebony sky that enveloped the lake in darkness. The only sound was the steady roar of a diesel engine that powered the 46-foot-long commercial fishing boat operated by members of the Sault Tribe of Chippewa Indians. Captain Brady Baker put the boat on autopilot, lit a cigarette, and focused his attention on navigational devices that cast a green shadow on his mustachioed face. It was 4:40 A.M. Fifty minutes passed before the sun began to rise, a pink hue absorbing the stars into the morning sky. To the south, a freighter was headed west through the Straits of Mackinac, into the treacherous, rock-infested waters that surrounded the Beaver Island archipelago.

Baker and three other members of the Sault tribe huddled in the tiny cabin of the *Thomas A.*, a working boat so Spartan it had no head. With Dave Frazier at his side, Baker steered the boat east toward the Mackinaw Bridge as the two youngest crewmembers, Geno Graphos and Kevin Frazier, curled up in the cramped bow to steal a nap before reaching the first of 10 fishing nets they would lift that day. Frazier and Graphos awoke an hour and 15 minutes later, put on yellow rubber chest waders and long rubber gloves, and prepared for the task at hand. Their mission was simple, repetitious, and arduous: Lift 10 trap nets scattered across a 200-square-mile area in a roughly eight-hour shift, and return the bounty—succulent lake whitefish, *Coregonus clupeaformis*—to the dock for processing at King's Fish Co.

Lake whitefish was the lobster of the Great Lakes commercial fishery. The fish were plentiful, loaded with beneficial omega-3 fatty acids, and had a light, almost sweet taste. They were among the least fishy-tasting of all Great Lakes fish. Native American and non-tribal commercial fishing operations caught 9,355,906 pounds of whitefish, valued at $6.7 million, in the five Great Lakes in 2005. Lake Michigan yielded the biggest catch, 4.5 million pounds. Commercial anglers caught 3.3 million pounds of whitefish in Lake Huron, 1.5 million pounds in Lake Superior, and 5,176 pounds in Lake Erie.[1] Called *Atikamig* by Native Americans, whitefish weighed an average of two to four pounds and grew to 22 inches in length. Only their fillets were white. The fish were greenish-brown along the spine and silver on the sides; their trademark feature was a small head and a downward-pointing snout. Whitefish were perhaps the most delicious ugly fish in North America. The 1918 edition of Fannie Farmer's Boston Cooking School cookbook called it "the finest fish found in the Great Lakes."[2]

Dining on Great Lakes whitefish was considered glamorous in culinary circles. The effort that went into catching the fish was far from elegant. Harvesting lake whitefish was backbreaking manual labor flavored with a heavy dose of fish blood and guts. Any notion that there was pleasure to be found in the work aboard the *Thomas A.* was dashed by a sign in the cabin window. The sign faced toward the rear deck, where workers hauled tons of fish out of nets each day and packed them in crates of ice. "Shut up and keep working," read the handwritten sign. Said Baker: "It's our company's motivational video."

On an average day, when the weather and fish cooperated, lifting 10 trap nets yielded 2,000 pounds of whitefish. The fillets sold for about a dollar per pound. But there was an air of uncertainty as the crew of the *Thomas A.* set out on their daily fishing trip. None of the fishermen could predict from one day to the next if the massive submerged nets they used to trap fish would be teeming with whitefish or blanketed with *Cladophora*, an algae which scared away fish and occasionally sank the nets. The algae flourished after zebra and quagga mussels invaded the lake and increased water clarity, which allowed light to penetrate deeper. The phenomenon essentially created a larger field on which *Cladophora* could grow. *Cladophora* blooms turned parts of the Great Lakes into underwater kelp forests. The algae fouled commercial fishing nets and blanketed shorelines when waves or currents dislodged it from the lake bottom.

The first net the crew of the *Thomas A.* lifted on that unusually warm June day showed signs of promise before it reached the surface. "I see bubbles," Baker said. Bubbles were good. Bubbles equated to fish. A few bubbles soon became hundreds, and the water boiled with fish as the net was hoisted to the surface. The crew would estimate the haul from the first net at about 400 pounds—a decent start after weeks of disappointing catches.

Catching whitefish with nets was a controversial practice that had been refined by the descendants of Native Americans who first arrived in the Great Lakes region several thousand years ago. An 1836 treaty between the Ottawa and Chippewa tribes and the U.S. government gave the tribes exclusive rights to catch fish with nets in portions of Lakes Michigan, Huron, and Superior. For decades, Native American fishermen in Michigan used gill nets to catch whitefish. Gill nets snagged and killed most fish that ran into the mesh. Following a lawsuit by fishing groups and the state of Michigan, the tribes agreed in the 1990s to switch from gill nets to trap nets in most of the Great Lakes. Trap nets captured whitefish and kept them alive until they could be retrieved; non-target fish, such as salmon and lake trout, could be released back into the lake. Gill nets, on the other hand, essentially strangled killed every fish that swam into the massive devices.

The design of a trap net was brilliant in its simplicity, yet it was based on a thorough understanding of fish behavior. It consisted of a large mesh box with three vertical mesh walls extending from one end. Two mesh walls, called wings, extended from the outer edge of the box. A third vertical wall, called a lead, extended as much as 2,000 feet from the trap. The lead guided fish from shallow water into deeper water, where the trap was set. When fish bumped into the lead, they instinctively headed for deeper water to avoid the obstruction. The lead guided them into a series of smaller mesh boxes until they arrived in the main trap net. Fish could swim into the trap net but couldn't swim out. Their fate was sealed.[3]

To retrieve fish from the nets, crews used a winch to lift part of the large polypropylene net over the low-slung deck of the fishing vessel. With the trap net at the surface, the crews used smaller dip nets on long poles to remove the fish. They then lowered the net back into the lake's depths, where it would trap fish for as long as a week before being hoisted again. Trap nets proved to be a successful tool in the battle against zebra mussels. But tribal fishermen soon discovered that the nets were magnets for *Cladophora*. Every time the wind picked up, algae broke loose from the lake bottom and drifted with the lake currents until it collided with something stationary. Often, that was a fishing net suspended in the lake.

THE DEVASTATING EFFECTS ALGAE HAD ON TRAP NETS WERE EVIDENT AT a site in Sturgeon Bay, offshore of tiny Cross Village in northern Michigan. The first hint of trouble was patches of algae floating on the lake's surface—thousands of clumps of dead *Cladophora* ranging in diameter from the size of a marble to that of a baseball. The algae resembled floating oatmeal and made the lake surface look like a kid with chickenpox—brown pockmarks scarring Lake Michigan's dazzling blue water. The scene was equal parts tragedy and metaphor. The Great Lakes in the early years of the twenty-first century were ecologically sick. The disease: a plague of shipborne invasive species.

As Baker slowed the *Thomas A.* to a crawl after an hour of cruising across the Straits of Mackinac at 13 mph, everyone on the boat could see the trap net that was supposed to be invisible 60 feet below the lake's surface in Sturgeon Bay. The water was remarkably clear, but the net was coated with a layer of *Cladophora* that looked like a blanket of lime green hair.

Baker hopped into a 12-foot skiff to get a closer look at the algae blanketing the trap net. He reached over the gunwale of the aluminum boat and lifted a small portion of the net out of the water. Much of the algae dripped off the net as it was lifted out of the water. That was good. It meant Brady's crew wouldn't have to pull the entire contraption out of the water, haul it back to land, and power wash it before replanting it in the lake. The bad news: The algae weighed down the net, collapsing the tunnel that directed unsuspecting whitefish into the large mesh box from which they could not escape. "If the fish can see the net or the net gets weighed down, it stops fishing," Baker said. "The algae gets so thick sometimes we can't see the webbing."

Lifting a net out of the lake and taking it back to land for cleaning was a pain in the back and hard on the company's bottom line. The net would be out of service for at least two days, which translated into hundreds, if not thousands, of pounds of fish that wouldn't be caught. The result: thousands of dollars in lost revenue.

The algae-coated net in Sturgeon Bay yielded few fish that June day. Baker blamed foreign mussels and the algae. "Algae gets in the nets, the fish see the net and avoid it . . . or find their way out," he said. But with 2,000 pounds of whitefish already in the boat, Baker could afford to call it a day. He steered the *Thomas A.* north for the two-hour ride back to Naubinway. En route, the crew gutted hundreds of whitefish on the deck. The bloody job attracted a white cloud of seagulls behind the boat. The squawking birds dove into the water every time one of the workers threw a bucket of fish entrails overboard. Cleaned fish were placed in crates of crushed ice and stored below deck. Once the boat returned to the dock, another crew ran the fish through a series of machines that cut them into fillets and plucked out the tiny bones. Workers inside the frigid processing plant then packaged the fish for shipment to restaurants in Michigan and fish markets in Chicago and New York. By the end of the day, some of the fish would be tickling the palates of tourists in the spectacular dining room at the famed Grand Hotel on nearby Mackinac Island.

BOB KING, A FOURTH-GENERATION TRIBAL FISHERMAN, STROLLED TOWARD the dock as Baker steered the *Thomas A.* into the slip. I knew it was King before I had a chance to step out of the boat and shake his hand. His first name adorned the large metal buckle on his leather belt.

"How'd you do?" King asked Baker. King no longer went out on the boat, but he remained a consummate fisherman.

"We got a ton," Baker replied. It was an average haul.

Jeff Alexander

Commercial fishermen Brady Baker (*left*) and Dave Frazier are pictured removing whitefish from a trap net in northern Lake Michigan in 2007. The men were members of the Sault Tribe of Chippewa Indians and employed by King's Fish Co. in Naubinway, Michigan. Zebra and quagga mussels fueled algae growth, which coated fishing nets and made it more difficult to catch whitefish.

With the boat secured to the dock, Baker and his crew began hoisting dozens of green plastic crates loaded with whitefish out of the *Thomas A.*'s belly and onto a forklift. That task finished, their work was done for the day. A second crew would fillet the fish and package them for shipment. Baker and his crew were headed home for naps. They would be back the next morning, before the crack of dawn, to repeat the drill. They did this every day from April to December, at least when the weather cooperated. They didn't go out when gales whipped the lake into a potentially deadly frenzy, and they didn't fish in November, when lake whitefish spawned.

King's family fished these waters for more than a century. Much had changed since the early 1900s, when King's great-grandfather fished for whitefish from a sailboat. That was before sea lamprey decimated lake trout and countless dead alewives coated beaches and parts of the lake's surface in the mid-1900s. King went to work for the family business in 1960 and stayed

Jeff Alexander

Lake whitefish are the prized catch for commercial anglers in the Great Lakes. Decreasing amounts of fish food caused whitefish to shrink in size after zebra and quagga mussels invaded the lakes.

until 1974. He grew tired of fishing and tried his hand at other jobs. That was when he discovered he was hard-wired, genetically or culturally, to make a living as a fisherman. Faced with the prospect of his family's fishing heritage fading away when his father retired, King bought the business in 1987. Whitefish provided well for King's family: He and his wife raised their children in a ranch-style house a stone's throw from the spectacular north shore of Lake Michigan. His family seemingly owned most of Naubinway—the fish market, the hotel, and one of the diners along the main thoroughfare, U.S. 2.

King had a few glorious years of fishing in the late 1980s and early '90s before he found his livelihood threatened by a proliferation of zebra mussels, quagga mussels, and Native American casinos. Changes wrought by the prolific mollusks—increased water clarity, huge algae blooms, decreasing amounts of plankton, and shrinking fish—complicated King's job. The 2007 fishing season was typical of what King and his crews had experienced in recent years:

decent fishing in the early spring, followed by algae blooms that collapsed nets and depressed their catch, followed by good catches later in the year, before autumn storms ripped loose more blankets of net-clogging algae.

"We were having a great spring until the algae showed up," King said. "The algae gets so thick on the nets sometimes that the nets will literally quit fishing. We'll go from getting half a ton of fish in one net to getting six or eight fish. When that happens you have to pull the net out of the water and wash it off. You might pull a net out to wash it and then put it back, there will be a big blow and it gets covered with algae. You have to pull the net and start all over again. It didn't use to be that way. You used to be able to set nets in the spring and leave them in the lake all summer."[4]

The near elimination of *Diporeia* from the lake, another effect of the *Dreissena* mussel invasion, was also taking a toll on whitefish and King's livelihood. "The whitefish are getting smaller," he said. "There's not enough food for all the fish."

The Great Lakes fish food chain was crumbling, algae blooms made the fishing difficult, a resurgent cormorant population feasted on whitefish, there were fewer men willing to work in the occasionally brutal conditions the lakes could produce, labor and fuel costs were rising, and aquaculture operations were becoming increasingly competitive. Those factors converged in the late 1990s to threaten the future of an $18-million-a-year fishery with cultural ties that dated back to the earliest days of human habitation in the Great Lakes region.

At the same time foreign mussels were dismantling Lake Michigan's food chain in the 1990s, Native American–owned casinos proliferated across northern Michigan. The casinos lured workers away from the commercial fishing industry with promises of better pay, better benefits, and safer working conditions. King's fleet of fishing boats shrank from thirteen in 1990 and to one in 2007. "Why fish when you can work in the casino?" he said. "You get a good wage at the casino, good benefits and work year round. Fishing is hard work and it's dangerous. Have we made money? Yeah, but we put it all back into equipment. We don't get subsidies like farmers do."[5]

King's two sons followed him into the fishing business. He said his grandson also planned to join the family business. As he neared retirement and the day when he turned over the business to his sons, Kenny and Theron, King said he had no idea what the future held for lake whitefish and the foreign species that jeopardized this sentinel Great Lakes species. "There are so many exotics coming in now that we have to compete with. The only way to stop this is to keep ocean freighters out of the lakes and that's not feasible," King said. "I don't know what's going to happen with the mussels. I don't think there is anybody who does know what's going to happen." Truer words were never spoken.

Who gets to decide what is a good fishery and what isn't a good fishery?

—Jim Seelye, retired U.S. Fish and Wildlife Service fishery biologist, 2008

17 PARADOX

Biologist Tammy Newcomb stood before a large group of anglers at a fisheries workshop in Michigan and asked a most provocative question: Was the changing character of Lake Huron's fishery in a state of pandemonium or promise? The response from the crowd of charter-boat captains and master anglers? Silence. No one dared to venture a guess as to which scenario most accurately described the hurricane of ecological changes swirling in the third largest of the five Great Lakes. The year was 2008. Over the course of a decade, a confluence of human-induced events—overstocking of salmon, decreased quantities of prey fish, zebra and quagga mussels hogging fish food—conspired to turn Lake Huron's fishery on its head.

The lake's thriving salmon fishery produced record catches in 2002, only to collapse the following year. Alewife, the invasive fish that supported the salmon fishery, vanished from the lake. That was bad for salmon but good for native fish species, including walleye, lake trout, and emerald shiners. Walleye made a dramatic recovery in the absence of alewife, and lake trout were showing signs of making a comeback. But the total volume of prey fish in the lake was dropping like a lead sinker, as was the abundance of *Diporeia*, an amphipod that was the most important source of fish food in the lake.

Confounding the situation was the conquest of the lake bottom by zebra and quagga mussels, which injected unprecedented chaos into the ecosystem. The mussels increased biological productivity near the shoreline, but left deeper areas of the lake devoid of aquatic life. *Dreissena* mussels transformed Lake Huron into a mirror image of Lake Superior, which had strikingly clear water but supported far less fish than its sister lakes to the south. So what was the state of Huron's fishery: pandemonium or promise? The answer depended on your perspective. Newcomb said there was a fair amount of both. Lake Huron, she told the anglers, faced "wickedly complex ecological resource issues." The operative word was complex. The lake's ecosystem was changing faster than the experts had believed possible. "Change is inherent in natural systems," Newcomb said. "What's surprising to us as managers is the rate of change in Lake Huron."[1]

I posed Newcomb's perplexing question to one of her peers at the fisheries workshop, which was held in the Fraternal Order of Eagles Lodge in Alpena, Michigan. Jim Johnson was a longtime fisheries biologist for the state of Michigan; he had spent most of his career studying Lake Huron's fishery. Much of Johnson's work in the first decade of the twenty-first century focused on finding new ways to manage fisheries in the face of an onslaught of foreign species that rewired the lake's ecosystem. "Lake Huron is not the same lake we knew just 10 years ago," Johnson said. "From the standpoint of the fishery in Lake Huron, I'd say we've experienced tradeoffs. Because invasive species contributed to the collapse of alewives, we've experienced a major rebound in two native species, lake trout and walleye. That's a good thing, but it's happening for the wrong reasons."[2]

Lake Huron was reverting to its presettlement condition, when native fish species were dominant and the water was clearer and cleaner. Its ecosystem was drifting away from the glorified fish farm Huron had become in the

David Jude, University of Michigan, School of Natural Resources and Environment

This emaciated Chinook salmon was caught in Lake Huron in 2005. The lake's salmon fishery was decimated by the disappearance of alewives, which was caused by overstocking of salmon and the invasion of foreign mussels that disrupted the fish food chain.

Davic Jude, University of Michigan, School of Natural Resources and Environment

As Lake Huron's imported salmon fishery collapsed, some native fish species rebounded. Shown here are walleye, which made a dramatic recovery after alewives virtually disappeared from the lake.

John Lyons, Wisconsin Department of Natural Resources

1960s, when Michigan imported a salmon fishery to rein in alewife and create a recreational economy. A more natural lake was desirable, but the changes would create winners and losers. What gnawed at Johnson was the knowledge that unnatural forces, primarily zebra and quagga mussels, were causing the most profound ecological shifts in the lake. It was hard to applaud change that was forged by invasive species and fraught with uncertainty. Having zebra and quagga mussels at the helm of the Lake Huron ecosystem was like allowing a blind man to drive a car and then applauding when he managed to avoid causing an accident. Odds were, the blind driver would eventually crash the vehicle. Would foreign mussels drive Lake Huron's fishery over the brink of an ecological cliff? No one knew the answer to that question.

Clues about the ecological future of Lake Huron, indeed the entire Great Lakes ecosystem, would be found swimming in the lakes and dangling from fishing poles. Fish were barometers of ecosystem health and, perhaps, a predictor of human health. Jim Seelye, a retired fishery biologist who supervised the sea lamprey control program for more than a decade, said all people—not just anglers—should take heed of the state of Great Lakes fisheries. "People used to ask me why I worked so hard to protect fish," Seelye said. "I'd tell them that as fish go, we go. Humans seem to have this infallibility gene. We think things will always get better and that bad things can't happen to us."[3]

EVERY JULY SINCE 1975, HUNDREDS OF ANGLERS HAVE CONVERGED ON THE Lake Huron port city of Alpena to participate in the Michigan Brown Trout Festival. The event that began by accident is, according to organizers, "the longest continuous fishing tournament on the Great Lakes." The birth of the brown-trout fishery in Lake Huron's Thunder Bay dates back to 1970. That year, after stocking several nearby rivers with brown trout—a species native to Asia and Europe that was imported to New York and Michigan waters in the 1880s—biologists discarded about a thousand excess fish in the bay. No one expected the foreign fish to survive, let alone become an icon for a community that was home to the world's largest cement plant. Yet, it wasn't long before catching brown trout became a favored pastime for local anglers. The fish's finicky nature made it difficult to catch. Anglers who did land one were rewarded with a fish that was beautiful to look at and a pleasure to eat. Known for black and red spots that lined their bodies, brown trout were close relatives of Atlantic salmon. Their size was unspectacular, averaging 8 pounds. But an older, lake-run brown could tip the scales at 25 pounds or more. The overnight popularity of brown-trout fishing led to the creation of the Michigan Brown Trout Festival. By its third decade, the festival had grown into a 10-day event featuring multiple fishing tournaments, beer tents with live bands, a boat raffle, art festival, sidewalk sales at the local stores, and games for children. Good times for all. But how long could a fishing tournament based on an introduced fish species last? Organizers of the event never asked that question, until forced to by a dramatic turn of events.

As the festival expanded and attracted more anglers in the late 1990s, there were hints of trouble in the brown-trout fishery. Anglers were catching fewer browns but more walleye, a native species. Unfazed, organizers adapted by adding a walleye-fishing contest to the tournament. Theirs was a fortuitous decision. For the first time in the tournament's history, anglers in 2007 caught more walleye than brown trout or any other species of fish. "We caught more walleye than salmon at the Brown Trout Festival in 2007—that was unheard of," Johnson said.[4] The future of the longest-running fishing tournament on the Great Lakes was suddenly endangered by a storm of ecological changes triggered by invasive species. There was irony in the fact that despised foreign species, zebra and quagga mussels, were causing problems for a desirable import, the brown trout. The rise of walleye in the wake of the foreign mussels' conquest of the lake raised another vexing question: Was it appropriate to continue calling Alpena's popular event the Brown Trout Festival if catching the namesake fish was no longer the rule, but the exception?

LAKE HURON'S SALMON FISHERY WAS LONG VIEWED AS A STEPCHILD TO the chinook fishery in Lake Michigan. Much of the stigma was due to family bloodlines. Lake Michigan, after all, was the birthplace of the modern Great Lakes salmon fishery. It also was better suited than Huron for large numbers of salmon. Michigan's waters contained more nutrients, phosphorus and chlorophyll—the building blocks of the fish food chain. But in the 1990s, when an epidemic of kidney disease decimated Lake Michigan's chinook fishery, anglers flocked to Lake Huron. The catch rate for chinook salmon in Lake Huron—the number of fish caught per five hours of angler efforts—exceeded Lake Michigan's for much of the '90s. The catch rate in Lake Huron in 1997 was nearly double that in Lake Michigan. Such a trend was unusual, almost insulting to Huron's neighboring Great Lake.

By 2002, a resurgent Lake Michigan salmon fishery was closing the gap on Huron's catch rate. The following year marked the beginning of a trend that saw the chinook catch rate in Lake Michigan increase markedly for seven consecutive years. Lake Huron still had a higher catch rate in 2002, when anglers hauled in a record number of chinook: 11,060. But the total number of king salmon caught in Lake Huron in 2002 was a fraction of the 46,031 chinook hauled out of Lake Michigan. At the peak of Lake Huron's salmon fishery, in 2002, anglers caught an average of one chinook for every five hours of effort. Anglers were happy, and charter-boat captains were pocketing record revenues. Bait shops and bars, party stores and hotels in port communities along the Michigan coast of Lake Huron—towns like Harbor Beach, Oscoda, and Grindstone City—cashed in on the salmon bonanza. The good times were rolling. There was no indication that the thriving salmon population was about to eat its way into near oblivion, taking brown trout and steelhead down with it.

Newcomb returned to her home state of Michigan in 2002 to take a job with the state's Department of Natural Resources. She figured her job as the DNR's coordinator of fish management programs in the Lake Huron basin was a plum assignment. It was, briefly. "In 2002, I thought this job was a piece of cake," Newcomb said. "That year was the best chinook salmon fishing in Lake Huron, ever. We were heroes."[5] The next year, the biological wheels flew off the salmon fishery. Anglers reported catching few salmon in the lake in 2003 and the following year. Many of the fish they did catch were emaciated. "The fish were starving," Newcomb said. Biologists and anglers were stunned and perplexed by the dramatic reversal of fortune.

Ed Retherford, a high school biology teacher who had operated a charter boat on Lake Huron since the early 1970s, recalled the astonishing turn of events. "In 2003 and 2004 we caught huge amounts of salmon but they had big heads and small bodies," he said. "The fish were in very poor condition."[6] The changes left anglers and biologists wondering what had happened and how it unfolded so quickly.

The problem was fairly simple: There were too many salmon in the lake in 2003 and 2004 and not enough alewives for the popular fish to eat. Figuring out why proved more difficult. Once biologists solved the riddle, it shook their faith in the ability of humans to manage a massive aquatic ecosystem where foreign species reigned supreme.

A convergence of factors caused the salmon fishery's sudden collapse. In the early 1990s, as Lake Huron's salmon population increased, zebra and quagga mussels were spreading rapidly across the lake bottom. The mussels siphoned tremendous quantities of nutrients and plankton out of the water, essentially stealing fish food. That set off a biological domino effect in the lake. The population of *Diporeia* on the lake bottom plummeted as the mussels spread. That contributed to a precipitous drop in the volume of prey fish in the lake—the small fish eaten by larger fish. Among the prey fish robbed of food by zebra and quagga mussels were alewives, the sole source of food for chinook salmon.

At the same time the foreign mussels were dismantling Lake Huron's food chain, the salmon population was rapidly increasing. Unbeknownst to biologists, the number of salmon reproducing naturally in rivers that flowed into the lake soared after 1995. Natural salmon reproduction was especially high in the vicinity of the North Channel and Georgian Bay, biologically fertile areas on the east side of Lake Huron. Natural reproduction of salmon was critically important. The state of Michigan and the province of Ontario stocked millions of salmon in Lake Huron each year. Figuring out how many fish to stock in the lake was a delicate balancing act. Too many salmon would eat all the alewives, causing the salmon population to crash. Too few salmon in the lake would hurt a popular fishery that generated millions of dollars in revenue for coastal communities where fishing was a popular recreational activity.

Fisheries biologists believed, prior to 2004, that 85 percent of the salmon in Lake Huron were stocked fish that had been raised in hatcheries; the remaining 15 percent were thought to be products of natural reproduction. Tests conducted the following year revealed just the opposite: The three million fish stocked in the lake represented just 15 percent of all salmon in the lake. The other 85 percent were naturally reproduced fish. It was a glaring miscalculation, due, in part, to Ontario's unwillingness to insert tiny microprocessors in hatchery-raised fish the province deposited in the lake. Without a way to document how many of the salmon that anglers caught were hatchery fish, biologists couldn't estimate the number of naturally produced fish in the lake. Fish management agencies had stocked the lake with far too many salmon. With too many chinooks feasting on a shrinking population of alewives, which were being starved by zebra and quagga mussels, salmon literally ate their way into a biological abyss. Alewives all but disappeared from Lake Huron in 2004; salmon followed suit the following year. The collapse of the salmon fishery was quick, extraordinary, and economically devastating.

The number of salmon caught in Michigan's portion of Lake Huron plummeted from 11,060 in 2002, to 1,861 in 2007. The fish were scarce and smaller. In 1997, the average chinook caught in Lake Huron weighed 17 pounds; that figure dropped to 8 pounds by 2004. The condition, or plumpness, of chinook caught in Lake Huron in 2004 was the lowest ever recorded in the Great Lakes.[7] With the number and size of salmon shrinking, anglers abandoned the lake in droves. Many headed back to Lake Michigan, where the salmon fishery was enjoying a resurgence that featured near-record catches in 2005, 2006, and 2007.

The exodus of anglers dealt a crippling economic blow to many of the small port communities along Lake Huron. Bait shops, hotels, general stores, and bars that counted salmon anglers among their best customers were left high and dry. Many closed. Marinas once filled with fishing boats were deserted. In 2007, the once-thriving port community of Harbor Beach, Michigan, had just one charter-fishing business left. "It was like a ghost town," Johnson said. "Anglers spent 1.3 million hours fishing for chinook salmon in Lake Huron in 2003. In 2006, that dropped to 303,000 hours—24 percent of the long-term average. That decrease in angler effort translated into a $19 million loss of economic activity in 10 port communities, just from the chinook decline."[8]

Researchers concluded in 2007 that zebra and quagga mussels had shifted much of the fish food in Lake Huron from the lake's water column, or pelagic zone, to its benthic zone, the lake bottom. Think of a giant vacuum cleaner sucking vast quantities of plankton out of the water column and depositing that material on the lake bottom, in the form of zebra and quagga mussel feces. That's precisely what the mussels were doing in Lakes Huron, Michigan, Erie, and Ontario. The shift in food energy from the open waters to the lake bottom was a profound change in the very essence of how the lakes' ecosystems functioned. "Clearly, Lake Huron is experiencing the full force of an invasive species storm, precipitated for the most part by stowaway creatures in the ballast water of salt-water freighters," concluded a study by the Michigan Department of Natural Resources.[9]

The state of Michigan made a desperate attempt to resurrect alewives in Lake Huron, which, if successful, would have provided more food for a shrinking salmon population. The state cut salmon stocking in half, from 3 million fish in 2005 to 1.5 million the following year. But that effort was too little, too late for the salmon fishery—alewives were history. Newcomb said biologists in 2007 resigned themselves to a "new reality" for the future of Lake Huron's fishery. One of the basic tenets of that new reality was that alewives were unlikely to return. "We don't believe alewife are coming back any time soon," she said. That, by extension, meant that the lake's once thriving salmon fishery was likely history as well.

Dozens of charter-boat operators and hundreds of other anglers had already moved across the state by 2007 to fish for chinook in Lake Michigan. A few hearty anglers who stayed put were rewarded for their stubborn belief that there were still salmon lurking in the depths of Lake Huron. Retherford was one who refused to give up on the fishery. "We still catch a lot of salmon but nowhere near the number we caught in the 1990s and the years prior to 2004," Retherford said. "There are still salmon out there; we catch them every day. But it's not the same. In the '90s, we used to joke about when we would get our limit of salmon and get back to the dock. The joke was whether we'd go to breakfast at 8 A.M. or 9 A.M. The days of that kind of salmon fishing have crashed and probably won't come back."[10]

Charter-boat operators could catch huge numbers of lake trout in Huron after the alewife and salmon populations crashed. But Retherford said most serious anglers wouldn't pay hundreds of dollars to catch lake trout. "Once someone hooks a salmon, they're not going to come back if all they're going to catch is lake trout," he said. Lake trout could grow to 30 pounds, but didn't put up much of a fight when caught by an angler's lure. Salmon anglers often referred to the fish as "grease balls," even though lake trout, not salmon, were the top native fish species in the Great Lakes. Catching lake trout also became more difficult in the *Dreissena* mussel era. The mussels fueled algae growth on the lake bottom, where lake trout lived. "We can't fish the bottom of the lake for lake trout because of all the green slime [algae]," Retherford said. "You can't let your hook touch the bottom or you're done—you'll pull up green slime or a bunch of zebra and quagga mussels."[11]

Scientists who devoted entire careers to studying the Great Lakes were astounded by the pace and magnitude of changes that zebra and quagga mussels caused in all but Lake Superior. "When they first arrived we wondered what kind of impact they might have and now we're starting to see it in Lakes Michigan and Huron; I didn't think it would happen this quickly," said David Jude, a University of Michigan research scientist.[12]

Retherford was one of the few charter-boat captains who held out hope that Lake Huron's salmon fishery would rebound someday. But he didn't deny the harsh reality that foreign mussels had reconfigured the lake ecosystem, crippled the mighty salmon fishery, and decimated a sport-fishing economy that once pumped millions of dollars into towns along the east coast of Michigan. "At one time we had 20 charter boats working out of Presque Isle and another 15 to 20 in Alpena. A lot of those guys have left," Retherford said. "I don't blame them for leaving. There are fewer fish and the fish are smaller." Among the charter-boat captains who moved to Lake Michigan were Retherford's two sons. Retherford considered moving his boat, the *Trout Scout*, from Alpena to Saginaw Bay, to cash in on the bay's booming walleye population. "Anybody can catch walleye," he said.[13]

BIOLOGISTS WHO STUDIED THE COLLAPSE OF LAKE HURON'S SALMON FISHery were surprised to learn how important a role alewives played in the lake's fish communities. It was well known that salmon dined almost exclusively on alewives, and that the invasive fish depressed populations of several native species, including lake trout, perch, walleye, and emerald shiners. The degree to which alewives affected native fish species came into focus after the invaders were virtually extirpated from the lake. "I don't think we could have designed a scientific experiment any better than what happened with the alewife," said James Baker, a fisheries manager for the Michigan Department of Natural Resources. "The collapse of alewife showed scientists how large a role alewife played in suppressing native species; we had no idea it was that profound."[14]

The lake trout fishing in Lake Huron in 2007 was as good as in Lake Superior, which supported the Great Lakes' healthiest lake trout population. "Our charter boat lake trout catch is as good as the lake trout catch in Lake Superior—that's damn good," Johnson said. "To say fishing has collapsed in Lake Huron would be totally wrong. The walleye and lake trout fishing has never been better. But people want chinook salmon. Probably three-fourths of the people who fish out of these ports on Lake Huron won't fish here if the salmon aren't here."[15]

Even more impressive than the resurrection of lake trout was the recovery of walleye, which struggled for decades in the presence of alewives that devoured their eggs. "Walleye are staging a major comeback and reproducing at 1940 levels," Johnson said. Michigan officials said walleye fishing in Lake Huron's Saginaw Bay in 2007 was nearly as good as western Lake Erie, which Ohio officials declared the "Walleye Capital of the World."

Despite the recovery of walleye, lake trout, perch, and emerald shiners, there was a growing body of evidence that zebra and quagga mussels were turning the deep waters of Lake Huron into a biological desert. Biologists from the U.S. Geological Survey's Great Lakes Science Center said they were amazed by what they didn't see when they dragged the lake bottom in 2007 to assess prey-fish abundance. "We didn't see any alewives," USGS biologist Jeff Schaefer said. "We saw large areas of the lake that are devoid of fish and the fish that are out there are going deeper. There seem to be very clear indications that the food web is changing. Food that used to go into fish is suspected of being shunted into dreissenids [zebra and quagga mussels] and *Bythotrephes* [spiny water fleas]." The upside was that changes caused by those invasive species had opened the door to restoration of native species, Schaeffer said. "With the decline of alewife, the door to restoration of native species that everyone thought was closed has been kicked open."[16]

Lake trout and walleye also were thriving, in part, because they took a liking to round gobies, the invasive fish from Europe. Whitefish, like salmon, didn't eat gobies. That put a pinch on commercial fishing operations, which could only harvest whitefish from Lake Huron. With less *Diporeia* to eat, whitefish were forced to eat quagga mussels just to survive. Quagga mussels had far less fatty tissue than *Diporeia*, which meant whitefish were eating more food with less nutritional benefit. The result: emaciated whitefish, some of which were too skinny and mushy to be sold. John Gauthier, a veteran commercial fisherman in Rogers City, Michigan, said the phenomenon drove the price of whitefish down by 30 cents per pound between 2002 and 2005. Gauthier sold whitefish for 50 cents per pound that year, the same price his father was paid for whitefish five decades earlier. "The people who smoke whitefish want plump fish, and that's the reason we're only getting 50-cents a pound," Gauthier said.[17]

LAKE MICHIGAN AND HURON ARE TECHNICALLY ONE LAKE, LINKED BY THE Straits of Mackinaw. The collapse of Huron's salmon fishery after 2004 had many Lake Michigan anglers wondering if the birthplace of Michigan's salmon fishery could withstand the insidious effects of a plague of zebra mussels, quagga mussels, and round gobies. Would the artificial salmon fishery created in 1966 survive to see its 50th anniversary in 2016? Scientists were divided on that question. Anglers, the ultimate optimists, were confident that Lake Michigan's mighty chinook fishery would avoid the catastrophic changes that brought down Lake Huron's salmon fishery.

There were legitimate reasons to be optimistic and pessimistic about the future of Lake Michigan's salmon fishery. Anglers caught near record numbers of chinook salmon from 2005 through 2007. But the fish were roughly half the size of those caught 15 years earlier. Fish managers reduced the number of chinook stocked in Lake Michigan from 9 million in 1995 to 3.5 million fish a decade later. It was all they could do to try and find a sustainable balance between the number of salmon and alewives in the lake.

A most troubling sign was the continued decline in the volume of prey fish in Lake Michigan. Prey-fish abundance in the lake hit record lows in 2006 and 2007. The volume of all prey fish in the lake—bloaters, alewife, and other small fish eaten by salmon, lake trout, and whitefish—dropped from 61 kilotons in 2006 to 30 kilotons in 2007, according to data compiled by the U.S. Geological Survey's Great Lakes Science Center. That was the lowest volume recorded since the government began tracking prey fish densities in 1973. Prey-fish abundance in 2007 was 92 percent below the record volume of 400 kilotons recorded in the late 1980s, said Chuck Madenjian, a USGS research fishery biologist.[18]

A shrinking food supply caused the average size of chinook in Lake Michigan to drop from 18 pounds in 1995 to 12 pounds in 2007. A similar trend played out on Lake Ontario, which traditionally supported the largest salmon among all the Great Lakes. The average size of chinook there increased from 23 pounds in 1992 to 26 pounds in 2000, before dropping to 18 pounds in 2007.[19] The rise of *Dreissena* mussels and decline of prey fish prompted an obvious question, one that many anglers were reluctant to discuss: Was Lake Michigan's salmon population destined to run out of food and crash, much like the fishery did in Lake Huron? No one knew the answer—there were too many variables to predict the fishery's future with any degree of certainty.

The extent of changes that *Dreissena* mussels inflicted on Lakes Michigan, Huron, Erie, and Ontario in the span of two decades were so profound as to be almost unbelievable. Trillions of the tiny mussels had effectively reprogrammed the biological mechanisms that governed the massive lakes. With fish populations suffering, beaches despoiled by algae blooms, and fish-eating birds dropping dead from food poisoning passed up the food chain by the foreign mussels, one could only wonder if an already dire situation could get any worse. It could—and did—grow more ominous. Algae blooms fueled by the foreign mussels also jeopardized safety at one of the most inherently dangerous industries operating near the Great Lakes: nuclear power plants. Giant mats of *Cladophora* fertilized by *Dreissena* mussels clogged water intakes in Lake Ontario, forcing several emergency shutdowns at nuclear power plants along the coast. Imagine that: Mussels native to Russian waters creating conditions that could, in the absolute worst-case scenario, cause a nuclear disaster capable of turning Great Lakes communities into mirror images of Chernobyl. In the annals of Great Lakes invasive species, that would be the sickest irony of all.

In the battle between thumbnail-sized mussels and an 850-megawatt nuclear power plant, you'd be smart to bet on the shellfish.

—Charles McChesney, journalist, 2007

18 FEAR THIS

The guard at the New York side of the border crossing in Niagara Falls stared me down with the requisite skepticism as my car inched toward his booth. He wasted no time with small talk, immediately popping the question that always made me nervous.

"Citizenship?"

"U.S."

"Where you from?"

"Michigan."

I had nothing to hide. Still, the possibility of the guard searching my vehicle was enough to make my heart race and cause beads of sweat to form on my brow. I presented my driver's license and birth certificate to the guard. He stepped into the booth to check my name against a database of known or suspected terrorists. Satisfied that I wasn't on a mission of mass destruction, the guard launched into a series of rapid-fire questions that produced an uneasy verbal dance between us.

"Do you have any liquor, firearms, or drugs?" he said.

"Nope." (I always wondered if drug smugglers or motorists packing heat answered that question in the affirmative.)

"How long were you in Canada?"

"Just passing through," I said.

"Where you goin' and what's the purpose of your visit?"

That turned out to be the pivotal question.

"I'm headed to Bailey's Harbor, New York, to do research for a book I'm writing," I said.

"What kind of book?"

"It's a book about how exotics, invasive species, have changed the Great Lakes."

"Exotic species, huh," the guard said. "Like zebra mussels?"

"Yep."

Suddenly, the guard who seconds earlier was giving me the third degree was more interested in discussing zebra mussels than determining whether I was a criminal or a terrorist.

"I used to have a jet ski and after those zebra mussels came in the lake was beautiful," the guard said, referring to Lake Erie. "You could see down 15 feet. I could see sandbars and everything. Sometimes I could see fish. It was amazing. Zebra mussels have been good for the lakes, right?"

"Sure," I lied.

Having driven seven hours from west Michigan, I was too exhausted to explain why a crystal clear lake was not necessarily a healthy lake. Besides, I learned long ago that it rarely made sense to argue with people who carried loaded guns.

To my surprise, the guard sent me on my way with a friendly greeting: "Good luck with your writing."

He was obviously unaware that the same foreign mussels that made parts of Lake Erie look like a swimming pool had caused an emergency shutdown three days earlier at a nuclear power plant 150 miles east of Buffalo, on the shores of Lake Ontario. Six years after terrorists flew airplanes into buildings in New York and Washington, D.C., one of the least-known but serious threats to homeland security in the Great Lakes region was algae blooms. Giant mats of *Cladophora*, the growth of which was fueled by zebra and quagga mussels, began washing onto Lake Ontario beaches in 2005. On several occasions, the algae clogged water intakes at nuclear power plants, forcing emergency shutdowns. Ponder that for a few seconds: the notion of dime-sized mussels crippling nuclear power plants stationed at the edge of the world's largest reservoir of fresh surface water. Who could have imagined that the unintended effects of two incredible engineering feats—nuclear fission and construction of the St. Lawrence Seaway—would collide on the shores of Lake Ontario 46 years after the Seaway opened? Certainly not the engineers who figured out how to generate electricity by splitting atoms, or the politicians who lobbied for construction of the Seaway.

As I drove toward Oswego, New York, the border guard's comments about zebra mussels echoed in my brain like an annoying jingle. What would it take, I wondered, for people in the United States and Canada to comprehend the dire threat zebra mussels and other invasive species posed to these incomparable lakes and many other freshwater ecosystems in North America? Would massive fish kills or waves of dead water birds washing up on beaches convince the skeptics, the ignorant, or those simply ambivalent about the immense risks foreign species posed to the environment and public health? Perhaps images of people being sickened, or dying, after coming in contact with toxic algae blooms triggered by the mussels would drive home the message. What if zebra and quagga mussels caused a meltdown at one of the 13 nuclear power plants that relied on the Great Lakes for cooling water? That seemingly absurd scenario was entirely plausible. In the first decade of the twenty-first century, foreign mussels created conditions that led to seven emergency shutdowns at three nuclear power plants on Lake Ontario. That was no trivial matter. Emergency shutdowns of nuclear reactors, if not conducted properly, could lead to a catastrophe capable of endangering human life and water quality in one of the world's largest lakes. The frightening possibility took me back to one of the scariest times of my childhood.

NUCLEAR FISSION WAS ONE OF THE MOST EFFICIENT AND CONTROVERSIAL methods for generating electricity. Few issues evoked more powerful emotions than nuclear power. For the most part, people either supported nuclear power, or opposed it with every ounce of their being. Rarely was there a middle ground. Developed during the construction of atomic bombs that were dropped on Hiroshima and Nagasaki, nuclear fission was perfected in December 1942 at the University of Chicago. The mastermind was an Italian physicist named Enrico Fermi. The process was first used to produce electricity in 1951, when an experimental reactor in Idaho became the world's first to generate power. The device illuminated four light bulbs. Six years passed before the first nuclear power plant, in Shippingport, Pennsylvania, was built and brought to full design capacity. The nuclear-power era was just getting cranked up when the Seaway opened in 1959.

Nuclear reactors were one of the world's cleanest and most dangerous sources of electricity. Properly operated reactors didn't pollute the air or contribute to global warming the way coal- and gas-fired power plants did. In fact, nuclear power was one of the cleanest ways to produce power, if you ignored the occasional spills of radioactive cooling water and the radioactive waste that remains highly toxic for thousands of years. Poorly operated power plants could suffer meltdowns capable of releasing enough radiation to instantly kill humans and poison vast areas of the landscape downwind. That was illustrated in 1986, when the Chernobyl nuclear power plant in the former Soviet Union suffered two explosions. The blasts leveled part of the facility and sent a cloud of radioactive dust across much of Europe. Investigators determined that the horrific accident was the result of poor design and a series of human errors that resulted in the world's worst nuclear disaster.

Among the human errors at Chernobyl was the disabling of a safety procedure known as SCRAM. SCRAM was an acronym for "Safety Control Rod Axe Man." It was not code for "Run for your life!" when problems arose in a nuclear reactor. SCRAM was developed by Fermi and his team of engineers who built the world's first nuclear reactor—known as Chicago Pile 1—on the floor of a squash court beneath the University of Chicago's athletic stadium. Fermi's team discovered that inserting cadmium rods into the reactor disrupted the fission process by absorbing neutrons. Halting fission was crucial if a crisis erupted—such as one caused by the sudden loss of reactor cooling water—that could trigger a meltdown or an explosion. Most modern nuclear power plants automatically launched a SCRAM if cooling-water levels reached dangerously low levels, or if the water became too hot. Modern versions of SCRAM were similar to the one Fermi developed, just more advanced and automated. The SCRAM process gave power plant operators a powerful tool to prevent a nuclear meltdown. The procedure also played a role in *The China Syndrome*, a 1979 American film that dramatized the risks of a nuclear disaster.

The movie's title referred to the baseless theory that a runaway nuclear reaction could cause a reactor to melt through the floor of its containment building and burn a hole through the Earth, all the way to China. It was a ridiculous notion, but one that made great fodder for Hollywood scriptwriters. To an impressionable teenager, which I was in 1979, the movie was terrifying. In the film, a television news crew was filming an upbeat story about energy production when the reporter—portrayed by Jane Fonda—saw employees panic in the control room of a nuclear power plant she was visiting. The plant's operators scrambled to regain control of the plant after cooling water in the reactor plunged to dangerously low levels. Operators initiated a SCRAM to control what managers at the power plant later described as a "swift containment of a potentially costly event." A suspicious Fonda persuaded a disgruntled plant operator, portrayed by actor Jack Lemmon, to blow the whistle on problems at the fictional nuclear power plant. Those problems, Fonda warned viewers in the fictional newscast, could lead to a nuclear meltdown that could "render an area the size of Pennsylvania permanently uninhabitable."[1]

The release date of *The China Syndrome*—March 16, 1979—was uncanny. Twelve days after the film opened, amid criticism from industry officials who characterized it as liberal fear-mongering, a nuclear power plant in Pennsylvania suffered a partial meltdown. The accident at the Three Mile Island facility near Middletown, Pennsylvania, was the worst nuclear power accident in U.S. history. Investigators concluded that the incident was caused by a series of malfunctions that resulted in a loss of cooling water in the reactor core. Though a SCRAM was initiated to prevent Three Mile Island's core from overheating and spewing radioactive gas into the air, the procedure was deployed too late. One reactor suffered a partial meltdown and released what government officials characterized as a small amount of radiation into the air. No one was killed or injured in the accident, but the reactor was severely damaged. Cleaning up radioactive water in the damaged reactor cost $1 billion. That reactor, one of two at Three Mile Island, was never put back into service.[2]

Government and industry officials tried to assure the public that the amount of radiation released at Three Mile Island was minimal and posed no threat to public health or the environment. Area residents and anti-nuclear-power activists would have none of it. Fear swept the nation as Three Mile Island became an icon for the dangers inherent in nuclear power production. The accident led to sweeping safety reforms. It also tarnished the nation's nuclear power industry and galvanized opposition to it. There wasn't a single nuclear power plant built in the United States for 30 years after the release of *The China Syndrome* and the Three Mile Island accident. Nor were there meltdowns at any other U.S. reactors during that period.

The risk of a meltdown, however slight, is omnipresent when operating a nuclear reactor—it increases dramatically when cooling water drops to dangerously low levels. That very scenario, the loss of cooling water, became a serious problem for nuclear reactors on Lake Ontario in 2005. Massive algae blooms stimulated by zebra and quagga mussels began clogging water intakes. The seriousness of the problem cannot be overstated: The sudden loss of cooling water is one of the worst things that can happen at a nuclear reactor. It is one of the quickest ways to cause a meltdown. The only thing that stood between mussel-fueled algae blooms and nuclear disaster at the edge of the world's largest source of fresh surface water was the ability to successfully initiate SCRAM procedures every time a slug of *Cladophora* plugged a cooling water intake. It was a perilous situation.

THE SUMMER OF 2005 WAS UNUSUALLY WARM IN SOUTHERN ONTARIO AND northern New York. The sweltering heat produced ideal growing conditions for *Cladophora* in the shallow coastal regions of Lake Ontario. In mid-August, powerful winds out of the southwest ripped huge quantities of *Cladophora* off the lake bottom and pushed it toward the Ontario shoreline. Tons of the filamentous green algae drifted into a canal where the massive Pickering nuclear power plant drew huge quantities of cooling water from Lake Ontario. Located 26 miles east of Toronto, and a stone's throw from the lake's edge, the Pickering facility is one of the world's largest nuclear power plants. Its six reactors crank out 3,100 megawatts of electricity, enough to power a city of 1.5 million people. Lake Ontario plays an integral role in power production at the Pickering facility. Pumping cooling water out of the lake is a vitally

Ontario Power Generation

Massive algae blooms in Lake Ontario, fueled by zebra and quagga mussels, clogged water intakes and forced emergency shutdowns at the Pickering Nuclear Power Plant (*shown here*), east of Toronto. The algae blooms forced emergency shutdowns at two other nuclear power plants on the shores of Lake Ontario.

important process that, in the complex nature of operating a nuclear plant, is considered fairly routine. That changed in August 2005—sucking water out of the lake became a vexing challenge with potentially deadly consequences.

On the morning of August 19, waves of *Cladophora* plugged filter screens that kept debris out of the Pickering power plant. The situation was analogous to someone jamming a giant cork in the water intake. Operators were forced to shut down three of Pickering's six nuclear reactors; two others were idled until the algae could be removed. Only one reactor was able to operate, due to the reduced flow of cooling water. The problem was quickly resolved, but soon returned.[3]

Six weeks after the Pickering incident, algae clogged the water intake at the Darlington nuclear power plant. That facility, located 21 miles east of the Pickering plant, features four massive reactors. Those reactors generate 3,524 megawatts of electricity—20 percent of all the power needed in the sprawling province of Ontario. Officials at Ontario Power Generation, which owns the Pickering and Darlington facilities, believed the Darlington facility was immune to floating mats of *Cladophora* because its intake drew water from the bottom of Lake Ontario. The Pickering facility inhaled water from the surface of the lake, where *Cladophora* dislodged from the lake bottom often accumulated until winds blew it ashore or out to sea. Like their colleagues at Pickering, plant operators at the Darlington facility received a rude awakening in 2005. Strong winds and high waves pushed a dense green mass of *Cladophora* along the lake bottom, where it was sucked into Darlington's intake pipe. The incident shut down one of the facility's four reactors. The problem returned two years later, forcing officials there to shut down one of six reactors until the algae could be removed from filters and screens.[4]

Ontario Power Generation officials noted in press releases that the algae blockages posed "no environmental or safety concerns for the public or staff."[5] True, the algae didn't cause a nuclear meltdown or an explosion in any of the reactors. But any one of those blockages could have caused a nuclear disaster, absent quick intervention by skilled power plant operators.

Beyond safety concerns, shutting down a nuclear reactor is a costly and complicated process. It isn't like turning off a vehicle's engine and then restarting it on a moment's notice. The SCRAM procedure is only used in emergency situations. In the United States, initiating a SCRAM is serious enough to require notification of the U.S. Nuclear Regulatory Commission.[6] The financial implications associated with taking a reactor off line also are severe. Down time means lost revenue. OPG officials estimated that the algae blockages cost the company $30 million in lost revenue between 1995—when *Cladophora* began to clog water intakes—and 2005, when the problem reached its apex.[7]

The recurring problem prompted Ontario Power Generation to install a huge net near the Pickering facility's water intake. The vertical mesh barrier was anchored to the lake bottom, near the end of the power plant's water-intake canal, from May through December. It was supposed to reduce the volume of *Cladophora* entering the intake by 30 percent. The net helped, but it wasn't foolproof. Enough algae slipped through the net on August 9, 2007, to clog a water intake and force OPG officials to shut down one of the reactors. That incident was less serious than the 2005 algae blockages at the Pickering and Darlington facilities. It was a minor disruption compared to a series of algae blockages that crippled a nuclear power plant on the New York side of Lake Ontario.

THE CONCRETE COOLING TOWER AT THE JAMES A. FITZPATRICK NUCLEAR power plant rises 543 feet above an otherwise scenic stretch of Lake Ontario shoreline. The massive structure, which resembles an inverted funnel, is the icon of an inherently risky industry situated adjacent to one of the world's largest lakes. It is a powerful reminder of how humans toy with the fate of the Great Lakes on a daily basis. That the FitzPatrick facility is located a few miles north of the Seaway Trail is ironic. The Seaway Trail, a two-lane highway that follows Lake Ontario's southern coast, is a tribute to the engineering marvel that gave ocean freighters access to the Great Lakes. Absent the Seaway, the FitzPatrick power plant likely would not have been faced with a critical cooling-water crisis that forced employees to SCRAM the facility twice in six weeks.

September 12, 2007, was a day like no other at the FitzPatrick nuclear power plant. It marked the first time *Cladophora* completely shut down the 850-megawatt facility. Unlike similar problems at the Pickering and Darlington facilities, the September 2007 incident brought all power production at the FitzPatrick facility to a screeching halt. Huge, dense mats of algae smothered filter screens near the power plant's water intake. The facility, which inhales 360 million gallons of Lake Ontario water daily to keep its nuclear reactors cool, was suddenly sucking air.[8] The *Cladophora* flowed in so rapidly, and was so dense, that power plant operators had to SCRAM the reactor to keep it from overheating.[9]

Officials at the FitzPatrick facility were still trying to figure out how to cope with the algae problem when *Cladophora* clogged the intake six weeks later, on October 13, 2007. High winds that day dislodged more *Cladophora* from the lake bottom. Wind-driven currents pushed the algae into the power plant's intake, where it quickly overwhelmed filter screens that kept debris out of the facility. The weight of the algae placed so much strain on the large revolving screens, the metal pins that secured the devices to motors sheared off. Workers used backup pumps to send cooling water into the reactor, but were soon forced to shut down the facility to avoid a potential disaster.

Fifteen days later, the menacing algae struck again. Employees knew from experience that high winds out of the west could send a torrent of *Cladophora* into the facility's water intake. Winds reached 30 mph on October 28, with gusts of 50 mph. The workers were prepared for the worst, or so they thought. They sped up the rotating filter screens with the hope of keeping *Cladophora* from encasing the devices in an impenetrable blanket of fetid algae. Novel as it was, the defensive procedure was for naught. A huge slug of *Cladophora* began flowing into the intake late on a Saturday afternoon. Shortly after midnight, at 12:48 A.M. to be precise, operators in the control room initiated "abnormal operating procedures" when the algae began to reduce the flow of cooling water. Down in the water-intake structure, crews attacked the algae-coated filter screens with fire hoses, to no avail. At 12:55 A.M., plant operators reduced power in the facility to decrease the need for cooling water. It was a desperate but short-lived effort to keep the plant operating. Four minutes later, at 12:59 A.M., the operators gave up and initiated a SCRAM procedure.

A spokeswoman at the FitzPatrick plant tried to downplay the algae problem in a 2007 newspaper interview. "It's a natural problem and we're talking to a marine biologist," spokeswoman Bonnie Bostian told the *Syracuse Post-Standard*.[10] Her answer was half correct. *Cladophora* was natural and native to the Great Lakes. But its growth rate in the era of *Dreissena* mussels, imported two decades earlier by ocean freighters, could only be described as supernatural.

After the second algae incident, crews at the Fitzpatrick facility strengthened the filter screens and took other steps to keep algae from clogging the water intake and shutting down the reactor. The stakes were huge. Every time the reactor was out of service cost its owner, Entergy Nuclear Operations Inc., upwards of $2 million a day in lost revenue. Valiant efforts to keep the algae at bay failed time and again. The *Cladophora* returned and shut down the power plant a third time, on October 28. Two weeks later, algae forced a fourth emergency shutdown. Officials at the U.S. Nuclear Regulatory Commission, who investigated the problem, praised the utility's efforts to deal with the algae blockages. In a 2008 report, federal investigators said:

> Over the last few months, FitzPatrick has taken significant steps, including changes in operating strategy and procedures, as well as equipment enhancements, to reduce vulnerability of the plant to this phenomenon. FitzPatrick has also taken steps to minimize *Cladophora* through use of divers harvesting the algae in areas of high concentration.[11]

Those efforts were no guarantee that the problem was solved. In reality, it probably wasn't. *Cladophora* would continue to thrive in the Great Lakes as long as zebra and quagga mussels were present. There was a chance the algae problem could worsen as quagga mussels, the larger and more durable cousins of zebra mussels, colonized more of the lake bottom. Quagga mussels provided more habitat and fertilizer for *Cladophora*.

That zebra and quagga mussels created conditions capable of crippling nuclear reactors on the shores of the Great Lakes was perhaps the most frightening twist in the strange saga of foreign species conquering the lakes. The environmental horror show seemed to get more bizarre with every passing year. And yet, the toll invasive species inflicted on the lakes, and all who relied on them for work and play, was only half the story. The rest of the story focused on two burning questions: Was this ecological mayhem preventable? And, if it wasn't preventable, could more have been done to slow the rate of invasion and the resulting devastation? The short answer to both questions: Absolutely. That led to an even more pointed query: If government officials in the United States and Canada could have kept *Dreissena* mussels and other invaders out of the Great Lakes, or slowed the rate of invasion, why didn't they? It was a simple question with a complex answer. Politics, scientific uncertainty, recalcitrant shipping interests, narrow-minded environmental activists, and dysfunctional government bureaucracies created a regulatory quagmire that produced much rhetoric but little meaningful action. Not only were the lakes' ecosystems wilting under the weight of foreign species, some of the invaders were conspiring to create a Great Lakes version of *The China Syndrome*. The scenario proved, once again, that the truth is often stranger than fiction.

Invasive species cost Canadian society billions of dollars each year and threaten Canada's ecosystems, native species and their habitats. They affect all of us, no matter where we live, but so far these invaders have met only limited resistance. The federal government's inaction, more than its ignorance of the problem, is leading to a loss of biodiversity and mounting costs to Canadians.

—Johanne Gelinas, Canada's Commissioner of the Environment and Sustainable Development, 2002

We were in a position to start taking action in the 1980s and early 1990s, and we could have stopped many of these species from coming in. Opponents to ballast regulations said there wasn't the technology. They said it was too expensive. They said it wasn't enforceable and that we needed more science. Bullshit. We don't need any more scientific certainty. We know how adversely these species have affected the Great Lakes. It's time to take action.

—U.S. Rep. James Oberstar of Duluth, Minn., 2006

04 BETRAYAL

The tragedy of this situation is that much of the invasive species problem could have been prevented.

—Gary Becker, mayor of Racine, Wisconsin, 2007

19

DIRTY SECRETS

William Jefferson Clinton was in the second half of his eight-year term as president of the United States when he waded into the quagmire of science and politics otherwise known as "the war on invasive species." After six years in office, Clinton finally got the message from scientists who for years had pleaded with the federal government to get serious about dealing with foreign organisms that endangered human health, ecosystems, and the nation's food supply. The president chose to attack the problem with one of the most powerful tools at his disposal: an executive order. Clinton directed officials in the Department of Interior and the Department of Agriculture to draft an executive order that would force the world's largest bureaucracy to focus like a laser beam on slowing the flood of invaders entering the United States on the wings of expanding global commerce. The order appeared to be a huge advance in the war on invasive species. But Clinton's order was far from novel—President Jimmy Carter had issued a similar directive in 1977. And the time it took government officials just to draft Clinton's order was a demoralizing reality check—two years to write a four-page document. At that pace, it would be years, if not decades, before the most powerful nation on Earth created an effective shield against the flood of foreign species pouring into the United States.

Secretary of the Interior Bruce Babbitt was designated to be the Clinton administration's general in the war on foreign species. One his first assignments was to rally the troops. On January 26, 1999, Babbitt strode to the podium at the Massachusetts Institute of Technology to deliver the keynote address at the First National Conference on Marine Bioinvasions. His speech came a decade after zebra and quagga mussels began laying siege to the Great Lakes. Babbitt's speech, titled "Launching a Counterattack against the Pathogens of Global Commerce," was designed to energize a growing army of scientists and policymakers battling invasive species. Instead, the former Arizona governor delivered a sobering message about the daunting challenge of controlling the spread of species in a global economy that moved largely on transoceanic ships. "In the time it takes me to deliver this speech, two million gallons of foreign plankton will have been discharged somewhere in American waters," Babbitt said. "We'd better get busy. And fast."[1]

Babbitt drew a parallel between the *Exxon Valdez* oil spill, which stained Alaska's Prince William Sound with 11 million gallons of crude oil, and foreign species invading the same biologically rich waters. "Television crews on the scene [after the *Valdez* accident] broadcast scenes of seabirds, otters and sea lions slicked black with oil. Those images fixated the world on the dangers of oil spills and led to many new laws and regulations designed to prevent another such tragedy. Yet the biological spills taking place in Prince William Sound from oil tankers go virtually unnoticed." Scientists discovered four foreign species of zooplankton in Prince William Sound in the decade after the *Valdez* oil spill. The invaders were delivered by ships from East Asia that came to Alaska after stopping in San Francisco Bay, according to Babbitt. "In the long run these zooplankton . . . may change the sound more extensively and permanently than any oil spill," he said.[2]

The fact that foreign zooplankton were invading Prince William Sound, and zebra mussels discovered in the Great Lakes were spreading across the eastern United States, was not surprising, Babbitt said. The nation's response to the threat of aquatic invasive species up to 1999 was, in his words, "pitiful. Frankly, in light of the economic and ecological devastation, we have been too timid. We restrain ourselves with voluntary guidelines, a scattered approach and limited, unenforced codes. No longer." Babbitt said the United States was going to take the offensive on invasive species, much as President Reagan did in the 1980s with his war on illicit drugs. President Clinton's executive order, signed the week after Babbitt's speech, proclaimed his administration's commitment to stemming the tide of foreign species.

Clinton's order established a National Invasive Species Council, sort of a joint chiefs of staff in the symbolic war on invasives. The council included many of the government's big hitters: the secretary of state, the secretary of defense, the secretaries of several other agencies—Interior, Agriculture, Commerce, Transportation—and the Environmental Protection Agency's administrator. Clinton's order said the council's charge was to "provide national leadership regarding invasive species." Its job was to keep the sprawling federal bureaucracy focused on the president's directive, which required the government to "prevent the introduction of invasive species and provide for their control and to minimize the economic, ecological and human health impacts that invasive species cause."[3]

The funny thing was that the federal government already had a powerful regulatory tool at its disposal. A coalition of environmental groups and attorneys general was about to bring that little known fact to the government's attention. The coalition's campaign to make the government enforce a law that had been on the books for three decades advanced the war on invasive species further, and faster, than Clinton's edict or countless other pledges to protect the Great Lakes from new invaders.

TWO WEEKS BEFORE BABBITT'S SPEECH AT MIT, A COALITION OF ENVIRONmental advocates, anglers, and California water utilities filed a petition with EPA administrator Carol Browner. The groups wanted Browner to repeal a provision in the Clean Water Act, 40 C.F.R. 122.3(a), that made ballast water discharges from ships exempt from the law. The notorious ballast water exemption, which allowed transoceanic freighters to dump billions of gallons of ballast water swarming with foreign species into the Great Lakes, was suddenly in the cross hairs of environmental lawyers. One of the most important battles in the environmental history of the Great Lakes would be fought in California, of all places. The reason: Invasive species were wreaking havoc on the marine ecosystem of San Francisco Bay in much the same way invaders were harming the Great Lakes. California would be the proving ground for a regulatory sea change on invasive species. The coalition's Clean Water Act petition had the potential to turn the nation's ballast water regulatory regime on its head. Craig N. Johnston, the attorney for the coalition called Northwest Environmental Advocates, drafted the petition that was submitted to Browner. Its argument was simple, direct. "This exclusion is illegal. It conflicts with the statute and runs counter to case law that is directly on point."[4]

To review, the EPA's ballast discharge exemption issued in 1973 read as follows: "The following discharges do not require NPDES permits: Any discharge of sewage from vessels, effluent from properly functioning marine engines, laundry, shower and galley sink wastes or any other discharge incidental to the normal operation of a vessel." The Clean Water Act defined the phrase "discharge incidental to the normal operation of a vessel" to include ballast water.

The exemption allowed ships to discharge unlimited quantities of polluted ballast water into the Great Lakes and other U.S. waters without having to obtain a discharge permit under the Clean Water Act. This was no trivial matter: Ships discharged more than 21 billion gallons of ballast water into the nation's oceans, lakes, and rivers every year. Transoceanic freighters dumped about a billion gallons of ballast water into the Great Lakes annually, roughly the equivalent of 1,500 Olympic-sized swimming pools teeming with trillions of plankton, invertebrates, cholera, and other potentially deadly pathogens.[5] Each ocean freighter could carry as much as 3 million gallons of ballast water into the lakes.[6]

The Northwest Environmental Advocates' petition challenging the ballast water exemption claimed the following:

> The Clean Water Act prohibits all point source discharges of pollutants into the waters of the United States unless a permit has been issued. . . . Nowhere does the statute exempt "discharges incidental to the normal operation of a vessel" from the requirement to obtain a permit. To the contrary, the Act specifies that vessels are point sources under the Clean Water Act. It is also clear that ballast water contains large numbers of nonindigenous species,

> which qualify as biological pollutants under the definitions of the Act, as well as other non-biological pollutants.[7]

The coalition wanted the EPA to regulate ships the same way it regulated thousands of factories and municipal sewage-treatment plants that discharged chemical and biological pollutants into surface waters. The stakes were extraordinarily high: Regulating ballast water under the Clean Water Act would require all ships to install ballast treatment systems. The law also allowed citizen lawsuits, meaning any individual or organization could sue a shipping company suspected of discharging biological pollutants in excess of its EPA-issued discharge permit. That scenario was one of the shipping industry's worst nightmares.

The EPA's initial response to the petition was a meek attempt to placate the environmental groups. The agency's assistant administrator, J. Charles Fox, sent the groups' lawyer a letter on April 6, 1999—he called it an unofficial response to their petition. Fox promised the EPA would release a draft report by September 1, 1999, on new and existing technologies to assess, control, and treat ballast water. Fox added this remarkable commentary:

"I share your view that nonindigenous species pose a serious threat to the ecological health of the nation's waters and the economies they support. I also agree that the Clean Water Act provides EPA with broad authority for controlling the discharge of pollutants, and that authority could be extended to the control of ballast water discharges from vessels in some cases."[8] In other words, Fox agreed that the EPA had the authority to regulate ballast water discharges. The agency simply chose not to do so.

Sandra Zellmer, an environmental-law expert at the University of Toledo, said in a 1999 essay that the EPA's refusal to regulate ballast water discharges under the Clean Water Act was legally inexcusable. "Establishing emission controls for ballast discharges under the Clean Water Act will, no doubt, pose a challenge, because ships, as mobile sources, are quite different than traditional sources of pollution—wastewater discharged from industrial pipes," Zellmer said in Michigan's 1999 State of the Great Lakes Report. "Nonetheless, courts have repeatedly held that administrative difficulties are no excuse for regulatory avoidance. In the end, EPA in cooperation with the Coast Guard and Canadian regulatory bodies . . . must regulate ballast water discharges under the Clean Water Act."[9]

Critics of the ballast water exemption were sure the EPA didn't have a legal leg to stand on. Despite Fox's assurance that the EPA was addressing the ballast water issue, the agency didn't formally respond to the Northwest Environmental Advocates' petition to eliminate the exemption. And the agency didn't provide the coalition with the ballast water treatment study Fox had promised to deliver by September 1, 1999. The coalition sent two more letters to the EPA demanding a formal response to their petition. The EPA was required by law to respond to such petitions, but the agency ignored both requests for a formal response. Frustrated by the EPA's stonewalling, the coalition sued the agency in U.S. District Court on April 2, 2001. The complaint alleged that the EPA was violating federal law by not regulating ballast water and by failing to formally respond to the petition that sought to revoke the 1973 exemption. The attorneys general from New York, Michigan, Illinois, and Minnesota joined the fight in 2002. They asked the federal court to force the EPA to regulate ballast water discharges as a pollutant under the Clean Water Act and require ship owners to apply for NPDES permits. "By failing to act, the EPA is allowing this serious problem to worsen," said Eliot Spitzer, then attorney general of New York. EPA officials remained silent on the issue, claiming the agency didn't comment on pending litigation.[10]

The EPA finally made its intentions clear on September 2, 2003. Top agency officials appointed by the administration of President George W. Bush denied the petition to repeal the ballast water exemption. EPA officials said the Coast Guard was the lead agency on ballast water regulations. They said the Coast Guard was given that responsibility as part of the Nonindigenous Aquatic Nuisance Prevention and Control Act of 1990 and a subsequent law, the National Invasive Species Act of 1996. The problem was that neither of those laws treated ballast water as a pollutant, established discharge limits, or required shipping companies to disinfect ballast water. Those laws only required ocean ships to flush ballast tanks with seawater before entering the Great Lakes. That requirement didn't protect the lakes—the number of shipborne invaders increased after the 1990 and 1996 laws took effect. The ballast exchange rule essentially required ships to replace just 85 percent of fresh water in their ballast tanks with seawater before entering the Great Lakes. And the NoBOB loophole quietly crafted by Coast Guard officials allowed between 80 and 85 percent of ocean freighters to enter the Great Lakes without flushing their ballast tanks with seawater. That ridiculous loophole remained in place for 18 years after the 1990 law took effect, despite studies that showed the average NoBOB ship had 157 metric tons of biologically rich mud in its ballast tanks. Much of that material ended up in the lakes when NoBOB ships took on cargo—and discharged ballast water—at Great Lakes ports.

Despite the obvious, gaping holes in the ballast water regulations, EPA officials refused to regulate ballast water discharges as a pollutant under the Clean Water Act. EPA officials, in several classic examples of passing the buck, repeatedly expressed concern about invasive species entering the Great Lakes via ballast water discharges, but never exercised their legal authority to address the problem. Time and again, agency officials deflected the issue to the Coast Guard, which had no expertise in regulating pollution discharges.

Marianne Lamont Horinko, then EPA administrator, explained the agency's rationale in 2003 when she rejected the petition to revoke the ballast water exemption. "EPA believes that regulation of all discharges incidental to the normal operation of a vessel, including discharges of ballast water, would be a massive undertaking,"

Jim Hanlon, then EPA director of wastewater management, added that "The U.S. Coast Guard is the arm of the government there. Vessels are part of their mission. Our feeling is that adding ballast water to their responsibilities is better than adding vessel management to the EPA's."[11]

Members of the coalition were livid. It was apparent that EPA officials had made a conscious decision to ignore the biological pollutants freighters were dumping into the nation's surface waters. Worse yet, the agency put the regulation of ballast water squarely in the Coast Guard's lap. The Coast Guard's primary missions were ensuring marine safety and facilitating the movement of commerce—it was ill-prepared to regulate an industry the agency was supposed to protect and promote. "EPA has completely abdicated its responsibility," said Linda Sheehan, regional director of the Ocean Conservancy's office in San Francisco, California.[12]

John Berge, vice president of the Pacific Merchant Shipping Association in San Francisco, told the *San Jose Mercury News* that the industry was working with environmental groups and state agencies to "tackle this difficult problem." "We're on top of it," Berge said. "Eventually we hope to have technology that will provide onboard treatment of ballast water of ships so there is zero discharge. We're confident that will happen."[13]

Having grown weary of the shipping industry's promises to address the ballast water problem, and the EPA's intransigence on the issue, the coalition forged ahead with its lawsuit against the EPA. The group was victorious.

U.S. District Judge Susan Illston, who decided the case in the court's Northern California District, ruled in 2005 that the EPA "acted in excess of its regulatory authority in exempting an entire category of discharges from the NPDES permit program and denying plaintiff's petition to rescind [the exemption]."[14] Judge Illston rejected the EPA's argument that Congress acquiesced to the agency's interpretation of the law because lawmakers had never challenged it in the three decades since the Clean Water Act was enacted. EPA lawyers claimed that the failure of Congress to overturn or revise the ballast water exemption meant that lawmakers supported it. The judge said agency officials presented no evidence that Congress knew about the exemption or ever debated its merits when lawmakers revised the Clean Water Act in 1981, 1987, and 1997.

Judge Illston also rejected a government claim that two laws that gave the Coast Guard authority to regulate freighters' ballast water—the Non-Indigenous Aquatic Nuisance Prevention and Control Act of 1990, and the National Invasive Species Act of 1996—excused the EPA from its regulatory responsibilities under the Clean Water Act. She said neither law prohibited using the Clean Water Act to regulate ballast water discharges. Perhaps most significantly, Judge Illston said that ballast water discharges constituted a discharge under the Clean Water Act's definitions. "Ballast water discharges clearly introduce biological materials from outside sources, as demonstrated in the introduction of the zebra mussel in the Great Lakes region."[15]

The ruling was a stunning, slam-dunk victory for environmentalists, fishing groups, and millions of Great Lakes residents who had watched the lakes deteriorate as EPA officials turned a blind eye to the problem of shipborne biological pollutants. A former top EPA administrator who worked for years on the invasive-species issue said Judge Illston's ruling was to be expected. "I think, legally, it was pretty clear that ballast water discharges were subject to the Clean Water Act, but that was a wrenching thought to a lot of people [in the EPA]," said Tracy Mehan, the EPA's assistant administrator for water from 2001 through 2003. "I figured we would lose the case but we had to fight it to avoid the appearance of being a pushover."[16]

Mehan had a unique perspective on the ballast water lawsuit. He was director of Michigan's Office of the Great Lakes for several years in the 1990s, a position that put him on the front lines of the war on invasive species. He left Michigan in 2001 to join the Bush Administration's EPA. Mehan said he had tried to persuade top EPA officials to develop ballast water regulations before the issue ended up in court, only to be rebuffed at every turn. "When I would go into the office of the head of the water division, or the general counsel's office, no one wanted to touch the issue," Mehan said. "They thought it was a political hot potato, which it was, and that it would be an administrative nightmare."[17] He said most top EPA officials, who were not from the Great Lakes region, didn't understand the severity of the invasive-species problem or the need to clamp down on ballast water discharges.

In the end, Judge Illston made the decision for the EPA. She ordered the agency to revoke the ballast water exemption and begin regulating ballast discharges from ships nationwide by September 30, 2008. "The court found that Congress had directly expressed its intention that discharges from vessels be regulated under the Clean Water Act, and that the regulation at issue contradicted that intention," Judge Illston wrote in her 2006 order. "The EPA regulation is plainly contrary to the congressional intent embodied in the Clean Water Act. For this reason, the court believes that it is appropriate to set aside the [exemption] at issue, and that the proposed remedies of the EPA and the Shipping Coalition, both of which would leave the regulation in place indefinitely, are inadequate."[18]

Judge Illston acknowledged that her ruling would have "dramatic effects" on the shipping industry and the EPA. But, she added, "The potential harm that ballast waters represent to our nation's ecosystems leads the court to conclude that there is an urgency to promulgating new regulations that EPA has not, to this point in the litigation, acknowledged."[19] The judge concluded that existing rules governing ballast water—those enforced by the Coast Guard, and global ballast water treatment standards proposed in 2004 by the International Maritime Organization—were inadequate to prevent new introductions of invasive species. Sixteen years after Congress passed the first law aimed at keeping other foreign species from invading the Great Lakes, Judge Illston made a pointed observation: "Many of the Coast Guard regulations remain voluntary. More importantly, the Coast Guard regulations do not relieve EPA of its duty to follow the mandates that Congress has established."

The EPA's response? It joined with the shipping industry to appeal Judge Illston's ruling—and lost again. EPA officials never wavered from their position that the Coast Guard should regulate ballast water discharges from ships. The agency stuck to that position despite overwhelming evidence that allowing the Coast Guard to regulate ballast water was tantamount to a fox guarding the chicken coop.

Any doubt that the Coast Guard had dropped the ball on protecting the Great Lakes from filthy ballast water discharges was erased in 2005. Coast Guard officials acknowledged that the 1993 ballast water exchange rules only applied to about 20 percent of the ocean freighters entering the lakes. The Coast Guard's regulatory loophole that allowed 80 percent of transoceanic freighters to avoid the ballast exchange regulations was laid bare. It was a low point in the nation's supposedly dogged efforts to protect the Great Lakes and other surface waters from shipborne invasive species. The 1993 ballast water exchange rules, it turned out, were little more than a finger in a porous regulatory dike that was supposed to shield the Great Lakes from new invaders.

DID IT MATTER THAT THE EPA CHOSE NOT TO REGULATE BALLAST WATER IN 1973, or that it left the job to the Coast Guard? Of course it did—legally, morally, ecologically, and economically. When the Clean Water Act was approved in 1972, there were 126 invasive species in the Great Lakes. That figure stood at 183 in 2005, when Judge Illston ordered the EPA to regulate ballast water discharges. Thirty-eight of the species discovered in the lakes between 1972 and 2005 were attributed to ballast water discharges from ocean freighters.[20] Among the most notorious invaders of that era were zebra mussels, quagga mussels, and the round goby. Coping with zebra mussels cost industries and municipalities in the United States and Canada more than $1.5 billion in the two decades after the mollusks invaded the Great Lakes. By 2005, zebra mussels had spread to lakes and rivers in 20 states in the United States and two Canadian provinces.

One of the most dramatic examples of how the Clean Water Act reduced chemical and biological pollution in the Great Lakes, only to have the EPA's ballast water exemption reverse some of that progress, was found in Lake Erie. In 1969, cities and farms dumped about 29,000 metric tons of phosphorus into the lake every year, fueling massive algae blooms. Passage of the Clean Water Act reduced phosphorus inputs to the lake to 11,000 metric tons by 1990. The results were amazing. The walleye harvest just from Ohio's portion of Lake Erie increased from 112,000 fish in 1976 to five million in 1988. The number of small businesses associated with charter fishing in Ohio soared from 34 in 1974 to 900 in 2002, according to Jeffrey Reutter, director of the Ohio Sea Grant College Program. "Lake Erie has gone from being the poster child for pollution problems in this country to being one of the best examples in the world of ecosystem recovery. Lake Erie became known as the 'Walleye Capital of the World,'" Reutter told the U.S. Commission on Ocean Policy in 2002.[21]

Then, zebra mussels invaded and crashed Lake Erie's environmental comeback. The mussels initially increased water clarity in some areas of the lake sixfold, which was great for aesthetics but terrible for walleye—the easily spooked fish preferred murky water. The sport-fishing industry suffered greatly as a result. The number of hours that Ohio anglers spent fishing for walleye plummeted from 12 million in 1987 to 2 million in 2005. The number of charter-boat captains fishing for walleye dropped from 1,207 in 1989 to 783 in 2006.[22] Water quality changes caused by zebra and quagga mussels shrank the walleye population by one-third; the fishery's economic value plummeted from more than $600 million in the late 1980s to about $250 million in 2002. "The walleye population is still huge," Reutter told me in 2007, "but it would be so much larger without these mussels."[23]

Perhaps the greatest tragedy was the realization that the government could have prevented much of the damage invasive species inflicted in the Great Lakes. But in a scheme to avoid new work and save money, EPA officials in the 1970s sold out the Great Lakes and the nation's other coastal waters to profit-driven shipping interests. Who was really to blame for the resulting ecological meltdown in the lakes—shippers, or government officials who had a legal responsibility to regulate the industry and a moral duty and to protect these freshwater seas?

Establishing a regime that addresses the introduction and transfers of potentially harmful aquatic organisms via ballast water is my highest marine environmental protection priority.

—U.S. Coast Guard Commandant James M. Loy, 2001

When it comes to invasive species, the Coast Guard has always been a shield for the shipping industry, not a regulator.

—Mark Coscarelli, former Michigan Office of the Great Lakes staffer, 2007

20
WHO'S IN CHARGE?

In the fall of 2007, I received a surprising e-mail from an official in Transport Canada's Marine Safety Division. I had been pestering Michel Boulianne for permission to observe how government officials inspected the ballast tanks of transoceanic freighters when the ships arrived in Montreal. You would have thought I was asking for a private meeting with the Canadian prime minister. After several e-mail exchanges, Boulianne finally relented. We agreed on a date in September when I would meet his crew at St. Lambert Lock. There, Transport Canada officials would allow me to board an ocean freighter with them and observe how they checked the salinity of ballast water in ships entering the St. Lawrence Seaway from distant waters. I packed my bags a few days later and began the 14-hour drive from my home in Michigan to Montreal. Boulianne sent another e-mail a few days before my scheduled visit. The tour was off—national security concerns, he said. My initial fears that my name was on some government watch list for potential terrorists turned out to be unfounded, thankfully. The problem, Boulianne said in his final e-mail, was that it would take weeks to get the proper security clearance needed to board a foreign freighter entering the Great Lakes. His advice? Come back the following year.

Undaunted, I reverted to Plan B. I telephoned the U.S. Coast Guard's office in Massena, New York, to see if officials there would allow me to tag along as they inspected an incoming ocean freighter. Lieutenant Commander Matt Edwards, who supervised the Coast Guard's marine-safety detachment in Massena, was very accommodating. Stop by the next day, Edwards said, and his staff would take me along as they performed a ballast water inspection. The U.S. Coast Guard inspected some of the so-called "salties" at St. Lambert Lock, near Montreal. The agency inspected all others when the ships traveled between the first two locks on the U.S. side of the Seaway, the Snell and Eisenhower locks. I looked forward to the trip with great anticipation, only to be waylaid by a logistical nightmare. What transpired over the next 24 hours could only be described as surreal.

I arrived the next morning at 180 Andrews Street, on the north side of Massena. That was the address of the Coast Guard's Massena ship-inspection detachment. But the building didn't look like a Coast Guard office—there was no Coast Guard sign or logo on the front of the building. The reason: The building was the New York office of the St. Lawrence Seaway Development Corp., the government agency that operated the U.S. portion of the Seaway. That was odd. I was sure I had the right address. I asked the secretary in the vintage 1950s office of the Seaway Development Corp. where I could find the Coast Guard office.

"It's downstairs," she said.

The regional office of the Coast Guard, the agency responsible for inspecting ocean freighters entering the Great Lakes, was located in the basement of the agency that owned and operated the Seaway. The arrangement seemed akin to having a branch office of the U.S. Environmental Protection Agency located in the basement of Dow Chemical Co. Allowing the Seaway Development Corp. to share office space with the Coast Guard may have saved U.S. taxpayers some money on rent. But it did nothing to alleviate widespread claims that the Coast Guard was a paper tiger when it came to regulating ships plying the Great Lakes.

Lt. Cmdr. Edwards assured me that the Coast Guard wasn't compromised by its ties to the shipping industry, but was on top of the ballast water situation. The only reason the Coast Guard shared office space with the Seaway Development Corp., he said, was that both were part of the U.S. Department of Transportation before the terrorist attacks on September 11, 2001. The Coast Guard in 2003 was shifted to the Department of Homeland Security. Edwards dismissed allegations by a chorus of critics—environmentalists, government officials from several Great Lakes states, and a former Coast Guard commander—who claimed the guard was too friendly with the shipping industry to effectively police it. "The Coast Guard is really serious about this ballast water inspection program," Edwards said. "I think we're doing a pretty good job of closing the door on exotics. We board a lot of vessels and check a lot of ballast tanks. We have a very high compliance rate."[1] All freighters carrying pumpable ballast water into the Great Lakes from other continents had to conduct a mid-ocean ballast exchange before entering the St. Lawrence Seaway. But the Coast Guard's twisted interpretation of the 1990 and 1996 federal laws allowed 80 percent of ocean freighters, those fully loaded with cargo, to enter the lakes without flushing their ballast tanks.

Edwards clearly was proud of the job his staff was doing and seemed eager to display how they went about inspecting ballast tanks. "Come back this afternoon and we'll get you on a ship," he said. It was 11 A.M. The next ship was due to arrive at Snell Lock at roughly 3:30 P.M. Edwards told me to return to his office an hour before the ship arrived to meet with inspectors who would board the freighter at Snell Lock and ride it to Eisenhower Lock as they tested the salinity of its ballast water. Flush with renewed anticipation, I spent three hours cooling my heels in the gritty community of Massena. The city that hoped to be transformed by the St. Lawrence Seaway into the Pittsburgh of the North was a rust-belt community of 11,000 people with a struggling downtown, the requisite Wal-Mart, strip malls, chain restaurants, and fast-food joints. The downtime gave me a chance to check out the Eisenhower Lock and the massive Moses-Saunders Dam, which generated phenomenal quantities of cheap electricity by harnessing the St. Lawrence River. My cell phone rang as I drove back to the Coast Guard office. It was Edwards.

"The trip is off," he said.

"Isn't the ship coming in?"

"It is, but we can't let you go [on the ship] because the inspection is coming during a shift change," Edwards said.

"A shift change? How does that interfere with my visit?" I asked.

"They said it would be too complicated. We can't do it," Edwards said.

He wouldn't budge and didn't offer to reschedule my visit. The trip was off. I was screwed. My plan to observe a ballast water inspection was history. I was suspicious of Edwards's explanation. But I later learned from other sources that boarding foreign freighters entering the Seaway became very difficult after the 9/11 terrorist attacks. I could live with that explanation. But canceling a ship-inspection tour that the Coast Guard had offered—because of a shift change? That explanation seemed fishy. It made me wonder if there was merit to the numerous allegations that the Coast Guard was in cahoots with the shipping industry. The time had come to explore that issue.

EVERY WINTER, WHEN ICE-COVERED CHANNELS FORCED THE ST. LAWRENCE Seaway to close for two to three months, shipping-industry executives and Coast Guard officials gathered in Cleveland, Ohio, for the Great Lakes Marine Community Day. The event, hosted by the Coast Guard and shipping interests, gave industry officials a chance to rub elbows and discuss regulations with high-ranking government officials. The Coast Guard made no secret of the event's mission. A press release issued by the agency spelled out the event's objective: "Marine Community Day is the largest annual event in the Great Lakes region which brings together maritime industry leaders and regulating authorities from both the U.S. and Canada. The event's objective is to enhance communication between the maritime community and those agencies which regulate them and to discuss selected issues concerning maritime safety, maritime security, waterways management and the marine transportation system."[2]

Some of the terminology used to promote the conference was interesting in its own right. The Coast Guard touted the 2005 conference as "An Industry Celebration." One of the highlights of the annual event was the Admiral's Dinner, a party that featured cocktails and a social hour followed by dinner with whoever was head of the Coast Guard that year. In 2008, the man of the

Muskegon Chronicle

The U.S. Coast Guard cutter *Woodrush* is shown clearing a path for a freighter on an icy Great Lake. Critics said the Coast Guard was so focused on supporting the shipping industry, the agency failed to deal effectively with invasive species imported by transoceanic freighters.

U.S. Coast Guard

A domestic freighter unloads gravel at a Great Lakes port. The Coast Guard allowed ships to wash coal, gravel, salt, iron ore, and other dry cargo residues into the lakes, even though the practice was contrary to federal environmental laws and international treaties.

night was Coast Guard Commandant Admiral Thad Allen. He spoke to more than 250 representatives of the shipping industry and government officials. Allen's speech was the climax of a day that featured an "executive session"—a closed-door meeting—with more than 25 industry officials to discuss "issues surrounding maritime transportation safety, security and protection of the environment," according to a Coast Guard press release. "During the session, Allen listened to the concerns of maritime industry leaders and reiterated the commitment he made in his State of the Coast Guard Address [the preceding week] to enhance the Coast Guard's marine safety program."[3]

The Great Lakes Maritime Community Day was open to the public, for a price. Attending the conference cost $60. Admission to the Admiral's Dinner was another $70. The fact that it was open to the public was progress from the event's early years. In the 1990s, the event was known as Industry Day and was not open to the public. "It was like Dick Cheney discussing the nation's energy policy behind closed doors with oil tycoons," said Eric Reeves, a retired Coast Guard commander who managed the agency's ballast water inspection program from 1993 to 1998. "We would get together to talk to shipping officials about what the industry didn't like about our regulations and how we [the Coast Guard] were going to deal with it."[4]

Reeves persuaded his superiors to change the name of the conference from Industry Day to Maritime Community Day and open it to the public. He hoped some environmental advocates would attend to become better educated about ballast water and other maritime issues. Few ever showed up. Reeves said he was bothered by the Coast Guard embracing its role as a "partner" with the shipping industry. "We were not supposed to be their enemies, but we were their regulators. We should not be their partners," he said.[5] When an agency's mission statement cited the promotion of maritime commerce as one of its primary responsibilities, which the Coast Guard's did, it made me wonder how the agency could promote *and* regulate the shipping industry. It was an untenable relationship.

U.S. Rep. Vernon Ehlers, a Republican from Michigan who worked for years on the ballast water issue, said the blame for failing to solve the problem rested solely with the Coast Guard. Ehlers, a physicist turned politician, lambasted the Coast Guard during a 2007 press conference to discuss ballast water regulations. "I personally think the Coast Guard really fell down on the job," Ehlers said. "Congress passed a law on invasive species in 1990 and the Coast

Guard did nothing. Congress passed another law in 1996 and the Coast Guard did nothing. They really didn't want anything to do with it—they didn't see it as part of their mission. And they tended to side with the shipping industry. The Coast Guard tried to talk me out of passing tougher regulations on the shipping industry."[6]

There is no doubt the Coast Guard fumbled the ballast water regulations. That does not mean the agency was completely inept. The Coast Guard does many things very well. It has done a remarkable job of saving lives and promoting safety on the Great Lakes and the nation's oceans. Nowhere was the skill and bravery of Coast Guard personnel more evident than in the aftermath of Hurricane Katrina, in 2005. Members of the Coast Guard rescued more than 33,000 people after the hurricane laid siege to the city of New Orleans. In 2007, on its 217th anniversary, the Coast Guard announced that it had saved the lives of 1,109,310 people since its inception. That broke down to an average of 5,112 rescues per year, or 14 lives saved every day, of every year, for more than two centuries. Countless documentaries have highlighted the heroic actions of Coast Guard personnel who pulled sailors off sinking ships and plucked others out of frigid waters or dangerously high seas.

The Coast Guard also does a fine job of maintaining safety on the Great Lakes, marking shipping lanes with buoys and freeing freighters that become trapped by ice. The agency also was given the responsibility of protecting the nation's shores from terrorists after the murderous attacks in 2001. In sum, the Coast Guard has a distinguished record in many of the areas it works. Ballast water management proved to be one of the agency's most difficult challenges, technically and politically. Perhaps the Coast Guard—in its role as a facilitator of maritime commerce—was too close to the shipping industry to effectively regulate it.

One sign of the fraternal relationship between the industry and its regulators was the number of Coast Guard officers who left their government jobs for high-ranking positions with the shipping industry. The late Admiral William M. Benkert retired as the Coast Guard's chief of the Office of Merchant Marine Safety in 1978 to become president of the American Institute of Merchant Shipping. The late Admiral John William Kime, who was commandant of the Coast Guard from 1990 to 1994, retired after 41 years of service to become chief executive officer of ship-management companies in the United States and three other countries. He also worked with the world's largest private shipping organization.[7] One of the most influential shipping organizations in the Great Lakes was led by a former Coast Guard engineer who took a job with a shipping company before becoming an industry lobbyist. "These guys are completely shameless about it," Reeves said. "This problem of the revolving door is, of course, nothing unique to the Coast Guard. It is a common practice in almost all regulatory agencies, most of which have been captured by the industries they regulate."[8]

That so-called revolving door also lured leaders of the shipping industry into regulatory agencies. One of the shipping industry's most outspoken advocates for ocean freighters having unfettered access to the Great Lakes accepted a high-ranking post in 2006 in the U.S. Department of Transportation, which managed and promoted the Seaway. Helen Brohl was something of a legend for her tense encounters with environmentalists and scientists who suggested banning ocean freighters from the Great Lakes until the vessels were equipped with ballast water treatment systems. She was named the first executive director of the Executive Secretariat to the Committee on the Marine Transportation System. The job put her in charge of an agency created by President George W. Bush to coordinate the work of several government agencies that regulated the shipping industry.

The movement of Coast Guard officials to companies they once regulated, and vice-versa, was not limited to the Great Lakes. A 2002 article in the *Seattle Post-Intelligencer* examined whether Coast Guard officials who took jobs in the shipping industry had a conflict of interest when lobbying their former employer, the federal government. The Seattle newspaper documented five cases of top Coast Guard officials on the West Coast taking jobs with shipping interests. It should be noted that many Coast Guard officials who wanted to continue working on shipping issues after leaving the guard had few options other than work for private industry. Still, it created the appearance of a good-old-boys network. Ed Wenk, a University of Washington professor and former presidential marine-safety advisor, told the Seattle newspaper that concerns about the government-to-industry revolving door creating conflicts of interest was "widespread among people who are concerned about lack of stronger regulation by the Coast Guard."[9]

The Coast Guard was not shy about its friendly relationship with the shipping industry. When longtime Duluth Port Authority director Davis Helberg retired in 2003, the Coast Guard pulled out all the stops to honor him.

The guard's top official, Commandant Admiral Thomas H. Collins, traveled from Washington, D.C., to Cleveland for a special ceremony recognizing Helberg's two decades of working to support "responsible, safe and secure maritime transportation." Collins presented Helberg with the Coast Guard's Distinguished Public Service Award, the highest award given to civilians other than those involved in a life-saving rescue.[10]

The Coast Guard's alleged bias toward the shipping industry was laid bare in 2006, when the agency was forced to deal with the problem of shippers washing cargo residues into the Great Lakes. Since the 1930s, shippers routinely washed cargo residuals—coal, iron ore, limestone, and salt—into the lakes after leaving port. The practice kept ships clean and prevented the remnants of one load from contaminating future cargo. Shippers washed two million pounds of cargo residue into the lakes in 2002, according to a Coast Guard study. Coast Guard officials were aware of the practice but turned a blind eye, even though it violated the U.S. Clean Water Act and an international shipping treaty called MARPOL Annex V. Congressional approval of the MARPOL treaty in 1990 effectively outlawed the practice of washing cargo residues into the lakes. Instead of banning the practice, the Coast Guard and Congress created an interim policy that amounted to a legal loophole. The interim policy allowed shippers to continue using the lakes as waste-disposal facilities for another 15 years, through 2008. The Coast Guard extended the interim policy indefinitely when it expired in 2008. Agency officials explained their rationale for the indefinite extension of the loophole in a notice in the *Federal Register*. It read as follows:

> Under Coast Guard regulations . . . dry cargo residue is an operational waste and constitutes garbage, the discharge of which into the navigable waters of the United States is prohibited. If these regulations were strictly enforced on the Great Lakes, they would put an end to the practice of cargo sweeping.[11]

Coast Guard officials said that banning dry cargo sweepings, and forcing ships to discard cargo residuals at disposal facilities on land, would cost the shipping industry $35 million annually while providing little environmental benefit. The agency's public notice acknowledged the possibility that cargo residues could harm water quality or aquatic life. Yet, the notice said more study was needed to determine whether the practice posed a significant threat to the environment. Shipping groups said that outlawing cargo sweeping would be financially catastrophic for the industry. "It would shut down power production, steel production and all kinds of construction activities in the region," said James Weakley, president of the Cleveland-based Lake Carriers Association. Discarding 1,000 tons of cargo residuals into the lakes annually was a small price to pay for ships carrying 165 million tons of cargo on the lakes each year, Weakley said in a newspaper interview. "It's the equivalent of sweeping out my garage. I'm pretty sure the dust and dirt I sweep out of my garage is nontoxic but I don't have any scientific data to back that up."[12]

Scientists who examined the potential risks associated with cargo sweeping concluded in 1999 that dumping thousands of tons of iron ore, coal, and other materials into the lakes could cause environmental problems. They issued a report summarizing their findings, which said the following: "Iron ore, coal, petroleum coke and slag were determined by the committee to have the potential for both acute and chronic environmental impacts and were worthy of more intense scrutiny. Of greatest concern to the committee, however, is the repetitive addition and probable buildup of these materials in bottom sediments and the potential chronic effects on both hard and soft bottom habitats." The scientists added that some of the chemicals found in coal, slag, and iron ore could be "quite toxic."[13]

Whether or not cargo sweeping was environmentally benign was largely irrelevant. The practice violated the U.S. Clean Water Act, the U.S.-Canada Great Lakes Water Quality Agreement, and MARPOL Annex V, the international treaty that prohibited ships from dumping waste in the world's oceans or the Great Lakes. The Coast Guard's policy also created a glaring double standard: Recreational boaters who tossed so much as a pop can into the Great Lakes were subject to fines of up to $50,000 under U.S. law. Freighters, on the other hand, could wash unlimited quantities of coal, iron ore, and other materials into the lakes with the blessing of the U.S. Coast Guard.

Arming the Great Lakes

The terrorist attacks of September 11, 2001, gave the U.S. Coast Guard a new assignment—protect America's coasts from terrorists. In the Great Lakes region, that charge shifted some of the guard's focus from battling invasive species to preventing terrorists from attacking coastal cities and nuclear power plants. The guard's blueprint for doing that produced one of the most controversial activities ever proposed for the Great Lakes.

The Coast Guard published a notice in the *Federal Register* in August 2006 announcing its plan to establish 34 "safety zones" on the Great Lakes. Those safety zones were actually unmarked firing ranges that the guard wanted to demarcate on all five lakes. The guard planned to conduct target practice periodically at each of the sites, using machine guns and real bullets to fire at floating targets. The plan wasn't disclosed to members of Congress, governors of Great Lakes states, or mayors of coastal communities. The guard only mentioned its plan in the *Federal Register*, an obscure publication federal agencies use to announce new policies. The target-practice plan initially generated little public interest. That changed on August 30, when the *Muskegon (Mich.) Chronicle* published a map showing where the proposed firing ranges were located on the lakes. Several were in areas that passenger ferries, freighters, and fishing boats traversed on a regular basis.

News of the Coast Guard's plan spread across the region with the speed of a celebrity rumor in Hollywood. The proposal ignited a firestorm of controversy. Many people wondered if leaders of the Coast Guard—a government agency known for engendering goodwill among boaters and the shipping industry—had lost their collective minds. "When I heard, I thought it was something from The Onion newspaper or an Internet hoax," said Sarnia, Ontario, mayor Mike Bradley in a *New York Times* article. "This whole thing was done way below the radar."

Coast Guard spokesman Brian Lanier told newspapers that the firing ranges were necessary to make sure the guard was "ready for anything." Lanier said the invisible boundaries of the so-called safety zones would be marked on navigational maps. Boaters were to be warned to stay out of the zones before Coast Guard soldiers began shooting guns capable of firing bullets a distance of two miles.

Some critics feared that stray bullets would kill or injure nearby anglers or sailors. Environmentalists said the 6,900 pounds of lead and 2,800 pounds of copper the guard would deposit in the lakes each year by firing 430,000 bullets could harm water quality and aquatic life. The plan would have made the Coast Guard the nation's largest source of lead entering the lakes. Some scientists said the bullets posed little risk to the environment because they would break down slowly in the water. Regarding safety concerns, Coast Guard officials said the 34 firing ranges collectively encompassed just 2.5 percent (2,376 square miles) of the lake's 94,488 square miles of surface area. Critics were not assuaged.

U.S. Coast Guard

A U.S. Coast Guard crew is shown conducting a harbor patrol at a Great Lakes port. The Coast Guard's 2006 plan to conduct live-fire weapons training on the Great Lakes was derailed by intense opposition in the United States and Canada. The proposed target practice would have deposited thousands of pounds of lead and copper bullets in the lakes each year.

Opponents said the plan violated the Rush-Bagot Treaty of 1917, an agreement between the United States and Canada that banned warships from the Great Lakes. "The proposal to do regular training with automatic weapons is totally contrary to the long history of peaceful relations and environmental cooperation between the United States and Canada on the Great Lakes," Toronto mayor David Miller said in a letter supported by the mayors of 80 Great Lakes communities. Coast Guard officials said Canadian leaders had approved an exception to the Rush-Bagot Treaty because the deadly, unpredictable nature of terrorism called for extraordinary measures to ensure homeland security.

What the Coast Guard couldn't successfully navigate was the 1972 Great Lakes Water Quality Agreement between the United States and Canada. That agreement called for both nations to eliminate, to the greatest extent possible, the discharge of toxic pollutants into the lakes. The Coast Guard's proposal to pump bullets containing 9,700 pounds of potentially harmful lead and copper into the lakes annually flew in the face of the Water Quality Agreement. Canadian officials refused to withdraw their environmental concerns. "Any activity that results in a net increase in pollution to the Great Lakes is counterproductive to both governments' commitments under the Canada-U.S. Great Lakes Water Quality Agreement," the Canadian government said in a letter to the Coast Guard.

The Coast Guard scrapped the weapons-training plan in January 2007 amid growing opposition. Rear Admiral John E. Crowley Jr., then commander of the guard's Great Lakes office, said the agency would conduct weapons training on the Atlantic Ocean, or use eye-safe lasers to carry out target practice on the lakes. That the Coast Guard even proposed to pump tons of lead bullets into the Great Lakes spoke volumes about its commitment to protecting the lakes' fragile ecosystems.

The Coast Guard took a similar hands-off approach to regulating ballast water. The agency was loath to penalize the owners of transoceanic ships that violated the 1993 ballast water exchange law. Though the rate of compliance with the law through 2005 was 90 percent, the Coast Guard didn't penalize the 10 percent of shippers who violated the law, according to Commander Kathy Moore, chief of the Coast Guard's environmental-standards division. Moore said that fines were impractical because ships that failed to conduct a mid-ocean ballast exchange before entering the St. Lawrence Seaway could retain all of their ballast water while traveling around the lakes. Those ships were checked when departing the Seaway to make sure they didn't illegally dump ballast water into the lakes. Moore explained the enforcement process to a congressional committee in 2005, when a lawmaker asked if the Coast Guard had ever fined a ship that violated the ballast exchange regulations. She said: "By virtue of the system that's been set up in terms of reporting, sampling and enforcement, we haven't had an opportunity, if you will, to actually exert a fine in the Great Lakes system."[14]

The shipping industry went ballistic in 2006, when U.S. District Judge Susan Illston changed the rules of the regulatory game. Judge Illston ordered the U.S. Environmental Protection Agency to regulate ballast water discharges the same way it regulated industrial and municipal wastewater discharges. Her ruling meant that the Coast Guard was no longer the lone regulator of ballast water in the United States. Shippers argued that the EPA couldn't regulate ballast water discharges—they said the agency knew nothing about ships. Shipping interests also argued that ballast water regulations on the books in 2005 were sufficient to protect the environment, despite scientific studies proving just the opposite. So fearful were shipping interests that they would have to deal with the EPA, instead of the Coast Guard, the industry took the remarkable step of joining the EPA in appealing Judge Illston's ruling. That appeal, which failed, was the shipping industry's last-ditch effort to protect the Coast Guard's authority as the United States' lone regulator of ballast water. Wonder why?

IN MAY 2008, EIGHT MONTHS AFTER OFFICIALS AT TRANSPORT CANADA AND the U.S. Coast Guard reneged on their promise to let me observe a ballast water inspection aboard an ocean freighter, another invitation arrived via e-mail. It came from the St. Lawrence Seaway Development Corp. The Seaway Development Corp. had organized a media day focusing on ballast water inspections at the St. Lambert Lock. Why the sudden desire to invite dozens of journalists and photographers to observe a process that had been so tightly guarded in the past? One motivation was the desire of Seaway officials to tout new rules requiring all ships, not just those hauling large quantities of ballast water, to flush ballast tanks with seawater before entering the Great Lakes. The event was a brilliant public relations move at a critical juncture in the Seaway's history. At the time, a growing chorus of scientists, politicians, and environmental advocates were calling for banning ocean freighters from the Great Lakes until all were equipped with ballast treatment systems. It was against that politically charged backdrop that the savvy new director of the Seaway Development Corporation, Collister "Terry" Johnson Jr., took the bold step of ordering all ocean freighters to flush their ballast tanks with seawater before entering the St. Lawrence River. Johnson took over as leader of the U.S. side of the Seaway in October 2006. He immediately went about searching for a way to close the gaping loophole in the 1993 ballast water regulations that allowed ships loaded with cargo, the so-called NoBOBs, to enter the lakes without flushing their ballast tanks with seawater. The Coast Guard studied ways to deal with NoBOBs for close to a decade, during which time the number of shipborne invaders in the lakes continued to escalate.

Johnson, a college roommate of former president George W. Bush when they were undergraduates at Yale, chose not to wait for the Coast Guard to close the NoBOB loophole. The Seaway Development Corporation in 2008 ordered all ships entering the Seaway to flush ballast tanks with seawater at least 200 miles offshore of the Atlantic Coast, in waters 2,000 feet deep.

Violators were subject to fines up to $36,625. The rule was similar to a Canadian mandate enacted in 2006. Johnson imposed the regulation after a 2007 study found that requiring the NoBOB ships to flush ballast tanks with seawater before entering the Great Lakes killed at least 95 percent of organisms living in the residual muck in the tanks.

Closing the NoBOB loophole was a significant advancement in efforts to stop the flow of shipborne invaders into the Great Lakes. Johnson wanted to promote the new rule, and deservedly so. After more than a decade of foot-dragging by the Coast Guard, Johnson's agency dealt with a major portion of a ballast water problem that had plagued the lakes since the Seaway opened in 1959. The ballast water exchange rule Johnson enacted in 2008 would remain in effect until the Coast Guard developed its own comprehensive ballast water management regulations.

On a glorious spring day in 2008, at a carefully organized event, reporters from the United States and Canada were shown how ballast water inspections were conducted aboard ocean freighters entering the Seaway. The demonstration took place on board the M/V *Federal Kivalina*, a 600-foot-long, Hong Kong–flagged freighter owned by Canada-based Fednav Ltd. Surrounded by two dozen newspaper and television reporters and photographers, Seaway officials and ship personnel demonstrated a ballast water inspection. The ship, by the way, passed its inspection. Seaway officials and scientists declared during press conferences that the battle to keep ocean ships from importing foreign species into the lakes via ballast was close to being won. "This step will almost shut the door [on new invasives], it will leave only a crack open," Johnson said at the media event.[15] Richard Corfe, Johnson's counterpart at Canada's St. Lawrence Seaway Management Corporation, said the mandatory ballast flushing for all ships was the best regulators could do until federal agencies in the United States or Canada established treatment standards for ballast water. "What we're doing now with saltwater flushing is as good [a ballast water treatment] as you can get in today's world," Corfe said.[16]

All that remained in the regulatory process was establishing ballast water treatment standards and setting deadlines for freighters to install systems capable of killing a broad spectrum of aquatic organisms. The EPA should have initiated the process in the early 1970s, under the auspices of the federal Clean Water Act. Reaching a consensus on ballast treatment standards 30 years later would prove to be more challenging than sending a man to the moon.

In the near future there must be complete elimination of aquatic nuisance species entry by all ships. This is an achievable goal.

—Canadian Shipowners Association, 2004

Clearly, zero or near zero discharge is the ultimate protection from aquatic invasive species, although it is simply not currently achievable.

—Joseph J. Cox, Shipping Industry Ballast Water Coalition, 2004

21 MISSION IMPOSSIBLE

A cadre of shipping industry officials and scientists gathered in Superior, Wisconsin, on a summer day in 2006 for a historic event: the christening of a $3.5 million facility to test the efficacy of ballast water treatment systems. The project, known as the Great Ships Initiative, was a collaboration between maritime interests, university scientists, and government agencies that regulated the shipping industry. Organizers hailed it as the world's first facility designed specifically for testing how well chemical biocides, filtration, ultraviolet light, and other technologies eradicated aquatic organisms in ballast water. Shipping officials called the endeavor an industry-led effort, though government agencies and publicly funded universities provided most of the startup money. To mark the occasion, the mastermind of the project—an invasive-species policy expert named Allegra

Cangelosi—clambered up the metal-grate stairs to a catwalk adjacent to one of four 50,000-gallon water-storage tanks at the facility. There, she swung a champagne bottle at the metal tank. The bottle didn't break. Cangelosi tried a second time. Nothing. The bottle bounced off the cylindrical tank, intact. Frustrated, she removed the bottle from its rope harness and smashed it on a pier that jutted into the Wisconsin side of Duluth-Superior Harbor. "This ballast water issue is a hard one to crack," Cangelosi said after the bottle shattered.[1] Her comment was a rare moment of levity in the prolonged, vexing campaign to equip freighters with treatment systems capable of disinfecting ballast tanks. It was also a classic understatement.

The Great Ships Initiative was the culmination of a grand experiment 10 years in the making. The idea was that government officials and shipping interests could solve the ballast water problem faster by working together, as "partners." That industry-friendly approach was an alternative to government agencies enforcing an existing federal law that could have required ocean ships to disinfect ballast water before discharging it into the Great Lakes. The government-industry partnership was an unconventional approach to a conventional pollution problem. Cangelosi was one of its most outspoken advocates. A biologist by training, Cangelosi was a staffer for U.S. Senator John Glenn in the 1990s; she helped write the world's first laws regulating ballast water. After Congress passed the National Invasive Species Act in 1996, Cangelosi left Sen. Glenn's staff to join the Northeast-Midwest Institute, a nonprofit think tank in Washington, D.C. There, she was named manager of the Ballast Water Technology Demonstration Project, an experiment mandated and funded by the 1996 law she helped write.

Cangelosi's project brought together dozens of bureaucrats, scientists, and shipping industry officials who conducted research on how to kill an array of aquatic organisms in ballast water tanks. Over the course of a decade, the program awarded $10 million for 50 different research projects. Those projects advanced scientific understanding of invasive species and produced useful data on how to eradicate live organisms in ballast water. But all that research failed to produce a magic bullet for the war on shipborne invaders infecting the Great Lakes. It wasn't for a lack of effort. Engineers in the United States and abroad developed several ballast treatment systems after 1996. Not one was deemed an acceptable solution. The reason: ballast water treatment was ensnared in a Catch-22, a quandary with no solution. There were no national

Photo by the Associated Press, courtesy of the *Muskegon Chronicle*

A crew is shown inspecting a massive ballast tank in a Great Lakes freighter. A single freighter can hold millions of gallons of ballast water, which can harbor billions of live organisms and potentially harmful pathogens. Disinfecting the tanks can be a complicated and costly process.

or international standards for treating ballast water prior to 2008. Absent standards, there was no way to assess treatment technologies or require shipping companies to sanitize ballast tanks. The result was a circular argument that left regulators, shippers, and engineering firms wondering which came first: ballast treatment standards, or technology capable of meeting standards that had yet to be written. Such was the state of ballast water regulations in 2006, when the Great Ships Initiative facility opened amid much hype.

Supporters claimed that the facility provided researchers with the platform needed to gauge how well different ballast treatment technologies worked. What it couldn't do was determine whether any given technology worked well enough—there was no legal standard for measuring success. "Ideally, we'd like to have the treatment standard first, but if you have no standard you can still test," Cangelosi said.[2] But did the tests mean anything? Without

Jeff Alexander

The $3.5 million Great Ships Initiative testing facility was built in Superior, Wisconsin, to test the efficacy of ballast water treatment systems used in freshwater ecosystems.

standards, the answer was a resounding "no." Until those benchmarks were established, the Great Ships Initiative was little more than a glorified academic exercise. It was the latest in a long line of ballast treatment research projects that paved the U.S. government's regulatory road to nowhere.

"CATCH-22" IS A TERM USED WIDELY—OFTEN INCORRECTLY. THE TITLE and legacy of Joseph Heller's satiric 1961 novel about war, Catch-22 is used to refer to inherently illogical rules, with arbitrary exceptions, that create circular arguments. The most dramatic example in Heller's classic book was when the main character, a fighter pilot named Yossarian, tried to get himself grounded for fear he would be shot down on his next mission. Yossarian sought to be declared insane—pilots declared insane were not allowed to fly combat missions. His ploy was clever. But his commanding officer ruled that Yossarian's concern for his own life proved that he was sane, and thus capable of flying more combat missions. Yossarian was trapped in a no-win situation, a Catch-22.

The Catch-22 that paralyzed ballast water regulations in the United States after passage of the 1996 National Invasive Species Act was the result of a regulatory blunder two decades earlier. The EPA's 1973 decision to exempt ballast water discharges from the federal Clean Water Act unleashed a sort of regulatory anarchy that crippled future efforts to regulate ballast water. One of the cornerstones of the Clean Water Act was the establishment of a regulatory regime that banned "the discharge of any pollutant by any person" unless it complied with the law. The law's definition of pollutants included biological wastes, such as the live organisms and pathogens found in ballast water. The Clean Water Act prohibited the discharge of chemical or biological wastes into surface waters unless permitted by the EPA; all permittees had to meet specific discharge limits. The law put the burden of proof on industries and municipalities to find treatment technologies that enabled them to meet their assigned discharge limits.[3]

Had the EPA applied the Clean Water Act to ballast water, the law would have established discharge standards and forced shippers to find treatment technologies to meet their assigned limits. Instead, the EPA abdicated its authority and handed the regulatory reins to the U.S. Coast Guard. When foreign species began streaming into the Great Lakes in the mid-1980s, the Coast Guard could have asked the EPA to rescind the ballast water exemption and regulate discharges from ocean freighters. That never happened. Nor did the Coast Guard direct ocean freighters to disinfect ballast water before pumping it into the Great Lakes. Coast Guard officials said they couldn't order the treatment of ballast water without treatment standards. That put the agency on a regulatory course of action that was diametrically opposed to the approach used in the Clean Water Act. Instead of putting the burden on the shipping industry to prove its ballast water discharges weren't harming water quality, the Coast Guard's actions put the burden of proof on the U.S. government.

The Coast Guard's regulatory regime was based on the premise that the government wouldn't require ballast water treatment until it established treatment standards. That approach might have worked if the Coast Guard had moved promptly to establish treatment standards. But that didn't happen, either. The agency plodded along for more than a decade, conducting endless research on ballast water treatment without proposing any treatment standards. All the while, the number of shipborne invaders entering the lakes via ballast water continued to increase. The Coast Guard created a Catch-22 situation that confounded, for more than a decade, all efforts to make transoceanic ships sterilize ballast tanks before entering the Great Lakes. The actions of the EPA and Coast Guard led scientists, shippers, and regulatory officials on a quixotic mission to find the perfect gadget for killing aquatic organisms in ballast tanks. The quest for treatment technology that could measure up to a nonexistent performance standard was analogous to running a race with no finish line.

BY THE MID-1990S, WITH THE NUMBER OF DOCUMENTED INVADERS IN THE lakes rising steadily, regulatory officials and politicians in the United States and Canada faced a difficult choice. Would the two nations charged with protecting the lakes get tough with the international shipping industry? Would the so-called guardians of the Great Lakes regulate biological pollution from ballast water the same way the U.S. EPA controlled the volume of biological and chemical wastes that cities and industries discharged into the lakes? Or would government agencies allow the shipping industry to dictate the rules of the game and delay the imposition of costly ballast water treatment standards? Here's what the U.S. EPA pledged to do in a 1994 report to Congress:

> The Fish and Wildlife Service, states, Coast Guard, NOAA, the Great Lakes Fisheries Commission and EPA will work together to prevent further introductions of nonnative species and to mitigate the harmful effects of ones that have already entered the Great Lakes. They will monitor the ecosystem for new nonnative species and conduct research on environmentally-kind control techniques for disruptive nonnative species. The Coast Guard will establish requirements governing ship ballast water, a common pathway for the introduction of nonnative species.[4]

Congress, two years later, passed the National Invasive Species Act of 1996. The law expanded the Great Lakes' ballast exchange requirements to transoceanic freighters entering all U.S. ports. It did not, however, require those ships to disinfect ballast tanks before entering U.S. waters. Rather than clamping down on ballast water discharges, Congress added language to the 1996 law that encouraged the Coast Guard and other federal agencies to work cooperatively with the shipping industry to solve the problem. That provision was a turning point in the development of ballast water regulations. Federal officials were expected to work cooperatively with the very industries they regulated to resolve the ballast water conundrum. Industry groups, government agencies, and environmental advocates became "partners" in the war on invasive species. It was a ludicrous proposition—regulatory officials and environmental advocates were sleeping with the proverbial enemy. Instead of forcing the transoceanic shipping industry to clean up its act, the United States and Canada created a small army of bureaucrats and scientists whose job was to help shippers figure out how to sanitize ballast water. Cangelosi was among those who believed ballast treatment technologies would develop more quickly if government and industry officials worked together, as allies instead of adversaries. Nicole Mays, one of Cangelosi's associates at the Northeast-Midwest Institute, summed up the rationale for the government-industry collaboration in one of the institute's numerous ballast water reports. "Conceivably, the more the associated parties are encouraged to fraternize and converge towards the ultimate goal, the more support there will be when the outcome to the problem is attained and action needs to be taken," Mays said. "The shipping industries' input is imperative in making the transition between voluntary and regulatory ballast management procedures more acceptable and comfortable."[5]

The shipping industry embraced that approach. Helen Brohl, executive director of the U.S. Great Lakes Shipping Association in 2000, said the government-industry partnership formed as part of the Ballast Technology Demonstration Project was "one of the great successes in the United States and Canada with regard to ballast water management." Specifically, Brohl said that the industry's voluntary installation of experimental ballast treatment systems on two Great Lakes freighters provided invaluable data that "provided a basis for further study."

"Further research is needed to develop additional and alternative measures to current prevention methods," Brohl said in a 2000 essay she coauthored with Ivan Lantz, then vice president of operations for the Shipping Federation of Canada. "Ballast water exchange is the only real prevention tool available to the commercial marine shipping vector at this time. We discourage the creation of regulations before proper criteria are developed, but we strongly encourage further joint participation and research into improved prevention techniques."[6] Their essay was a complicated way of saying that the shipping industry was satisfied with the ballast water exchange requirement; shippers didn't want new rules that required the installation of costly ballast treatment systems on ocean freighters. By 2008, the industry had kept ballast treatment standards at bay for fourteen years . . . and counting.

The shipping industry embraced ballast water exchange because it was relatively inexpensive. Before entering the Gulf of St. Lawrence, transoceanic vessels discharged freshwater ballast taken on while in ports overseas and replaced it with seawater. The theory was that salt water would purge freshwater organisms from ballast tanks. Any marine species in the saltwater ballast would die instantly when discharged into the fresh waters of the Great Lakes. The trouble with ballast water exchange was that it only eliminated, at best, 85 percent of live freshwater organisms in ships' ballast tanks. Chris Wiley, who managed Canada's ballast water program for the Department of Fisheries and Oceans, said ballast water exchange was "woefully inadequate" as a treatment standard. His comments came in an essay that accompanied one written by Brohl and Lantz, who claimed that ballast water exchange was the only viable treatment method in 2000.

"Realistically," Wiley said, "a long-term goal would be the development of ballast water discharge standards similar to those utilized in the wastewater industry."[7] That's precisely what would have happened if the U.S. EPA had applied the Clean Water Act to ballast water when the agency began cracking down on industrial and municipal wastewater discharges in the early 1970s. From my vantage point, it seemed evident that government agencies in the United States and Canada were determined to remain allies with the shipping industry while trying to develop new ballast water regulations. It was a questionable arrangement.

Eric Reeves, a retired U.S. Coast Guard commander who managed the Great Lakes ship-inspection program in the 1990s, said ballast water management was an issue where the government needed to be heavy-handed and force ocean ships to disinfect their ballast tanks before entering the lakes. He believed efforts to solve the ballast water problem were derailed by shipping interests that co-opted regulatory agencies and environmental advocates, seized control of the debate and, ultimately, delayed meaningful rules for more than 14 years. Reeves elaborated in an essay, which appeared with those by Brohl, Lantz, and Wiley in the Spring 2000 edition of the *Toledo Journal of Great Lakes' Law, Science & Policy*. Reeves said the following:

> I am a strong proponent of the idea that the environmental community has to make deals with the business community, but I am advocating a settlement of a conflict, not a pretense that does not exist. The shipping and aquaculture industries, good people though they are, are not "partners" with the environmental community. They have conflicting interests. It is no more legitimate for government bureaucrats or environmental representatives to enter into "partnerships" with them than it would be for a lawyer to go into business with the opposing counsel. Besides, it is almost always an unequal partnership. Business people play the political game with much more skill than do environmentalists, and the environment loses almost every time. Open and honest disagreement, or real politics, is a much better basis for a final settlement.[8]

Reeves knew from experience how difficult it was to implement new ballast treatment techniques. In 1996, when it became clear that Congress was content to fund years of research before forcing ships to sanitize ballast water, Reeves suggested an interim solution until acceptable technologies emerged. His proposal: pour chlorine into ocean freighters' ballast tanks. The chemical, though highly toxic, had been used for decades to purify drinking water and disinfect sewage in thousands of U.S. and Canadian communities. Chlorine was inexpensive, and its ability to kill aquatic organisms was well documented. Reeves figured that chlorinating ballast tanks would slow the flow of foreign species entering the lakes. But it wasn't to be. Shippers, environmentalists, and government officials were united in their opposition to using chlorine. How they reacted to Reeves's chlorine proposal spoke volumes about the difficulty of developing a treatment system that was suitable—even on an interim basis—to all parties wallowing in the political quagmire of ballast water policy. The mere suggestion of using chlorine to slow an emerging biological catastrophe in the Great Lakes was an object lesson in ballast water politics.

Reeves unveiled his plan at a meeting of U.S. and Canadian officials monitoring how well the two nations were upholding terms of the Great

Lakes Water Quality Agreement. Stopping shipborne invasive species from entering the lakes was one of the issues discussed at the meeting. "Silly me, I said, 'Why don't we just put some chlorine in these ballast tanks?'" Reeves said. "There was this absolute, stone silence. One of the state agency officials at the meeting came up to me after the meeting and said, 'You may be absolutely right, but politically we can't talk about any new uses of chlorine.' She told me I might as well stop talking about it because it would never be allowed. So I stopped talking about it."[9]

His suggestion may have made sense, but his timing was awful. Four years earlier, the International Joint Commission—the U.S.-Canadian panel that moderated Great Lakes issues—had called on both nations to phase out the use of chlorine and chlorinated compounds in all manufacturing processes.[10] The IJC's stance was a response to environmental advocates who argued, correctly, that chlorine and chlorinated chemicals were at the core of many of the Great Lakes' worst environmental problems. Chlorinated chemicals contaminated fish, caused deformities in birds, and poisoned the bottoms of many harbors around the lakes. Chlorine was the building block for a class of chemicals that drove the bald eagle to the brink of extinction in the 1960s, killed countless other birds, and prompted Rachel Carson to write *Silent Spring*. Despite its use in a myriad of common consumer products, environmental advocates in the early 1990s declared chlorine a chemical pariah, particularly in the Great Lakes. Lost in the debate over the proposed chlorine ban were several facts: Chlorination was the most common method of purifying municipal drinking water, disinfecting municipal wastewater, and sanitizing freighters' on-board sewage-treatment systems. Ironically, chlorine was the primary weapon that hundreds of cities and industries around the Great Lakes used to keep zebra mussels from clogging water intakes. Environmental advocates said that using chlorine to kill zebra mussels in water intakes was okay. But deploying the chemical to kill organisms in the vessels that dumped zebra mussels in the lakes was out of the question. Their rationale amounted to an absurd double standard.

U.S. and Canadian officials knew that chlorine killed a wide variety of organisms in ballast tanks. On at least three occasions in the early 1990s, the U.S. Coast Guard and Transport Canada poured chlorine into the ballast tanks of transoceanic freighters that entered the St. Lawrence Seaway without performing a proper ballast water exchange in the North Atlantic. The use of chlorine was a precautionary measure to prevent those ships from dumping millions of gallons of ballast water from overseas ports into the Great Lakes.[11] Reeves said the chlorinated ballast water was discharged into the lakes several hours later, after much of the chlorine had evaporated; the ballast water was devoid of aquatic life.

Marine engineers and EPA officials also knew that chlorine worked well in sanitizing ballast water. For years, cruise ships used sodium hypochlorite (household bleach) to kill organisms in seawater that the vessels used to cool engines. Some cruise liners also used sodium hypochlorite to disinfect ballast water. Those systems worked so well, Congress exempted cruise ships from the ballast water exchange requirements contained in the National Invasive Species Act of 1996.[12] Congress did not, however, require freighters to follow the lead of cruise ships and disinfect ballast tanks with chlorine. Shipping officials were dead set against it. They said chlorine would corrode ship hulls and expose crews to potentially deadly fumes. Those arguments prevailed in Congress, despite the fact that Argentina had long required ships to disinfect ballast tanks with chlorine before entering its ports. Freighters managed to comply with Argentina's requirement without chlorine eating holes in ship hulls or chemical fumes killing sailors. So did cruise liners that carried thousands of people at a time, human cargo that was far more valuable than the steel and grain most transoceanic freighters hauled in and out of the Great Lakes. It was also important to remember that saltwater from the ocean corroded ships, just like chlorine did.

Impenetrable opposition to disinfecting ballast water with chlorine—a stance that marked an odd alliance between the shipping industry and environmental advocates—was another victory for shipping officials determined to keep new regulations at bay as long as possible. Who could blame them? No successful business invited more regulation. It was the government's responsibility to adopt rules that put the public interest—in this case the health of the Great Lakes—above private interests. Alas, the U.S. Coast Guard and Congress sided with industry and opted for years of research when swift, dramatic action was needed to stop a tsunami of foreign species flooding the lakes.

OVER THE COURSE OF A DECADE, SCIENTISTS INVOLVED WITH CANGELOSI'S Ballast Technology Demonstration Project experimented with a variety of treatment systems on ships, on barges, and in laboratories. Researchers tested the ability of fine mesh filters, pressure chambers, and ultraviolet light to kill organisms in ballast water. Those tests provided useful data on which types of organisms the treatment systems killed, or didn't. But in the end, the project did not reveal a suitable treatment technology. It was evident by 2005 that involving dozens of government agencies and industry groups in ballast water research caused the very delays Cangelosi had warned about in a 1999 report she authored. Or perhaps the business of treating ballast water in freighters was too complicated to be solved in, say, 14 years.

The technical difficulties of sanitizing ballast water could fill an entire book—but not this one. What gnawed at me as I researched this topic was how many years scientists, shippers, and government officials had worked on the issue without finding a solution. A sense of urgency warranted by the threat invasive species posed to the lakes seemed to be lacking in the research programs. Any foreign species that took hold in the lakes would likely remain forever. Given that, I wondered if 14 years was a long period of time to develop a new pollution-treatment technology and the associated government regulations. That was debatable. But when measured against one of the world's greatest technological achievements, the hunt for a ballast water solution clearly took an inordinate amount of time. The United States had needed just 11 years—from the initiation of research and development until successful completion of the task—to put a man on the moon. Apparently, rocket science was less complicated than disinfecting freighters' ballast water. Either that, or the U.S. and Canadian governments were co-opted by the shipping industry or horribly dysfunctional when it came to protecting what was arguably their greatest natural asset: the Great Lakes.

Treating Ballast Water

Was disinfecting ballast water as perplexing as the shipping industry claimed? Or did shippers drag their collective feet to avoid the expense of installing treatment systems? The short answers to those questions: Yes and yes. The shipping industry lobbied in the 1990s to slow the development of rules that would have required the installation of ballast treatment systems, which could cost up to $1.5 million per ship. But it was also true that killing all critters in ballast water tanks was a daunting task that, if conducted improperly, could endanger ship crews and harm the environment. The challenges included:

- VOLUME: A typical ocean freighter plying the Great Lakes carried up to 3.2 million gallons of ballast water and could inhale ballast water at a rate of 528,000 gallons per hour. The larger lake freighters, which never left the Great Lakes, carried up to 13 million gallons of ballast water and inhaled as much as 1.2 million gallons of ballast water per hour, according to data compiled for the Northeast Midwest Institute's Great Ships Initiative. Dave Knight, transportation manager for the Great Lakes Commission, said treatment systems had to be able to sanitize huge volumes of water without slowing the rate at which freighters took on or unloaded ballast water.
- MUDDY WATER: Disinfecting ballast water meant killing aquatic life and bacteria in millions of gallons of water and tons of muddy sediment in the ballast tanks. Ocean freighters entering the Great Lakes fully loaded with cargo carried an average of 157 metric tons of sediment, muddy slop, in the bottom of ballast tanks, according to a Canadian government study. That sediment teemed with aquatic organisms, particularly the resting spores of bacteria. Those spores could hibernate for months inside hard outer shells that made some pathogens nearly indestructible.
- SIZE MATTERED: Shippers wanted treatment systems that occupied as little space as possible on freighters. Ships maximized profits by hauling as much cargo as possible; adding large ballast treatment systems could cut into cargo space and reduce revenue.
- RULES: The International Maritime Organization ruled that ballast treatment systems had to meet four criteria. The systems could not jeopardize crew safety; could not harm the environment; had to be "reasonably affordable"; and had to meet discharge standards (which were not in place as of 2008).

"From the time the first onboard ballast treatment technology experiments were carried out in the Great Lakes . . . in 1996, it became evident that a two-stage process was necessary: A primary treatment [filtration] to eliminate larger organisms and a secondary, disinfectant treatment to kill microorganisms, including bacteria and viruses," Knight said in a 2007 essay.

The ideal chemical biocide was one that killed all organisms in ballast water without endangering crew safety or corroding ship hulls. The chemical also had to work quickly and degrade quickly so that ships wouldn't discharge toxic chemicals into the lakes. That was a tall order, considering it took large doses of chemicals to kill viruses and bacteria. Was there a chemical or other treatment technology that could meet such demanding criteria? That question divided researchers who studied invasive species, and engineers who designed systems to kill the unwanted stowaways. Among the more popular approaches to sanitizing ballast tanks were filtration and high-pressure chambers in conjunction with ultraviolet light, heat, or chemical biocides—chlorine, ozone, or hydrogen peroxide.

Fednav Ltd., the largest shipping company on the Great Lakes, installed the OceanSaver system on one of its freighters in 2006. It used filtration to keep larger organisms from getting into ballast tanks. Those creatures that slipped through the filters were suffocated with nitrogen or crushed in a chamber that sent pulses of intense pressure into the water, rupturing the organisms' cell walls.

OceanSaver was one of 27 ballast treatment systems that Lloyd's Register—an independent risk-management organization—identified as "generally effective" in a 2007 report. As of May 2007, 28 ballast treatment systems had been installed on ships worldwide, according to Lloyd's Register. Over half of those systems used ultraviolet light to disinfect ballast water.

The economic and environmental stakes associated with ballast water treatment were huge. A Dutch consulting firm estimated in 2001 that the market potential for ballast treatment systems could reach $1 billion per year once the devices were perfected and required.

But there were credible skeptics who claimed it was foolish to believe ballast water could be completely sanitized. Among them was David Reid, an invasive-species expert at the National Oceanic and Atmospheric Administration's Great Lakes Laboratory in Michigan. "We can't sterilize a ballast tank," Reid told the *Milwaukee Journal-Sentinel* in 2005. "Heck, we can't even sterilize a hospital room."

Reeves said the hundreds of individuals involved in the cooperative effort to solve the ballast water problem—shippers, bureaucrats, scientists, policy wonks, and environmental advocates—weren't bad people. Rather, he said, the process that governed their work was flawed to the point of preventing a solution. Reeves wrote a blistering assessment of the government-industry collaboration on ballast water management in a 2000 article in the *Toledo Journal of Great Lakes Law, Science & Policy*. He said:

> After ten years of special committees, workshops, scoping studies and continued efforts to work out "cooperative solutions" with industry, the only answer which has been produced is, in essence, "it needs more work." There is always some other interesting technological advance to look at. However, no option is ever deemed acceptable. Government-industry collaboration, which has lately been called "partnering," may sound like a nice idea. Partnership means an identity of interest, in both law and politics, and a government partnership with industry in that sense is a violation of the fiduciary responsibility to the general public.[13]

Reeves compared the hunt for the perfect ballast treatment system to a cruel prank in Charles Schulz's famous cartoon strip, "Peanuts." In the cartoon, the character Lucy would hold a football upright on the ground and invite her friend and dupe, Charlie Brown, to kick it. As Charlie Brown sprinted toward the ball and thrust his kicking foot forward, Lucy would pull the ball away at the last second. The boy would fly into the air and land on his butt, humiliated. Lucy played that trick on Charlie Brown time after time. The lovable fool always fell for it. So it was with efforts to find the ideal widget for sanitizing ballast water, according to Reeves.

Twenty years after zebra mussels were discovered in the Great Lakes, neither the United States nor Canada had certified a ballast water treatment system as suitable for protecting the world's largest freshwater ecosystem. Every time a government agency or researcher proposed a method for disinfecting ballast tanks, Reeves said someone from the shipping industry, an environmental advocacy group, or a government agency would object on grounds that the proposed solution was an unproven technology, threatened water quality or wildlife, was potentially dangerous to ships' crews, impractical, or too expensive. "In all the years since I left the Coast Guard [in 1998] nothing has changed," Reeves said in a 2008 interview. "It's still Lucy and Charlie Brown and the football. That's what it is with this never-ending ballast water research project. We're always almost there with some wonderful new technology that will be acceptable to the shipping industry. But we fall on our ass every year and never seem to connect with the football."[14]

STATE OFFICIALS IN MICHIGAN, A PENINSULA SURROUNDED BY FOUR OF THE five Great Lakes, were growing restless by the mid-1990s. They wanted federal officials to require ocean ships to install ballast treatment systems. That clearly was not in the works at the time, said Mark Coscarelli, an environmental specialist in Michigan's office of the Great Lakes in the 1990s. Coscarelli said that involving dozens of industry groups and government agencies in the ballast water debate paralyzed efforts to develop meaningful regulations. "There was a lack of vision from the EPA, Coast Guard, NOAA and the Fish and Wildlife Service on the issue," he said. "There was a lot of finger-pointing. They all had a role to play but no one wanted to emerge as a leader. It was an alphabet soup of federal agencies involved but not one agency was singled out to be the driver of this issue."[15]

Coscarelli said Reeves's failed campaign to get the Coast Guard to move quickly on ballast treatment standards reflected the difficulty of fighting a war against invasive species with an army of bureaucrats that had many soldiers, but no commander in chief. Frustrated by the lack of progress, Michigan officials resurrected the idea of treating ballast water with chlorine. In April 2000, shortly after a Michigan state senator proposed a law that would have required visiting ocean ships to sterilize ballast tanks with chlorine, 27 members of the Ballast Water Working Group met at Detroit Metropolitan Airport to discuss the issue. Many of the heavy hitters from industry and government were present. Michigan officials viewed the meeting as an ideal setting in which to throw down the gauntlet. Near the end of the meeting, a Michigan Department of Quality official named Steve Casey set a chlorine generator on the conference table. Tension immediately filled the air. Coscarelli recounted the scene: "Steve told the group, 'By this time next year, all your ships could be equipped with these pumps.'" Industry's reaction? "I think it scared a lot of people," Coscarelli said.[16]

Michigan officials told shippers they could treat the ballast water of every ocean freighter operating on the Great Lakes with 63 tons of chlorine per year. That was a fraction of the 1.8 million tons of chlorine used annually in the Great Lakes basin. Water filtration facilities and wastewater treatment plants used 73,400 tons of chlorine annually, and 38,600 tons of chlorine were used annually in swimming pools that were eventually drained into the Great Lakes basin. About 500 tons of chlorine were used annually to keep zebra mussels from clogging municipal and industrial water intakes in the lakes. Michigan officials claimed that each ship could be equipped with a chlorine treatment system for about $125,000. Each time the ship used the system to treat ballast water would cost about $756 for the chemicals.[17] Using 63 tons of chlorine annually to treat the ballast water of every ocean ship entering the lakes, and then dechlorinating ballast water with Vitamin C or other compounds before discharging it into the lakes, seemed like a no-brainer. "One of the industry representatives accused us of wanting a quick fix," Coscarelli said. "That's exactly what we wanted."[18]

Coscarelli said industry officials agreed at the conclusion of the Detroit meeting to study the feasibility of treating ballast water with chlorine. "True to the shipping industry's form, they would nod their willingness to study something in a meeting and then fight it behind the scenes," Coscarelli said. Michigan's chlorine treatment proposal—and the state legislation—were eventually defeated. Both were vilified by the shipping industry and an unusual ally: environmental groups that were steadfast in opposing the use of more chlorine in the Great Lakes, even if it could prevent ocean ships from importing more foreign species. At the time, many environmental groups were fixated on ridding the lakes of toxic chemicals; most were clueless about the ecological significance of invasive species. Their opposition to using chlorine helped the shipping industry prevail, again. Coscarelli and other Michigan officials were furious. But the issue was squarely in the hands of federal officials and their partners in the shipping industry. One shipping association went so far as to suggest that the United States and all other nations leave ballast water regulations to the International Maritime Organization, a branch of the United Nations.

The Shipping Industry Ballast Water Coalition, a group representing some of the world's largest and most influential shipping interests, urged Congress to let the IMO regulate ballast water treatment. The IMO in 2004 proposed the first global ballast water treatment standards. Those standards, which took a decade to negotiate, required the following: ballast water exchange would ensure that 95 percent of the original ballast water was replaced with seawater, and ships treating ballast water could not discharge more than ten viable organisms larger than 50 microns (about the width of a human hair) per cubic meter of water. Under the IMO standard, all ships would be required to treat ballast water or perform acceptable ballast water exchange by 2016. But there was a catch. The IMO regulations wouldn't take effect until one year after 30 of its member nations, representing 35 percent of all cargo moved by sea, ratified the deal. As of May 2008, just 14 nations, representing 3.5 percent of all merchant shipping tonnage, had endorsed the regulations. The United States was not among them. U.S. officials wanted discharge standards that were 100 times more restrictive than the IMO standard.[19]

Testifying before a congressional subcommittee one month after the IMO standards were announced, the director of the Shipping Industry Ballast Water Coalition urged federal lawmakers to ratify the IMO convention on ballast water treatment. Joseph J. Cox told lawmakers the following: "Since maritime shipping is an international business, it must be regulated by consistent international and domestic requirements. . . . The coalition believes that a carefully crafted and internationally consistent national ballast water management program should be the exclusive method of compliance for vessels trading in the U.S."[20]

Cox's comments made it clear that the shipping industry's goal was to keep individual nations, and their states, from developing ballast water treatment standards more restrictive than the proposed IMO regulations. That position reflected a classic strategy, which held that it was easier to delay new regulations, and secure less stringent rules, if more parties were represented at the bargaining table. It was the polar opposite of the divide and conquer strategy. Allowing the IMO to be the sole regulator of ballast water had the potential to unite the world around lax ballast water treatment standards. U.S. officials refused to surrender their authority to regulate ballast water; several U.S. states also wanted to retain their authority to regulate ballast water, a right guaranteed under the federal Clean Water Act.

With shipping officials pushing for adoption of the IMO standards, Michigan officials in 2005 resurrected the chlorination treatment option a second time. Michigan made the move after the U.S. Coast Guard failed to propose ballast treatment standards after years of study. That same year, the state legislature passed a law banning ocean ships from discharging ballast water in Michigan waters unless equipped with one of four state-approved treatment systems, one of which was chlorination. Michigan's law took effect in January 2007. The shipping industry filed a lawsuit in federal court challenging Michigan's ballast treatment regulations. One of North America's largest shipping companies, Fednav Ltd., joined with several other global shipping companies and international trade organizations to challenge Michigan's law in U.S. District Court. The shipping interests claimed that Michigan's ballast water regulations violated the Commerce Clause of the U.S. Constitution by placing "unreasonable burdens on interstate commerce," were prohibited by federal law, and conflicted with U.S. Coast Guard regulations.[21] U.S. District Judge John Feikens dismissed the industry's lawsuit and a federal appeals court later upheld Michigan's law.

Shipping officials said treating ballast water was far more complicated than the public or Michigan officials realized. They also maintained that the cost of installing ballast treatment systems on every transoceanic freighter would cripple global shipping in the Great Lakes. It was a bogus argument that was debunked by Cangelosi, one of the industry's staunchest allies. In her 2004 report outlining the Great Ships Initiative, Cangelosi said ballast treatment systems could cost up to $1.5 million per ship, or about $330 million for all transoceanic freighters that visited the Great Lakes. Because moving cargo on water was so much cheaper than hauling it by truck or train, between $8 and $21 less per metric ton, Cangelosi said the shipping industry "could absorb some additional costs without the industry or the regional economy suffering serious economic hardship." She calculated that the global shipping fleet that visited the Great Lakes could recover the cost of installing treatment systems in as little as two years.[22]

Cangelosi said the effectiveness of those treatment systems could be tested at the Great Ships Initiative facility on the Wisconsin shoreline of Duluth-Superior Harbor. It was her fervent desire that the testing facility would hasten the day when all ocean freighters plying the Great Lakes were equipped with ballast treatment systems. At the unveiling of the Great Ships Initiative facility, Cangelosi made a telling comment. "There's a sense of urgency because this has been a very confusing landscape for people to try to operate in," she told the *Duluth News Tribune*. "Industry's normal mode of operation is to wait for government to regulate them."[23] That's precisely what should have happened. But the EPA dropped the regulatory ball in 1973, and the Coast Guard was still trying to pick it up in 2008—fourteen years after promising to develop ballast water treatment rules. As the EPA and Coast Guard fiddled, the number of invasive species in the lakes increased from 166 in 1994, to 185 in 2008.[24]

Invader no. 183 was discovered in late 2006 by anglers fishing at night on Muskegon Lake, a major tributary to Lake Michigan in western Michigan. An

National Oceanic and Atmospheric Administration's Great Lakes Environmental Research Laboratory

A scientist collects muddy sediment found on the bottom of a supposedly empty ballast water tank in a transoceanic freighter. Ballast water residuals often harbor large quantities of aquatic organisms. Removing or killing all organisms in ballast water tanks can be difficult, dangerous, and expensive.

angler shining a light into the lake's dark, murky water saw a cloud of what appeared to be thousands of tiny red shrimp swarming below the surface. He reported his observation to scientists at the National Oceanic and Atmospheric Administration's Lake Michigan Field Station in Muskegon. A few days later, Steve Pothoven, a fisheries biologist at the NOAA research station, aimed a high-powered spotlight directly down into the waters of Muskegon Lake and saw the same thing. Pothoven could hardly believe his eyes. Just below the water's surface were thousands, if not millions, of bloody red shrimp, *Hemimysis anomala*. It was the latest in a long line of creatures from eastern European waters to invade the Great Lakes. Like its Caspian Sea relatives, bloody red shrimp brought with it the potential to take another link out of the lakes' already crumbling fish food chain.[25]

More disturbing was the fact that Canadian biologists Anthony Ricciardi and Joseph Rasmussen had predicted in 1998 that the bloody red shrimp would invade the Great Lakes. Their prophecy was based on the creature's spread

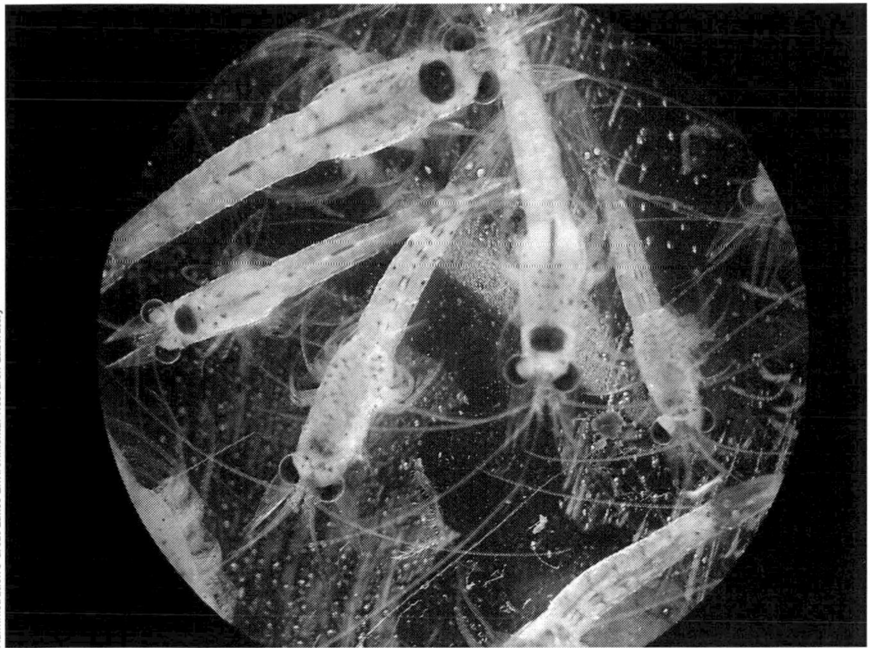

Microphotograph by Steve Pothoven, National Oceanic and Atmospheric Administration's Great Lakes Environmental Research Laboratory

Bloody red shrimp, *Hemimysis anomala*, captured in a tributary of Lake Michigan, are shown under a microscope.

across Europe and its ability to survive in ballast water for long periods of time. Ricciardi and Rasmussen predicted that 16 other species from Europe's Ponto-Caspian basin—which surrounded the Black, Caspian, and Azov seas—were likely to hopscotch the Atlantic Ocean in the ballast tanks of transoceanic freighters and invade the lakes.[26]

The November 2006 discovery of bloody red shrimp in Lake Michigan, and soon thereafter in Lake Ontario, was further proof that the lakes were paying a terrible price for endless efforts to develop suitable ballast treatment systems. Residents and politicians in the region were growing increasingly angry over the inability of federal agencies to halt the global shipping industry's relentless, albeit unintentional, ecological assault on the lakes. The depth of that outrage became clear one month after bloody red shrimp were discovered in the lakes. That's when one of the shipping industry's most loyal supporters became one of its harshest critics.

U.S. Rep. James Oberstar, a Democrat from the Great Lakes shipping mecca of Duluth, Minnesota, was a strong ally of the shipping industry. Among Oberstar's many industry-friendly actions on Capitol Hill was the securing of federal funds to build a second lock at the Soo Locks, a project that would allow more 1,000-foot-long lake freighters to safely travel between Lake Superior and Lake Huron. The Great Lakes Maritime Task Force in 2006 bestowed its "Iron Man Award" on Oberstar for his years of work on behalf of the shipping industry. Oberstar was a friend of the shipping industry, but he wasn't its pawn. He wasn't shy about expressing his displeasure over the industry's failure to solve the ballast water problem. At a 2000 meeting with European shippers, Oberstar told them, "Your ballast water from the Black Sea is destroying our Great Lakes! It's that simple."[27]

Oberstar stepped up his campaign for ballast water standards after Democrats regained control of Congress in the 2006 national election. The Democrats' victory put Oberstar in charge of the congressional committee that monitored shipping and other transportation industries. He made it clear in a newspaper interview that he aimed to force the transcontinental shipping industry to clean up its act. The time for studies had passed, Oberstar said. He and other Great Lakes lawmakers in Congress believed that the solution was passage of a federal law establishing ballast water treatment standards. That strategy allowed Congress to make an end run around the Coast Guard and force the shipping industry to install ballast treatment systems on all transoceanic freighters entering the Great Lakes.

Adolph Ojard, executive director of the Duluth Seaway Port Authority, said the international shipping industry also wanted national ballast water standards. He wrote an intriguing op-ed piece in the *Duluth News Tribune* in August 2007 expressing the shipping industry's concern about the invasive-species issue and its resolve to address the problem. Ojard said:

> The emergence of aquatic invasive species has become our industry's Achilles heel. . . . We stand ready to solve this—and let me assure you we will solve it. Technology vendors have already developed a host of products to treat ballast water. But absent federal standards, they are reluctant to make the investment necessary to bring these products to market. With the work that has been done developing the new Great Ships Initiative, we are prepared to test and certify

> current and developing technology. . . . I believe we can protect the aquatic environment and maintain a healthy environment. It is possible, but waiting for perfect won't get us there.[28]

Ojard's reference to "Achilles' heel" made me wonder if the shipping industry's years of intransigence on ballast water treatment standards might sink global shipping in the Great Lakes. Would powerful international shipping interests that violated the lakes as they fattened their wallets be brought down by poorly funded environmental groups—much as the seemingly indomitable Greek warrior Achilles was felled by an arrow that pierced his only weak spot? Such an outcome was hard to fathom. While visiting the Port of Duluth, I asked Ojard about his oft-repeated comment that invasive species were the shipping industry's Achilles' heel. Ojard changed his tune after being reminded that Achilles did not survive the arrow that pierced his heel. Ocean freighters, he said, would never be barred from the Great Lakes. There were forces at work, he said, that would never stand for closing the Seaway and severing the Great Lakes' connection to commerce from the world's seven seas—the U.S. and Canadian governments, to name two.

The improbable closure of the Seaway did not deter critics of the global shipping industry. They pressed their case that the only sure way to stop the flow of foreign species into the lakes via ballast water was to keep ocean freighters out of the freshwater seas. That stance was grounded in science. Even Cangelosi, a loyal ally to the shipping industry, reached that conclusion in her 2004 report proposing the Great Ships Initiative. "In reality," she said, "[ballast water] treatment generally will never be 100 percent effective."[29]

Unlike Cangelosi, who believed ocean freighters should be allowed to continue hauling cargo in and out of the Great Lakes provided they installed ballast treatment systems, industry critics demanded a fail-safe, 100-percent-effective solution. There was only one way to achieve that: ban transoceanic freighters from the lakes. It wouldn't be difficult, technically speaking. The St. Lambert Lock near Montreal was a chokepoint through which all ships entering the lakes from the Atlantic Ocean had to pass en route to Lake Ontario. The seemingly radical suggestion to close the Seaway evolved into a rallying cry as economists discovered that the global shipping trade caused far more environmental problems than economic benefits in the Great Lakes region. As it approached its 50th anniversary, the Seaway was developing an identity crisis. No longer was it known solely as an engineering marvel that linked shipping communities on the freshwater seas to seaports around the world. In some circles, the Seaway was viewed as a symbol of ecological ruin in the Great Lakes. Critics had their sights set on shutting it down, at least to ocean freighters. The battle lines would soon be drawn.

The shipping industry has evaded controls which would have come long ago in another context. Ask yourselves this question: If there were an oil or chemical spill on the Great Lakes that caused even a fraction of the harm caused by exotics, how would federal regulators and Congress have responded?

—Aaron Schlehuber, attorney for the Sault Ste. Marie Tribe of Chippewa Indians, 2001

22 SEAWAY HERETICS

The workday was in its first hour when my office telephone rang, delivering what would prove to be one of the most disturbing twists in the sordid tale of Great Lakes invasive species. I hadn't finished uttering the requisite one-word telephone greeting when the caller's staccato words burst into my ear. "Hey, it's Gary. You gotta get over here. Hurry. You won't believe your eyes." I didn't need caller identification to know, within a nanosecond of taking the call, that the person on the other end was Gary Fahnenstiel. The senior ecologist at the National Oceanic and Atmospheric Administration's Great Lakes Environmental Research Laboratory was one of the most animated, outspoken, and zealous scientists in his field. He was also one of the most distinguished of Great Lakes researchers, an expert on how the lakes' ecosystems functioned and responded to natural and human-induced changes. For him, studying the lakes was more than a paycheck, more than a career. It was a calling. He was passionate about protecting the lakes he had come to know and love as a child growing up in Saginaw, Michigan, and as a burgeoning scientist who earned all of his degrees—a bachelor's, master's, and PhD—at Michigan universities. His expertise could have taken him anywhere in the world, but he chose to remain in Michigan, near the lakes. "All I ever wanted to be was a Great Lakes scientist. I'm living my dream," he said.[1] The lakes were in his blood. He was consumed by their beauty, majesty, and tragedy.

On days when the weather was favorable, Fahnenstiel was known to drive a small boat across Muskegon Lake to his office on the Lake Michigan shoreline. Many of his summer vacations were spent cruising Lake Huron's Georgian Bay, with his wife and two sons, in a 35-foot cabin cruiser. His ideal vacation was laying claim to one of Georgian Bay's 30,000 islands and spending several days fishing, swimming, and reflecting on Lake Huron's beauty, tranquility, and awesome power. Huron, like the other four Great Lakes, was so large it altered weather patterns and was known to swallow ships when gales whipped it into a frothing menace. Fahnenstiel's research took him to all five Great Lakes and often entailed being on the lakes in bone-chilling cold, rolling with waves that caused more than a few sailors to puke. That didn't matter. Fahnenstiel couldn't get enough of the lakes. Superior, Michigan, Huron, Erie, and Ontario—all were part of him. And he was resolute about protecting them.

Fahnenstiel's zeal for safeguarding the lakes, and outrage over the damage caused by invasive species, prompted him to call me that morning in 2004. He wanted it known that zebra mussels, the scourge of power plants and municipal water facilities around the Great Lakes, had created a potentially deadly new phenomenon on inland lakes—toxic algae blooms. Some of the algae blooms released naturally occurring, but potentially deadly, compounds that were 100 times more toxic than cyanide. The rise of mussel-fueled algae blooms, on the heels of myriad other ecological problems that invasive species caused, fueled a white-hot rage in Fahnenstiel. He wondered how government agencies in the United States and Canada could allow ocean vessels to continue operating on the lakes in light of all the ecological damage their ballast water discharges had caused. Fahnenstiel's righteous indignation soon made him a lightning rod for controversy. The day after Christmas 2004, in the 45th year of the St. Lawrence Seaway's celebrated history, Fahnenstiel went public with a demand that was immediately denounced as heresy. The time had come, he told newspapers in Wisconsin and Michigan, to ban ocean freighters from the Great Lakes.[2]

Fahnenstiel was the first American scientist to call for closing the St. Lawrence Seaway. In his view, banning transoceanic freighters from the Great Lakes was the only sure way to keep the ships from importing more foreign species. His remarks were the product of frustration. For nearly two decades, he had witnessed zebra and quagga mussels, round gobies, and other invaders systematically deconstruct the lakes' ecosystems. He and his peers documented a litany of serious problems caused by shipborne invaders. The trouble was that their research had minimal effect on public policy. Time and again, Fahnenstiel and other scientists detailed how foreign species were dismantling fragile Great Lakes ecosystems. Then, nothing. Their findings were routinely ignored by government officials in positions of power—those with the authority to protect the lakes by forcing ocean freighters to sanitize their ballast tanks.

The last straw for Fahnenstiel was a study to determine whether foreign species lived in the muddy slop in the ballast tanks of the NoBOB ships—ocean freighters that reported having no ballast on board upon entering the Seaway. He was part of a research team in 2004 that discovered the NoBOB ships carried billions of foreign organisms and deadly pathogens in the muddy water that sloshed around in their supposedly empty ballast tanks. The NoBOB vessels were still exempt from ballast water regulations at the time, which meant that foreign organisms and pathogens lurking in their ballast tanks were routinely dumped in the lakes. The NoBOB loophole was an outrageous failure of public policy that threatened the environment and the health of millions of people who relied on the Great Lakes for drinking water. After 16 years of studies and rhetoric that failed to stop the flow of shipborne foreign species into the lakes, Fahnenstiel decided to express his views on what he considered an environmental scandal of the highest order. He believed the government's willingness to allow transoceanic freighters to play ecological Russian roulette with the world's largest freshwater ecosystem was a crime against nature and all who relied on the lakes. It was clear by 2004 that government agencies in the United States and Canada were unable, or unwilling, to stop the carnage. Tired of empty promises to stop the flood of invasives, Fahnenstiel drew a battle line.

"Let's close the Welland Canal. Let's start there. This is ridiculous," he told the *Milwaukee Journal-Sentinel*.[3] He struck the same day in an article published in the *Muskegon (Mich.) Chronicle*. "This is a simple problem with a simple solution," he said. "It's time to close the Welland Canal. We have a natural choke point and we can shut off the flow of exotics into the Great Lakes."[4] Technically, closing the Welland Canal would only prevent ocean freighters from reaching the four upper Great Lakes—the ships could still enter the St. Lawrence Seaway and proceed into Lake Ontario. Still, the intent of Fahnenstiel's comments was unmistakable.

The newspaper articles quoting Fahnenstiel were picked up by the Associated Press and distributed to newspapers across the United States and Canada. His comments sent shock waves through the vast community of scientists, government bureaucrats, and shipping officials working on the invasive species–ballast water issue. The notion of banning ocean ships from the lakes went against the conventional wisdom, which held that the Seaway was an environmentally benign marvel of modern engineering and a vital economic resource. Fahnenstiel's comments were a clarion call that changed the nature of the debate about invasive species, the future of the Seaway, and the ocean ships that used, and abused, the world's largest freshwater seas. He became somewhat of a folk hero among scientists and environmental activists for standing up for the beleaguered lakes at the risk of losing his well-paying government job. Shipping-industry officials had a different view. They considered Fahnenstiel a heretic, a loose cannon who had to be silenced.

HERETIC IS A LABEL OFTEN SLAPPED ON INDIVIDUALS WHO DISAVOW mainstream religion. It can be the most searing of insults, one usually reserved for pagans and atheists. That narrow definition of the word is unfortunate and incomplete. Heretics aren't necessarily pagans—some are simply nonconformists, free thinkers who dare to challenge conventional wisdom. That is not a revisionist definition on my part. No less an authority than *Merriam-Webster* offers two definitions for the word. A heretic, according to the dictionary, can be one of two things: "a dissenter from established religious dogma"; or "one who dissents from an accepted belief or doctrine: nonconformist." Some of the world's most brilliant and courageous individuals were declared

David Jude, University of Michigan, School of Natural Resources and Environment

A bloom of potentially toxic algae is shown on the surface of an inland lake. Toxic algae blooms are more common in lakes infested with zebra and quagga mussels.

heretics before their actions were later redefined as heroic. Joan of Arc, who led France to a military victory over England, was captured by her foes during the conflict, convicted of heresy, and burned at the stake in 1431. Five centuries later, she was canonized as a Catholic saint. In 1633, the Catholic Church found Galileo guilty of heresy for teaching that the Earth revolved around the sun. Galileo's theory proved correct, and he was posthumously deemed one of the great scientific minds of all time.

Fahnenstiel clearly was a Seaway heretic. He had the audacity to challenge the widely held belief that the massive public-works project was a sacred cow and, thus, immune to criticism. Initially a lone voice in the wilderness of invasive-species politics, he was soon joined by a growing chorus of Seaway heretics. They, like Fahnenstiel, dared to weigh the value of the Great Lakes against a man-made structure that invited ecological ruin into the heart of the freshwater universe.

THE SHIPPING INDUSTRY'S QUICK RESPONSE TO FAHNENSTIEL'S COMMENTS made it clear that the Seaway debate would be intense, bitter, and, at times, personal. Industry officials returned the verbal fire within a week after his comments were published. Helen Brohl, then executive director of the U.S. Great Lakes Shipping Association, and Michael H. Broad, president of the Shipping Federation of Canada, sent a strongly worded letter to Fahnenstiel's bosses at NOAA. They were irate over his comments about the Welland Canal, and his willingness to discuss preliminary results of a NOAA study that found that the majority of NoBOB ships entering the Great Lakes carried billions of viable organisms and deadly pathogens in their ballast tanks. Fahnenstiel said the study proved that the only sure way to keep ocean freighters from importing more foreign species into the lakes was to keep ocean freighters out of the lakes. Brohl and Broad demanded that Fahnenstiel be directed to write a letter to every newspaper in the United States that published his comments. They wanted Fahnenstiel's letter to "clarify how his comments were wrong." In their letter to Vice Admiral Conrad J. Lautenbacher Jr., NOAA's undersecretary of commerce for oceans, the shipping officials said the following:

> Dr. Fahnenstiel clearly has an agenda regarding Great Lakes shipping and is using NOAA and the Great Lakes Environmental Research Laboratory study to further it. This is, in the least, inappropriate and misleading and, at the most, unethical and libelous in its approach to address an issue we all take very seriously regionally, nationally and internationally.[5]

Fahnenstiel kept his job, and he never wrote letters to newspapers clarifying his comments, as the shipping industry demanded. But he never again spoke publicly about the ballast water issue. Brohl and Broad's letter had effectively silenced one of the global shipping industry's most vocal, and credible, critics. They could not, however, silence mounting concern about the biological chaos that ocean freighters turned loose in the lakes. Within a week of Fahnenstiel's incendiary comments, the debate about the costs and benefits of ocean shipping in the lakes erupted into a verbal conflagration.

The *Detroit Free Press* published an article on January 2, 2005, that quoted U.S. and Canadian scientists calling for closing the Seaway to ocean freighters. "Non-Great Lakes boats frankly should be banned from entering the Great Lakes," said Milt Clark, a senior health and research advisor at the U.S.

Environmental Protection Agency's Chicago office. The newspaper quoted a similar comment by Gail Krantzberg, then director of the Detroit regional office of the International Joint Commission—which mediated Great Lakes and border issues involving the United States and Canada. "We shouldn't be having the salties coming into the Great Lakes," Krantzberg was quoted as saying.[6]

Clark and Krantzberg had made their comments at a journalism conference in Pittsburgh three months earlier. The Detroit newspaper didn't publish their remarks until January 2005, after Fahnenstiel's remarks appeared in the Milwaukee and Muskegon newspapers. Krantzberg later said she had never called for banning ocean freighters from the Great Lakes. Rather, Krantzberg said she told journalists at the conference that banning transoceanic ships was the only guaranteed way to keep those vessels from importing more foreign species. That subtle distinction mattered little to her bosses at the IJC. Hours after the Detroit newspaper story hit the streets, Krantzberg received a call from an assistant to Herb Gray, a Canadian member of the IJC.

"I was in Ottawa celebrating the holidays with my family when I got the phone call," Krantzberg said. "Commissioner Gray's assistant asked me if my talk [at the Pittsburgh conference] called for shutting down the Seaway. I told him that what I said was the only sure way to keep invasives from getting into the lakes on ships was to close the Seaway. I called for a rational discourse on the issue. I thought it was a fair statement."[7]

Krantzberg's explanation failed to extinguish what was clearly a politically risky debate about the future of ocean shipping in the lakes. "It just got worse and worse after that," she said. "I was told the U.S. chairman of the IJC thought I exercised extremely poor judgment." A few hours after that heated discussion with the IJC official, Krantzberg received an e-mail from the *Detroit Free Press* reporter who wrote the article. "He said, 'I hope this didn't cause you any grief,'" she said. "I e-mailed him back and said, 'You know, your story is getting me fired.'"[8]

Three days after publication of the Detroit newspaper article, the IJC issued a statement from Lisa Bourget and Murray Clamen, who chaired the U.S. and Canadian sections of the commission, disavowing Krantzberg's comments. It read: "We would like to make clear that any such statement is not the position of the commission. The IJC has made a series of recommendations to prevent further introductions of aquatic invasive species in the Great

Lakes. Specifically, with regard to invasive species delivered via ballast water, we have recommended that the U.S. and Canada adopt protective measures more stringent than the International Maritime Organization Convention and implemented sooner. The commission also has recommended further study and assessment of management practices to minimize the threat of new invasions."[9]

Krantzberg said some IJC commissioners wanted her fired immediately after the newspaper article was published. She kept her critics at bay for six months, but relented in June 2005, when she agreed to resign. Her term as director of the commission's regional office was set to expire three months later, so her ouster was largely symbolic. Shortly thereafter, she landed a job as a professor and director at the Centre for Engineering and Public Policy, School of Engineering Practice at McMaster University in Hamilton, Ontario. Unlike the IJC, the university did not police her comments about the Seaway. She reflected on her ordeal in a 2007 interview:

"I found the whole process humorous, almost pathetically humorous, that I was getting fired for speaking truth to power. I certainly have no regrets. My point was that the Great Lakes are at tremendous risk from aquatic invasive species. How can we not have a study and a rational discourse about banning ocean ships from the Seaway?" Krantzberg said. "Technology, up until now, has not stopped the problem. I think there's a real fear on the part of the shipping industry that if we have a rational discussion about the economics of closing the Seaway we will come to the conclusion that keeping the Seaway open is not a viable way to go. The fishery of the Great Lakes alone is worth $7 billion; I doubt ocean shipping [in the lakes] is worth that much."

Krantzberg said she was never contacted by anyone associated with the shipping industry after the furor over the Detroit newspaper article. Still, she believed her comments caused a commotion among shippers. Though she paid a price for the remarks, Krantzberg—like Fahnenstiel—helped forge a serious review of the economic implications of banning ocean ships from the Great Lakes. The politically charged question of whether it made economic sense to keep the Seaway open—in light of the environmental damage it invited into the lakes—would finally be answered by a pair of economists commissioned to study the issue. Their conclusions struck like lightning, electrifying people on both sides of the increasingly contentious ballast-water debate.

JOHN TAYLOR AND JAMES ROACH WERE NOT THE TYPES OF ACADEMICS normally associated with divisive environmental issues. Taylor was an associate professor of marketing and logistics at Grand Valley State University in Michigan. Roach was a transportation consultant with 35 years of experience working with companies that moved cargo by land, sea, and rail. They were economists who studied traffic patterns and the movement of cargo, not radical environmentalists determined to ban ocean freighters from the Great Lakes. Roach was an advocate for the movement of cargo by ships when he worked for the Michigan Department of Transportation; he helped develop long-term goals for the Detroit–Wayne County Port Authority. A wild-eyed tree hugger he was not. Nor was Taylor, who served on several government-chartered transportation commissions for federal agencies in the United States and Canada. Taylor and Roach were experts in the economics of moving cargo—how to move the greatest amount of freight at the least expense. Overnight, they were thrust into the debate over the future of ocean shipping in the Great Lakes.

Taylor and Roach issued a report in August 2005 titled "Ocean Shipping in the Great Lakes: Transportation Cost Increases That Would Result from a Cessation of Ocean Vessel Shipping." The title of the report wasn't the stuff of eye-catching newspaper headlines. The executive summary was. It said: "The principle conclusion of this study is that a cessation of ocean shipping on the Great Lakes would result in a transportation cost penalty of $54.9 million [in U.S. dollars] per year." In other words, banning ocean ships from the lakes would add $55 million to the annual cost of shipping international cargo in and out of the region on trucks or trains. It was a paltry sum compared to the economic damage shipborne invasive species caused in the lakes, estimated at somewhere between $200 million and $500 million annually. The environmental costs of ocean ships plying the Great Lakes were far greater than the economic benefits—between four and ten times higher, according to the study.[10]

Freighters moved about 220 million tons of cargo on the Great Lakes annually, but the vast majority was shipped between ports within the lakes and along the St. Lawrence River. Transoceanic freighters accounted for about 5 percent of all cargo shipped on the lakes each year. In 2002, an average of two ocean freighters passed through the Montreal section of the St. Lawrence Seaway each day during the nine-month Great Lakes shipping season, according to the

Taylor-Roach study. Relatively few ocean freighters visited the lakes, but they left a massive trail of biological chaos in their collective wake. "Ocean vessels on the lakes make only a modest contribution to transportation cost savings for users of the system," the Taylor-Roach report said. "By way of comparison, we estimate the costs of existing invasive species on Great Lakes utilities at $200–$500 million [U.S. dollars] per year. It should be noted, however, that the costs of future invasions that might be introduced by continued ocean shipping is not known."[11]

Taylor and Roach concluded that keeping ocean freighters out of the lakes, and forcing them to off-load cargo to trucks and trains in Montreal or at ports along the Atlantic seaboard or Gulf of Mexico, would add 6 percent to the cost of moving international cargo in and out of the Great Lakes region. In purely economic terms, allowing ocean ships to continue entering the lakes made no sense. The environmental costs greatly exceeded the economic benefits. The economists didn't advocate banning ocean ships from the lakes by closing the St. Lawrence Seaway. But the implication of their report was clear: Closing the Seaway was a no-brainer, economically and environmentally. Politically, however, the Seaway was untouchable. There was no other explanation for maintaining a structure that never turned a profit or carried the volume of international cargo that was predicted when it opened in 1959.

When construction began on the Seaway in 1954, one economist projected that transoceanic ships traveling in and out of the Great Lakes would carry 50 million metric tons of cargo annually through the locks and canals in the St. Lawrence River.[12] The volume of transcontinental freight shipped through the Montreal section of the Seaway peaked at 23 million metric tons in 1978. By 2007, that figure dropped to 9 million metric tons.[13]

Shipping officials called the Taylor-Roach study simplistic and biased. Yet, they never produced data to discredit their findings. "The logic of the report is seriously flawed," said Albert Jacquez, administrator of the St. Lawrence Seaway Development Corp. in a 2006 article.[14] Jacquez said there was no credible evidence that shipborne invasive species were causing the amount of damage cited in the Taylor-Roach study. Jacquez's claim was nonsense. Conservative estimates put the damage from shipborne invasive species at $200 million to $500 million annually. Zebra mussels alone caused $1.5 billion damage in the two decades after they invaded the lakes, which broke down to $75 million annually. Jacquez noted, correctly, that ocean ships weren't the only source of invasive species in the lakes. Ballast water from ships accounted for roughly one-third of all invaders in the lakes. Bait dealers and aquarium owners also introduced foreign species into the lakes. The difference was that ocean freighters accounted for many of the most destructive invaders. Zebra and quagga mussels, the round goby, Eurasian ruffe, and spiny water flea—all were imported by ocean freighters after the Seaway opened in 1959. Other harmful invaders, such as sea lamprey, alewife, and white perch, snuck into the lakes through the Erie and Welland canals, both of which were built to increase the flow of global commerce in and out of the Great Lakes.

Shipping officials argued vehemently that closing the Seaway would create a logistical nightmare and inflict serious economic pain on the region. They said trucks and trains could not handle the surge of grain that was loaded onto freighters in Duluth and Thunder Bay, Ontario, each fall and shipped to Europe, Asia, and Africa. Richard Corfe, director of Canada's St. Lawrence Seaway Management Corp., said the Seaway could not survive financially on domestic shipping alone. Ocean freighters accounted for about 25 percent of the Seaway's revenue from tolls. "Now if that's not there, what do I have to do? I have to increase the cargo rate for what's remaining," Corfe told the *Milwaukee Journal-Sentinel*. "I don't believe we can operate the Seaway on domestic traffic. . . . I don't believe it's sustainable."[15]

Shippers attacked Taylor's credentials and questioned his motives. They claimed he knew little about moving freight by sea. One shipping-industry lobbyist said Taylor was a closet tree hugger funded by the pro-environment Joyce Foundation in Chicago. Nonsense, Taylor replied. The Joyce Foundation had a history of funding environmental projects, but Taylor insisted that his study was objective and would withstand the toughest scrutiny by other economists. Taylor, in fact, said that if he had any professional bias, it was in favor of the shipping industry. He was, after all, a protégé of the late John L. Hazard, a Michigan State University economist who supported construction of the Seaway. As industry officials tried, without success, to discredit the Taylor-Roach study, environmental activists seized upon its findings. Representatives of Great Lakes United and the Alliance for the Great Lakes said it was time to consider whether it made sense—economically and environmentally—to close the Seaway. That sentiment was shared by Michigan congressman Vernon Ehlers, a physicist and one of the most respected voices on Great Lakes issues in the U.S. Congress. The pro-business, conservative Republican told the *Milwaukee Journal-Sentinel* in 2005 that perhaps the Seaway had run its course.

"The reason for opening [the Seaway] was to improve the economy of the Great Lakes region," Ehlers said. "And if the economic advantages aren't great enough to justify the environmental costs . . . then I think both the state and federal government have to take a long, hard look at this and say, 'Does it make sense to keep operating this way?'"[16]

THE SEAWAY'S FINANCIAL STRUGGLES AND THE ENVIRONMENTAL PROBLEMS it invited into the Great Lakes weren't the only reasons to question its future. Security experts said the Seaway was vulnerable to a terrorist attack. The costs of preventing such an attack and dealing with the environmental aftermath of shipborne invasive species were "enormous," said Terry A. Breese, the U.S. State Department's director of Canadian Affairs. Speaking at a 2005 conference on Seaway security issues, Breese said, "I think we need to question whether we want to have a Seaway anymore."[17] Others at that conference said a terrorist attack on the Seaway could cause tremendous damage to economies in the Midwest and parts of Ontario and Quebec. They pointed to a 1985 lock failure in the Welland Canal—which trapped 53 freighters in Lake Erie for 24 days and cost shippers $24 million—as an example of the economic havoc terrorists could inflict by attacking one of the Seaway's canals.

The United States and Canada tightened security in the wake of the 9/11 terrorist attacks, erecting fences along many of the Seaway's locks and canals. Still, there were areas along the Seaway in 2008 where a person could stand on the edge of a canal, a stone's throw away from ships—some of which carried thousands of tons of oil and hazardous chemicals. There was one site along the Seaway where motorists routinely drove vehicles beneath one of the canals that carried ships. There were no checkpoints or security guards protecting that tunnel. A car bomb like the one Terry McVeigh used in 1995 to destroy the Murrah Federal Building in Oklahoma City could be stationed beneath the Seaway canal. An explosion there could collapse the concrete canal and halt ship traffic between Lake Ontario and the Atlantic

Ocean. I mention this not to give demented terrorists ideas, but to point out a gaping hole in the Seaway's porous security. If the Seaway is as valuable to the region's economy as U.S. and Canadian officials claim, it only seems prudent that they would do more to protect it from terrorists.

THE 2005 TAYLOR-ROACH STUDY CAME AS CONGRESS WAS CONSIDERING legislation to reauthorize the National Invasive Species Act of 1996. Environmental activists and Great Lakes politicians viewed the reauthorization of the law as a golden opportunity to strengthen ballast water regulations. The National Aquatic Invasive Species Act, known as NAISA, was introduced in the U.S. House of Representatives and the Senate in 2003. It enjoyed bipartisan support from lawmakers in the eight Great Lakes states, but ran into a brick wall of opposition among lawmakers from other parts of the country. Republican lawmakers who controlled both chambers of Congress from 2002 to 2006 wouldn't even afford their political peers the courtesy of holding hearings on the legislation. Conservative Republicans viewed NAISA as a new version of the federal Endangered Species Act, which many conservatives considered the epitome of excessive government regulation. NAISA died when the congressional term expired at the end of 2004. The legislation was reintroduced in 2005, but went nowhere. It wasn't until 2007, after Democrats won control of the House of Representatives and Senate, that Congress began to seriously consider regulating ballast water. By then, the ballast water regulations that the U.S. Coast Guard had repeatedly promised, but failed to produce, were long overdue.

Changes wrought by shipborne invasive species—coupled with lingering toxic chemicals, excessive nutrient loadings, and the destruction of coastal wetlands—were threatening to drive Great Lakes ecosystems into an ecological abyss. A group of distinguished scientists warned in a 2005 report that four centuries of human abuse of the lakes had dismantled their natural immune system and pushed the lakes to the brink of ecological collapse. "The Great Lakes are deteriorating at a rate unprecedented in their recorded history and are nearing the tipping point of ecosystem-wide breakdown. If we want to restore this resource, the time to act is now," said Alfred Beeton, one of the report's lead authors and the former director of the U.S. government's Great Lakes Environmental Research Laboratory.[18]

Several of the most dramatic, rapid changes in the lakes occurred in the last two decades of the twentieth century. Most were caused by a handful of foreign species—zebra and quagga mussels, the round goby, and foreign zooplankton—imported by ocean freighters from Europe. Zebra and quagga mussels reconfigured ecosystems in all of the lakes except Superior. Just those two species caused profound, devastating changes: the near elimination of vital fish food sources, which threatened a $7 billion fishery; noxious and toxic algae blooms that fouled beaches and threatened human health; the return of a biological dead zone in Lake Erie that hadn't been seen since the lake was gasping for life in the 1960s; the liberation of botulism spores that killed more than 70,000 fish-eating birds and countless fish; and the elimination of entire species of native mussels.

A rising tide of outrage over the problems caused by shipborne invaders—and the failure of the United States and Canada to stop the great invasion—boiled over at a 2007 congressional hearing on the NAISA legislation. The hearing was an effort to get Congress moving on new ballast water regulations, which had languished for five years. The hearing featured a mix of shipping officials and environmental activists, Native American tribal leaders and small-town politicians. Most were there to plead with Congress to circumvent the U.S. Coast Guard and promptly enact ballast water treatment standards. One of the most eloquent testimonials was delivered by Frank Ettawageshik, tribal chairman of the Little Traverse Bay Bands of Ottawa Indians in northern Michigan. Speaking to the House Transportation and Infrastructure Subcommittee on Water Resources and Environment, Ettawageshik said biological pollutants imported by ocean freighters threatened sacred food sources—whitefish and lake trout—that had sustained Native Americans in the Great Lakes region for thousands of years. He said:

> The tribes have been forced to helplessly watch the Great Lakes resource, and their treaty-based fishing industry being shamefully attacked, eroded and diminished by aquatic invasive species, particularly species that entered via the ballast water vector. We are appalled as to how such an obvious and destructive activity is allowed to continue, virtually unabated, while the federal government stands idle. To state it bluntly, the transoceanic shipping industry, through ballast water exchange practices and construction of canals, has severely impaired, and threatens to destroy, the treaty-based commercial

> and subsistence fishing industry. The ballast water vector is readily curable; it will simply require oceangoing ships to spend the necessary funds to fix their problem. Unfortunately, the shipping industry has stubbornly and shamefully refused to act, despite two decades of damage awareness. We believe they have had ample time to meet their responsibilities.[19]

Adolph Ojard, executive director of the Duluth Seaway Port Authority and president of the American Great Lakes Ports Association, provided the shipping industry's perspective. He repeated many of the industry's claims that shipping cargo was more efficient, safer, and caused less air pollution, than moving cargo by truck or train. He said the shipping industry had worked with the U.S. and Canadian governments to enact the world's first, albeit inadequate, ballast water regulations. The industry was committed to finding suitable ballast water treatment solutions to kill invasive species, which Ojard again called the industry's "Achilles' heel." Then he made one of the most remarkable comments ever uttered by a shipping industry official during a public discussion of the invasive-species issue. "The shipping industry—like any industry—operates under the terms of an unwritten social contract with the public. That is, our industry should add value to society and do no harm. Indeed, maritime commerce offers numerous benefits. Studies have shown that waterborne transportation is widely regarded as the fastest, safest, cleanest and least costly mode of commercial transportation."[20]

Global shipping certainly added economic value to the Great Lakes region, but at what cost? According to the Taylor-Roach study, ocean freighters reduced the cost of moving global commerce in and out of the lakes by $55 million annually, a savings of 6 percent. The resulting biological pollution cost the region a minimum of $75 million annually, likely much more. Taylor and University of Notre Dame biologist David Lodge put the annual damage caused by shipborne invaders at $200 million annually; other researchers claimed the cost was higher yet. Ojard's brazen claim that the shipping industry had lived up to its unwritten social contract to "do no harm" didn't hold water. There was abundant evidence that ocean freighters had caused widespread, profound harm to the Great Lakes.

Two months after the 2007 congressional hearing, a coalition of 90 environmental groups called for closing the Seaway until ocean freighters were equipped with treatment systems that sterilized ballast water. The proposal elicited a heated response from the U.S. boss of the Seaway. Closing the Seaway, said Collister Johnson Jr., was "a nice political statement, but it's completely impractical and impossible."[21] Johnson's comments seemed a bit over the top. Was closing the Seaway impossible? Technically, it would be quite easy—simply ban ocean freighters from entering the St. Lambert Lock near Montreal. That would keep all ocean ships from reaching the lakes. Legally, closing the Seaway was a tricky proposition. Politically, it seemed virtually impossible. The U.S. and Canadian officials had no desire to close the Seaway. No one in the Great Lakes region who understood political realities believed the United States and Canada would ever consider closing a massive public-works project they had built, operated, regulated, and subsidized. That would be akin to asking Americans to break their addiction to oil—it wouldn't happen anytime soon, if ever.

Still, for the Seaway administrator to say that closing the artificial canals that invited foreign commerce and ecological havoc into the lakes was impossible seemed like a bit of an overreaction. It reminded me of a famous line in Shakespeare's "Hamlet." At one point in the play, the queen was asked for her assessment of one of the characters. Her royal response: "The lady doth protest too much, methinks." That comment became synonymous with people who protested an idea too vigorously because, deep down, they feared it might become reality.

TAYLOR AND ROACH ADDED MORE FUEL TO THE RAGING DEBATE OVER THE future of the Seaway in 2007, when they released an update of their 2005 study. The second study enshrined them in the hall of Seaway heretics. In it, the economists concluded that a more detailed analysis of closing the Seaway to ocean ships showed that such a move would create more than 1,000 new jobs in the United States and Canada. Forcing transcoceanic freighters to unload cargo in Montreal, or at ports along the Atlantic Coast or Gulf of Mexico, would result in a net gain of 1,319 jobs for domestic dock workers, truckers, and railroad employees in the United States and Canada, according to their analysis. The Great Lakes' global shipping trade directly employed 1,001 people in the United States and Canada, and another 891 workers overseas. Banning ocean freighters from the lakes might eliminate the 1,001 North American jobs, but would replace them with 2,320 new jobs—a net gain of 1,319 jobs, according to Taylor and Roach.[22] As in their earlier study, the economists did not advocate closing the Seaway to ocean freighters. They simply analyzed whether such a move made economic sense. That was little consolation to shippers, who were infuriated by the study.

To even raise the possibility of barring ocean ships from the Great Lakes was "absurd," Johnson said. "That's just not going to happen. This [study] is a paper exercise."[23]

Steve Fisher, executive director of the American Great Lakes Ports Association, refused to even discuss the possibility of closing the Seaway. "It's a false idea," he said. "No one except these two professors is even talking about it."[24] Fisher's assertion may have been accurate if he was referring only to officials in the shipping industry. In other circles, the future of the Seaway was a hot topic of conversation and research. No less than the National Research Council, an advisory panel to the U.S. Congress, explored the merits of closing the Seaway. The council's 2008 report reached the following conclusions:

> The only way to eliminate all further aquatic invasive species introductions into the Great Lakes by vessels transiting the Seaway would be to close the waterway to all vessel traffic. Such action, which appears unlikely from a political perspective, would eliminate a trade route into and out of the Great Lakes and would not, therefore enhance the region's potential for global trade.[25]

The council concluded that it made more sense to keep the Seaway open but immediately require all ocean freighters to disinfect ballast tanks before entering the St. Lawrence River. The council was skeptical of the Taylor-Roach studies, but provided no scientific data to discredit their findings. That said, the council's report stated in unambiguous terms that the Seaway was not essential to the North American economy. "Available data suggest it would be hard to posit the continued use of the Seaway as vital to the economic health of North America," the council said in its report. "However, the Seaway may be critical for the continued operations of certain industries in the Great Lakes region, particularly in a time of heightened international competition."[26]

Taylor's response to the National Research Council report was blunt. "The Seaway is an inconsequential, irrelevant player in terms of global trade," he told the *Muskegon Chronicle*.[27]

Shipping-industry officials used the National Research Council report to bolster their argument that the Seaway should remain open. On the other side of the issue, the study did nothing to quiet a growing chorus of critics who said ocean freighters should be banned from the Great Lakes until shippers could prove their vessels were no longer importing foreign species. Krantzberg was among those who continued to speak out on the issue. She made no secret of her disdain for the global shipping industry's reluctance to make the changes needed to end their biological siege of the lakes. Krantzberg said ocean freighters would continue to import foreign species into the Great Lakes until there were laws that mandated zero discharge of biological pollutants from ships. To her, that was unacceptable.

"We need the political will to develop a discharge standard of zero and tell the industry it has five years to figure this out or we will close the Seaway," Krantzberg said. "We can only expect things to get worse. As climate change increases water temperatures in the Great Lakes, the lakes will be more susceptible to invasions by species from other parts of the world. I ask my students, 'What if the next thing to come into the lakes in an ocean freighter is cholera or some other human pathogen? What if it kills humans? Do we have to wait for that to happen before we force change?' If that's what it takes, shame on us."[28]

Unbeknownst to Krantzberg in 2007, one of the invasive species that transoceanic freighters imported into the Great Lakes had already claimed human lives. A foreign mussel discovered 18 years earlier in Lake Erie migrated west, triggering a series of events 2,000 miles away that culminated in the deaths of two men in California. They were the first human casualties of a biological plague that began in the Great Lakes and spread across the North American continent. Would more people die before the United States and Canada forced transoceanic freighters plying the Great Lakes to eliminate the potentially deadly byproducts of global commerce? Time would tell.

Frankly, Californians should be scared to death.

—Tina Swanson, San Francisco Bay Institute scientist, referring in 2007 to quagga mussels invading the West.

23 WESTWARD HO!

Two scuba divers with the California Department of Water Resources entered a canal at the Dos Amigos Pumping Plant on a mild winter morning in 2007. The facility, about 100 miles southeast of San Francisco, was one of 17 massive pumping plants in California's sprawling water collection and distribution system. California siphoned water from mountains and rivers along the Nevada and Arizona borders and transported it through hundreds of miles of canals to slake the thirst of enormous farms and star-crossed cities along the Pacific Coast. Transporting water from the Sierra Nevada mountains and the Colorado River to Los Angeles and other points west was terrifically complex and expensive. But in the West, it was often said that water flowed uphill toward money. Nowhere was that more evident than in Southern California, where water from adjacent states allowed land speculators and water barons to convert a semiarid desert into the metropolis of Los Angeles, entertainment capital of the Western world. The determination of California officials to keep water flowing to bustling cities and massive farms knew no bounds. In February 2007, that resolve collided with human error and the western migration of quagga mussels. The result was a horrifying tragedy.

Divers Tim Crawford and Martin Alvarado entered the water at the base of the Dos Amigos pumping facility at 10:10 A.M. on February 7. Each diver had 30 minutes of oxygen in their air tanks—it was all they needed for what was supposed to be a brief, routine job. Crawford, who had 18 years experience as a scuba diver, led the mission that day. Alvarado, who had been diving for just a year, followed Crawford's lead. The 44-year-old Alvarado, the father of seven children, became a scuba diver to supplement his $9-per-hour salary as a maintenance worker.[1] Their assignment that day was straightforward and relatively simple: Remove quagga mussels from the bars of submerged trash racks that kept debris from slowing the flow of water through the facility's massive pumps. The mussels, which had made their North American debut in Lake Erie in 1989, turned up in 2007 in the 444-mile-long California Aqueduct. Researchers believed that recreational boaters who towed boats from the Great Lakes to western states inadvertently carried quagga mussels over the Continental Divide and deposited them in Nevada reservoirs that supplied California with drinking water. Crawford and Alvarado's job was to prevent quagga mussels from crippling the Dos Amigos facility in the same manner that the invader's cousins, zebra mussels, had shut down the city of Monroe's Lake Erie water intake two decades earlier.

Scuba diving in the California Aqueduct was inherently dangerous—there were strong currents beneath the water's glassy surface. Each of the six pumps at the Dos Amigos facility pumped 2,600 cubic feet of water per second—a volume roughly equivalent to two Olympic swimming pools—up a 130-foot-high hill every minute. All of the pumps were supposed to be shut off before the divers descended to clean the trash racks. The divers were supposed to be equipped with two-way radios that allowed them to communicate with a dive tender at the surface. But on that fateful day, a series of mistakes turned deadly: One of the pumps was mistakenly left operating at full capacity; neither diver had a two-way radio; and their dive tender was not properly trained to ensure the safety of the divers. Ten minutes into the job, the dive tender watched helplessly as Crawford and Alvarado's bubble trails were swept toward the pump that was still operating, sucking water through the trash rack at a rate of five feet per second. There, the bubbles disappeared. The dive tender struck the trash rack with a metal rake to get the divers' attention, but there was no response.[2]

Crawford and Alvarado didn't stand a chance against the fierce underwater current the pump created in the canal. Both men were likely pinned against the trash rack. A rescue diver sent into the canal found Crawford and Alvarado lying face down, five feet apart, on the bottom of the concrete channel. Both were dead. They were pulled from the water two hours after their air bubbles stopped floating to the surface. California's Division of Occupational Safety and Health, Cal/OSHA, cited the Department of Water Resources for six safety violations and fined the agency $15,370.[3] It was a paltry sum considering the magnitude of the situation. The deaths of Tim Crawford and Martin Alvarado were significant for reasons beyond the obvious human tragedy. They were the first Americans to lose their lives to the plague of *Dreissena* mussels—quagga and zebra mussels—swarming North America's lakes, rivers, and water-distribution systems. Their deaths added a new element of horror to the scourge of shipborne invaders spreading across the continent. Nearly two centuries after European immigrants flocked to California to capitalize on the Gold Rush of 1848, another wave of European creatures was invading the West. The potential ramifications were ominous for a state that relied on a vast web of canals and pumps to keep its economy humming and provide drinking water to tens of millions of people. The same concrete canals that irrigated California provided an ideal conduit for the rapid spread of zebra and quagga mussels. The invaders were about to test the mettle of a state defined by its ability to ship water over mountains and across deserts.

EMPLOYEES AT THE LAS VEGAS BOAT HARBOR SHOWED UP FOR WORK ON A Saturday morning in January 2007 with expectations of an unusually busy day. High winds that buffeted the reservoir behind Hoover Dam a day earlier had damaged a floating breakwater that shielded the marina from Lake Mead's fury. Employee Eric Virgin was given the task of repairing cables that anchored floating boat docks to the lake bottom. Virgin wasn't far into the job when he spotted something unusual on one of the cables: a tan mussel, about the size of a fingernail, with stripes on its shell. That was strange—Lake Mead didn't have mussels, at least not of that variety. The harbor staff taped the tiny mollusk to a "Zap the Zebra" brochure, one of those colorful educational pamphlets that warned about the possibility of zebra mussels infesting western lakes. The irony was uncanny. One of the marina employees telephoned Wen Baldwin, a retiree who presided over the Lake Mead Boat Owners Association. Baldwin had spent countless hours of his own time leading a crusade to keep zebra mussels out of the lake. As soon as he saw the mussel, Baldwin was certain he was looking at Lake Mead's worst nightmare: a zebra mussel.[4]

Divers sent down later that day found more mussels clinging to docks, cables, and other hard surfaces in the lake. The evidence was overwhelming: *Dreissena* mussels that had colonized the Great Lakes in the late 1980s had somehow migrated nearly 2,000 miles west, jumped the Continental Divide, and, for the first time, invaded a lake west of Oklahoma. Baldwin and employees at the National Park Service, which managed the Lake Mead National Recreation Area, were defeated, dejected. They had lost the battle to keep zebra mussels out of Lake Mead. Kent Turner, resource-management chief at the National Park Service's Lake Mead office, told the *Las Vegas Sun*: "This is almost heartbreaking news that the mussels are here."[5] It was a very bad day.

The next week was worse. Scientists determined that something worse than zebra mussels had invaded Lake Mead: quagga mussels. Quaggas and zebra mussels were part of the same genus of *Dreissena* mussels, but quaggas could live in a wider variety of conditions—in warm or cold water, and on hard or soft surfaces. Quagga mussels were like zebra mussels on steroids. Quaggas had displaced zebra mussels from much of the Great Lakes by 2007 and intensified the profound ecological changes that zebra mussels had triggered in the lakes. By the time quaggas were discovered in Lake Mead, the mussels had colonized one third of the 225-square-mile reservoir, an area of lake bottom spanning about 30,000 acres.[6] Government officials in the West charged with managing fisheries and providing drinking water suddenly found themselves looking down the barrel of a cannon loaded with a biological bomb.

Biologists had long feared the prospect of zebra or quagga mussels invading Lake Mead or Lake Powell, the reservoir lakes created by damming the mighty Colorado River. The warm waters were ideal habitat for the European mussels. But there was a greater concern—the Colorado River was a primary source of drinking water for 38 million people in southern California, Nevada, and Arizona. Conquering the reservoirs would give the mussels a western beachhead from which they could spread to surface waters across the West, to the shores of the Pacific Ocean, from Mexico to Canada. The aqueducts and reservoirs that transported water from the Colorado River to southern California provided the means by which the mussels could rapidly spread. Managers of southern California's tentacled system of aqueducts and reservoirs knew it would cost millions of dollars annually to keep the mussels from clogging pipes and causing taste and odor problems in drinking water. How did government officials know the foreign mussels posed such a monumental threat to the waters of the West? They had spent a decade trying to keep the menacing invaders out.

WHEN CONGRESS AMENDED THE NONINDIGENOUS AQUATIC NUISANCE Prevention and Control Act in 1996, the changes expanded the scope of the nation's first federal law aimed at preventing the spread of invasive species. The first law had focused largely on regulating ballast water in ocean freighters that entered the Great Lakes. The 1996 version made the law national in scope and placed a greater emphasis on preventing future invasions. The law also established a Western Regional Panel on the Aquatic Nuisance Species Task Force. That panel was given a simple, daunting directive: Keep zebra mussels from spreading west of the 100th meridian, a longitudinal line that passed vertically through the middle of the United States. The 100th Meridian Initiative was an attempt to keep zebra and quagga mussels from invading the great rivers of the West—the Colorado, Yellowstone, and Columbia—and reservoirs that supplied water for tens of millions of people, thousands of industries, and America's most productive agricultural region.

The Western Regional Panel, comprised of some of the nation's top scientists, spent two years drawing up a plan to stop the spread of zebra mussels. It was called "The 100th Meridian Initiative: A Strategic Approach to Prevent the Westward Spread of Zebra Mussels and Other Aquatic Nuisance Species." At its core was a proverbial line in the sand—the 100th meridian. The imaginary line stretched from the southern tip of Texas, through the panhandle of Oklahoma, north through western Kansas, central Nebraska, the Dakotas, and into the Canadian province of Manitoba. It was the line beyond which zebra and quagga mussels could not be allowed to pass—the potential economic and environmental consequences were too great. The panel was resolute on that issue. It said the following:

> The establishment of zebra mussels West of the 100th meridian could devastate water resource projects, raw water users and aquatic ecosystems. If zebra mussels become established in headwater reservoirs, they would likely inhabit/colonize thousands of canals used to transport this water. This infested water would also be pumped to agricultural and municipal areas, thus spreading the mussels over large areas in a relatively short time. In lower elevations and wetter climates, surface water could become contaminated from waters originating at higher elevations. Once this water is transported to downstream receiving areas, they too would become infested.[7]

The 100th Meridian Initiative was the first comprehensive effort by state and federal agencies, Native American tribes, and Canadian officials to stop the western migration of *Dreissena* mussels. The major threat to lakes in the West was not the ocean freighters that infected the Great Lakes with zebra mussels and 60 other invasive species. No, the threat was pleasure boats that could carry the mollusks from infested lakes east of the Mississippi River to mussel-free lakes in the West. It was a formidable task, making sure every boat that was hauled by trailer from east to west was free of zebra and quagga mussels. A boat that spent as little as two days in a lake infested with *Dreissena* mussels could become a carrier of what amounted to a biological plague. Government officials believed that teaching boaters how to keep their boats from spreading *Dreissena* mussels from one lake to another would keep the plague of zebra and quagga mussels from spreading west of the Rocky Mountains. It was a noble but quixotic effort.

The authors of the 100th Meridian Initiative knew before the ink had dried on the 1998 document that zebra mussels, in all likelihood, were already in some lakes in Nevada and Manitoba. A 1994 survey of boats at the U.S.-Canada border revealed that 93 percent of those entering Manitoba came from lakes infested with zebra mussels. A separate study by the U.S. National Park Service concluded that "numerous boats entering Lake Powell were from areas infested with zebra mussels." Between 1993 and 1999, California Department of Agriculture inspectors found live or dead zebra mussels on

The 100th Meridian Initiative

This billboard in Missouri was part of a massive effort to prevent zebra and quagga mussels in the Great Lakes from spreading west of the 100th Meridian and into the western United States.

David Jude, University of Michigan, School of Natural Resources and Environment

Quagga mussels (*shown here*) crossed the Continental Divide on the hulls of trailered boats. The mussels were discovered in Nevada and California lakes in 2007, seventeen years after invading the Great Lakes.

18 boats that were trailered into the state.[8] Still, zebra mussels had not been detected in any lake or river west of the Rocky Mountains prior to 2007. The 100th Meridian Initiative report warned that an invasion was likely if boaters did not clean up their act. "Except for Oklahoma, zebra mussels have not been detected in open waters of the West," the report said. "However, without effective prevention measures, their invasion into the West is a real and imminent possibility."[9]

How did quagga mussels travel 1,700 miles from the Great Lakes to Lake Mead? Simple: The menacing pests could live out of water for up to 10 days, provided they were in a shady, damp area. They also fared well in bilge water, engine cooling systems, and live wells used to store fish. Zebra and quagga mussels also could be stealth invaders. Newborn mussels, called veligers, weren't visible to the naked eye. And since it took less than a week to trailer a boat from the Midwest to anywhere in the continental United States, it was only a matter of time before *Dreissena* mussels hitched a ride over the Continental Divide and invaded western lakes. Stopping their spread was almost beyond comprehension: Each female mussel produced up to one million offspring, or veligers, per year. The veligers didn't swim, but could spread rapidly by drifting with water currents. Once attached to a stable surface, the mussels grew for 12 to 18 months—filtering vast quantities of water and hogging valuable plankton that fish and other aquatic organisms need to survive—before reproducing. It didn't take a rocket scientist to figure out how a single boat infected with *Dreissena* mussels, or veligers, could infect a lake with the seeds of ecological devastation.

In a desperate attempt to prevent that very scenario, authors of the 100th Meridian Initiative drafted a plan of attack that included the following: inspecting all pleasure boats headed west from the Great Lakes region; distributing thousands of informational flyers warning about the problems caused by zebra mussels; and keeping infested boats out of places like Lake Mead. Inspectors fanned out to rest areas along 11 east-west highways that crossed the 100th Meridian. There, they conducted voluntary inspections of pleasure boats being towed west. Inspectors also were stationed at several boat launches on western lakes and given strict orders to thoroughly examine boats from mussel-infested lakes. The effort paid some dividends. Several mussel-encrusted boats were kept out of western lakes, and others were intercepted before crossing the 100th Meridian. In one of the most famous cases, a fisheries biologist from the Oklahoma Department of Wildlife Conservation was traveling west on I-40, in eastern Oklahoma, when he passed a semi-truck hauling a pleasure boat. The biologist, Hutchie Weeks, was shocked by what he saw: the boat and other equipment on the trailer were covered with zebra mussels. A subsequent report on the incident called it "the 100th Meridian Initiative's worst nightmare."[10]

The 100th Meridian Initiative also established rapid-response teams that would be sent into areas where zebra or quagga mussels had been discovered. Their job was straightforward, according to the plan: "Eradicate or contain zebra mussels immediately following detection."[11] It was an ambitious, slightly naive plan that would be tested in the waters of Lake Mead.

THERE WERE OMINOUS SIGNS IN 2004 THAT *DREISSENA* MUSSELS WERE on the verge of infesting western lakes, if they hadn't already. Inspectors checking a boat that year at a rest area on I-90, along the Idaho-Washington border, found live zebra mussels on the trim tabs of a 38-foot cabin cruiser being hauled from Tennessee to Washington's coast. Adult zebra mussels were found 17 days later on the hull of a 54-foot houseboat docked at Lake Mead's Temple Bar. That boat came from a marina on the Ohio River in Kentucky, a known hotspot for zebra mussels. Two months later, in July 2004, mechanics at a repair shop on Lake Mead were working on a boat from Chicago when they found zebra mussels on both outdrives.[12]

There was little doubt that it was only a matter of time before the *Dreissena* mussels hit the fan in waters of the West. The discovery of quagga mussels at the Las Vegas Boat Harbor in January 2007 merely confirmed the inevitable. The finding also changed the focus of the battle against zebra and quagga mussels. The campaign to deny the invaders access to western lakes morphed overnight into a frantic effort to keep quagga mussels from clogging water intakes and disrupting lake ecosystems in one of the most populous, water-starved regions of North America. Stunned by the Lake Mead discovery, government officials scoured reservoirs, water aqueducts, and fish hatcheries in Nevada, Utah, Arizona, and California looking for *Dreissena* mussels. They found the unwelcome guests nearly everywhere they looked. Controlling their spread was going to be an uphill race, and the mussels had a huge head start.

Within three weeks of the first quagga-mussel sighting in Lake Mead, the mollusks were found in other areas of the lake and further downstream—at Hoover Dam, and in Lakes Mohave and Havasu. The discoveries sent shock waves of concern through the halls of water-management agencies across the West. "The implications of this showing up in Lake Mead and the Colorado River system are enormous," said Allen Biaggi, director of the Nevada Conservation and Natural Resources Department.[13] Six weeks after Biaggi offered that assessment, quagga mussels were found at the Willow Beach National Fish Hatchery in Arizona, where the federal government reared trout and salmon 10 miles downstream of Hoover Dam. The quagga mussel invasion had erupted into a biological wildfire, sweeping across the West faster than anyone had ever thought possible.

A California science advisory panel released a report assessing the threats *Dreissena* mussels posed to the nation's most populous state, which also supported the world's eighth largest economy. The panel said:

> The significance and potential impact of this event cannot be overstated. Zebra mussels are harmful fouling organisms: they attach by the millions to submerged objects, fill and block water pipes and clog protective screens. Zebra mussels are efficient filter feeders: they strip food from the water that is needed to sustain other aquatic life. Direct economic costs are on the order of $100 million a year in eastern North America; unquantified secondary and environmental costs could be substantially larger. Impacts in California and the West could be as great or greater than those in the East.[14]

The advisory panel's report added that quagga mussels would most certainly spread to lakes and municipal water systems across the West if the pests weren't promptly eradicated from Lake Mead. "If not eradicated or contained, these populations will seed secondary invasions across western North America," the panel said. "The potential impacts include hundreds of millions to billions of dollars in direct economic costs."

Scientists concluded that the chemistry of surface waters in much of the West was more hospitable to zebra and quagga mussels than the Great Lakes and other lakes in the eastern United Sates. They also warned that California's massive economy was at greater risk of disruption because it was more dependent on water piped long distances than industries around the Great Lakes. The Bonneville Power Administration, which operated 13 hydroelectric dams in the Columbia River in Washington, warned in 2005 that a zebra mussel infestation could cost the utility $23 million at the outset and another $100,000 annually in maintenance costs.[15]

California officials faced a potential environmental and economic calamity. Knowing that, the state's Science Advisory Panel in May 2007 urged federal officials to ban boats from Lakes Mead, Mojave, and Havasu, all of which were infested with quagga mussels. The panel urged an aggressive counterattack. Its members recommended draining most of Lake Mead and pumping chemicals into parts of the lake—potassium chloride, copper sulfate, or sodium hypochlorite—to kill the quaggas. The panel warned that treating the infested lakes with chemicals and banning boats "may cause unacceptable short term interferences and interruptions to local economic activities, including tourism, fishing and recreation." But doing nothing "would inevitably result in large, long term environmental and economic costs across a large swath of western North America. That is the stark choice we face," the panel said in its May 2007 report.[16]

The moment for difficult decisions was at hand. The situation was eerily similar to one that unfolded two decades earlier, when biologists recommended using chemicals to eradicate Eurasian ruffe that had invaded Lake Superior's tributaries. The responses in both cases also were strikingly similar: government agencies did next to nothing.

Instead of attacking the quaggas with chemicals, state and federal officials ignored the California Science Advisory Panel's recommendations. Boats were not banned from Lakes Mead, Mojave, or Havasu, and there were no immediate efforts to drive quagga mussels out of those lakes. Instead of taking bold, decisive action, state and federal officials took the path of least resistance. They increased boat inspections at the infested lakes and ramped-up monitoring of a mussel invasion that was rapidly spreading beyond the point of control. It was a timid response to a clear and present danger that was bearing down on the environment and water-dependent economies of the West. Andrew N. Cohen, one of the scientists on the California Science Advisory Panel, said he was disappointed that federal officials did not move quickly to attack quagga mussels in Lake Mead. Cohen said the infestation was such a low priority among federal officials in Washington, D.C., that National Park Service officials in Nevada never had a chance to strike back at the mussels.

"If we wanted to badly enough, there are things we could do to get mussels out of Lake Mead," Cohen said in 2007. "But the people responsible were never given an opportunity to do that. I wouldn't suggest that they did nothing, but I'm baffled as to why the initial response wasn't more aggressive than it was. The first thing you do is contain the invasion and then figure out what to do after that, but none of that happened in the case of Lake Mead."[17]

Cohen's comments weren't the rantings of an environmental zealot. The senior scientist at the San Francisco Estuary Institute knew well the risks quagga mussels posed to aquatic environments and water distribution systems. Cohen was formerly director of the East Bay Municipal Water District, on the east shore of San Francisco Bay. "Quagga mussels and zebra mussels are quite a nightmare for people who run agencies that deliver water to people," he said.

The state of California tried, on its own, in early 2007 to slow the spread of quagga mussels. Agricultural inspectors began checking for *Dreissena* mussels on all trailered boats entering California. Similar inspection programs were initiated at several lakes and boat launches along the California-Nevada border. But, faced with limited funds and insufficient personnel to inspect all the places quaggas could have colonized, government agencies instead turned to the public for help. Officials in California and Arizona, along with the National Park Service, urged boaters to clean boats after taking them out of lakes. They asked scuba divers to report possible mussel sightings and established a toll-free phone number people could call to report mussel sightings. "We are already casting as wide a net as possible to stop the spread of quagga mussels into California waterways," said Ryan Broddrick, director of the state's Department of Fish and Game, in a January 2007 press release.[18] His was tough talk that made for good newspaper copy and sound bites on the nightly news. But his comments were, at best, ill-informed. The history of quagga mussel invasions in the Great Lakes suggested that it was impossible to keep the mollusks from spreading once they had invaded a large body of water. The best anyone could expect to achieve was slowing the rate at which quaggas spread from one waterway to another. Still, state and federal bureaucrats projected a strong, united front in the emerging war on quagga mussels. Bureaucrats and politicians wanted to show the public that they were on top of a problem that, in all likelihood, could not be controlled by human intervention.

There were only two documented cases, prior to 2009, of invasive mussels being purged from an aquatic ecosystem. Australia successfully eradicated the black-striped mussel—a species similar to the zebra mussel—after it was discovered in Darwin Harbor in 1999. The Australian government dumped 6,000 metric tons of an aquatic pesticide into the harbor eight days after discovering the foreign mussel. The treatment worked; the mussel was eliminated.[19] Australia took that bold step to keep the mussels from clogging industrial and municipal water intakes and threatening the nation's $225 million pearl industry. Canadian biologist Anthony Ricciardi, an invasive-species expert, said in a 2003 interview that the United States and Canada could have learned much from Australia's rapid, aggressive response to finding the invader. "This is an example of something we can emulate," Ricciardi told the *St. Catherines Standard*. "But as it stands now in Canada, we couldn't get a meeting together in eight days."

The only successful eradication of zebra or quagga mussels in North America was in Virginia. There, the state pumped 174,000 gallons of potassium chloride into the 12-acre Millbrook Quarry to kill zebra mussels. The treatment succeeded; the mussels were eliminated. Whether the Millbrook Quarry project could be replicated in Lake Mead was questionable. The quarry was a drop in the bucket compared to Lake Mead, a sprawling, deep reservoir that spanned 225 square miles of surface area when full.

The notion that tough-talking bureaucrats could control quagga mussels was laid bare a week after Broddrick's bold pronouncement, when the mollusks were found in Lake Havasu. Five months later, divers discovered that quagga mussels were multiplying at a phenomenal rate in a water-intake structure near the Arizona border. The density of mussels on equipment at the water intake increased a hundredfold between February and June of 2007.[20] That same month, officials shut down the Colorado River Aqueduct, which carries water to Southern California, after quaggas were found inside pumps at the Eagle Mountain and Hinds pumping plants. California officials idled the aqueduct for 10 days to inspect the concrete and steel canal. They found mussels throughout the 242-mile length of the aqueduct, which prompted chlorine treatments at selected locations to kill the invaders. Quagga mussels were at California's doorstep, and it seemed there was little anyone could do to keep the invaders from busting down the door.

The pernicious mussels turned up in two Southern California reservoirs in October 2007, nine months after they were discovered in Lake Mead. The city of San Diego purchased a remote-controlled submarine to inspect its reservoirs and water-intake facilities; the city also banned recreational boats from some reservoirs. There were no signs that the quaggas were retreating. It was painfully obvious one year after the mussels were discovered in Lake Mead that western states were headed for major challenges as the mollusks spread from one lake to another like a fast-moving infection.

There was more bad news in January 2008, when scientists discovered zebra mussels in two lakes in Colorado and California. The mussels were found in Colorado's Lake Pueblo, and in a reservoir in Hollister, California, that served huge farms. California officials expressed dismay at finding zebra mussels in the San Justo Reservoir, an artificial lake that was used to irrigate the lucrative fruit and vegetable farms south of San Jose. State officials acknowledged that there was little they could do to eradicate the mussels. Arman Nazemi, assistant San Benito County public-works director, told the *Los Angeles Times*: "It's not good news. If they're as invasive as they say, it could be a nightmare for our infrastructure."[21]

THE RUNAWAY MUSSEL INVASION MADE ME WONDER WHETHER THE GOVERNMENT's 100th Meridian Initiative, the most ambitious effort ever launched to slow the spread of zebra and quagga mussels, was a failure. David Britton, an aquatic invasive-species coordinator for the U.S. Fish and Wildlife Service who worked on the initiative, said it would be wrong to call the program a failure just because *Dreissena* mussels had successfully invaded the West. Keeping the mussels out of the West was not the program's only goal, he said. The initiative also aimed to keep other invasive species in the Great Lakes and other waters, including Asian carp in the Mississippi River, from spreading to the West. Britton said: "I absolutely do not consider the 100th Meridian Initiative a failure. The second goal clearly indicates that 100th Meridian Initiative partners were well aware that zebra mussels [and other aquatic nuisance species] could make it into the West. Our hopes were, of course, to prevent any invasion in the West, but such ideals are rarely achieved. For years, we have been interdicting boats from launching at ramps throughout the Western U.S. Nevertheless, there is still plenty of West left to protect."[22]

Britton said the *Dreissena* mussel invasion of the West illustrated the difficulty of getting government bureaucracies, which were structured to react to crises, to shift gears and focus on preventing biological pollution before the train of invasive species became a runaway locomotive. "Many understand the potential issues but have not taken our warnings as seriously as some of us might have wished," he said. "Part of the problem may be that we are flooded daily with countless impending crises (e.g. SARS, terrorists, West Nile virus, melting polar ice caps, out of control gas prices, *E. coli*, AIDS, obesity, bird flu, etc.) It is difficult for any of us to weed through the ever-growing list of agents of impending doom. Thus, prioritization is difficult and errors are often made. Many states are willing to roll the dice if there is even the smallest chance that they won't have to spend money on aquatic nuisance species issues in a particular fiscal year."[23]

Enemies at the Gate

Ocean freighters weren't the only source of invasive species in the Great Lakes. Sometimes, foreign creatures were dumped in the lakes—accidentally or intentionally—by anglers or aquarium owners. Occasionally, government agencies were the culprits. Asian carp, which were on the brink of invading Lake Michigan in 2008, were a classic case of good intentions gone awry.

The United States Fish and Wildlife Service imported Asian carp to Arkansas in 1963 to control algae in fish-farm ponds. The project was a direct response to the 1962 publication of *Silent Spring*, Rachel Carson's landmark book about the many dangers of rampant chemical use. Andrew Mitchell, a government researcher, told the *Milwaukee Journal-Sentinel* in 2006 that federal officials who imported Asian carp to the United States "were motivated by the greatest environmental concerns. They were acting in a very aggressive and appropriate manner for the time."

Two decades later, Asian carp escaped from private fish farms in Arkansas when the Mississippi River flooded. Thus began the northern migration of two species of Asian carp—bighead and silver carp—up the Mississippi and Illinois rivers. The fish were within 40 miles of Lake Michigan and, by extension, all five Great Lakes in 2008. Bighead carp grew to 150 pounds and hogged vast quantities of plankton that other fish species needed to survive. Silver carp, which could reach 40 pounds, leapt out of the water when boats passed by. Several boaters on the Mississippi and

U.S. Fish and Wildlife Service

Silver carp leap out of the water behind a boat on the Illinois River. The fish were one of five species of Asian carp imported to the southern United States in the 1960s. They escaped into the Mississippi River in the 1980s and began migrating north, toward the Great Lakes. The fish were within fifty miles of Lake Michigan in 2008.

Illinois rivers were injured when Asian carp rocketed out of the water. Some victims suffered concussions; others sustained broken bones.

A 2005 report by the Fish and Wildlife Service warned of more casualties as silver carp colonized more lakes and rivers. The agency said: "Injuries to humans from jumping fish will continue and may increase with silver carp populations, and human deaths may possibly occur."

The four species of Asian carp—bighead, silver, grass, and black—could pose a serious threat to Great Lakes fisheries if they invaded. The fish consume huge quantities of plankton at the base of the food chain. "The worst case scenario for the Great Lakes, if you compare it to what's happened in the Mississippi River, is pretty frightening," said Marc Gaden, a spokesman for the Great Lakes Fishery Commission. "This is a species that spreads quickly and is a feeding machine at the lower end of the food web—they take out all the other things fish eat."

The only thing standing between Asian carp and the Great Lakes was an electric fish barrier in the Chicago Sanitary and Ship Canal, 30 miles south of Chicago. The man-made canal, built in 1900 to reverse the flow of the Illinois River so Chicago's sewage would flow toward the Mississippi River instead of Lake Michigan, was the most likely conduit for Asian carp to invade the Great Lakes.

The U.S. Army Corps of Engineers installed the electric barrier in the canal in 2002. Its original function was to keep round gobies in Lake Michigan from migrating into the Illinois and Mississippi rivers. But gobies moved faster than the federal bureaucracy. The invasive fish escaped Lake Michigan and were 170 miles down in the Illinois River, halfway to the Mississippi, by 2007. Asian carp threatened to return the favor. The only obstacle was the electric fish barrier, which was experimental and designed to last three years. That barrier was still in service, barely, five years later. Only half of its electrodes worked. The Corps of Engineers built a new fish barrier in 2006, but the device failed safety tests. It remained idle in 2008 as Asian carp surged toward Lake Michigan like an approaching army.

The Asian carp's conquest of the Mississippi and Illinois rivers prompted a small-town bar manager in Bath, Illinois, to organize a fishing tournament in 2004 to kill as many silver carp as possible. Betty DeFord called the whimsical event the Redneck Fishing Tournament. By its third year, the event was attracting hundreds of anglers and national media attention. The tournament also produced hilarious videos, many of which ended up on the Internet, of anglers walloped by jumping fish.

Biologists said there was no hope of eradicating Asian carp from lakes and rivers it invaded. A single bighead carp could produce two million eggs per year. Experts predicted it was just a matter of time before Asian carp snuck into Lake Michigan and infested the many warm, murky rivers that flowed into the Great Lakes. A Canadian commercial fishing crew working on Lake Huron in July 2007 brought up a three-foot-long grass carp in one of its nets. "It's a big thing; it's very scary," said Tim Purdy, vice president of Purdy Fisheries, to the *London (Ontario) Free Press*. The fourth-generation commercial fisherman predicted that Asian carp would someday wreak havoc on Great Lakes fisheries. "They're devastating. Once they go through an area, there's nothing left," Purdy told the newspaper. "There's no perch left, there's no pickerel, there's nothing."

A 2001 study by scientists at the University of Texas-Arlington questioned the effectiveness of the 100th Meridian Initiative's massive boat-inspection program. Between 1998 and 2000, crews inspected 724 boats, but found only one obvious case of live zebra mussels being transported on a vessel. "Survey results suggested that the transport of zebra mussels by recreational boaters may occur unpredictably and at low levels, and that programs relying heavily on inspections and removal to control aquatic nuisance species dispersal into western states may prove ineffective," according to the study.[24] Authors of the study said it might have made more sense to simply prohibit people from hauling boats from one state to another. Boaters and anglers would likely view such a suggestion as heresy, an intrusion on their right to pursue happiness wherever and however they saw fit. Respected scientists suggested privately that the only sure way to prevent other invasive species from spreading across North America was to restrict the movement of their primary carriers: boats.

Given the widespread movement of boats and commerce across the continent and around the globe, scientists believed other invasive species in the Great Lakes would eventually spread to the West Coast. Chuck O'Neill, an invasive-species expert with New York Sea Grant, made a famous prediction about the fate of zebra mussels. O'Neill told a newspaper reporter he believed the mussels would eventually spread across the North American continent. "Ultimately, I think it will happen. I don't think there is any way to contain them in areas they started out in." He made that comment in 1991, sixteen years before quagga mussels turned up in Nevada and California.[25]

The cold, hard reality was that the natural barriers that kept species in their native environments for millennia were bypassed by the instruments of modern society: vehicles, ships, pleasure boats, airplanes, dams, canals, and bridges. The Rocky Mountains were no longer a barrier to *Dreissena* mussels moving from the Great Lakes to California, just as the Atlantic Ocean was no barrier to the mussels migrating from Europe to North America. Human ingenuity created this unnatural menagerie, and we are now suffering its consequences. University of Notre Dame biologist David Lodge told the *Paonia (Colo.) High Country News* in 2007 that Westerners "could be forgiven for thinking that they don't need to care about invasions in the Great Lakes, but it's clear that the desert is no barrier, and the Great Plains are no barrier, at least not an insurmountable one. The invasions from the Great Lakes are not going to stop with quagga mussels, unless effective policies are put in place."[26]

Cohen said the 100th Meridian Initiative helped educate boaters about the dangers of spreading zebra and quagga mussels, launched some useful monitoring programs, and possibly slowed the quagga's spread over the Rocky Mountains. But the initiative ultimately failed to meet its goal of keeping the mussels out of the West. And Cohen said the public, media, and politicians in

California and federal bureaucrats had not, as of 2008, grasped the significance of quagga mussels colonizing the West.

"This issue hasn't risen to a high enough level in the federal government because it's only received light press coverage," Cohen said. "If the press isn't paying attention, the politicians don't pay attention and if the politicians don't pay attention the government agencies don't pay attention. In all the articles I've seen, not one has quoted a representative of an environmental group saying this is a major environmental issue. I don't think the environmental community [in the West] is paying attention to this issue, for some reason."[27]

Perhaps the *Dreissena* mussel invasion didn't resonate in California or Nevada because the mussels had not, as of 2008, caused a catastrophic water loss as they had in Monroe, Michigan, two decades earlier. Cohen figured it was only a matter of time before quagga mussels caused a water crisis in the West. Quaggas also had the potential to eliminate many species of native mussels and fish in lakes and streams across much of California. The only area of the state immune to a quagga mussel invasion was the Sierra Nevada Mountains, where the water lacked adequate amounts of calcium to support the formation of mussel shells. In most of California's waters, the floodgates were wide open to the looming quagga mussel invasion.

The Great Lakes faced a similar situation in 2008. Two decades after zebra and quagga mussels invaded the lakes, the U.S. and Canadian governments still had not closed the door on invasive species sneaking into the lakes in the bellies of ocean freighters. Their failure to solve the ballast water problem put the entire continent, not just the Great Lakes, at risk for future invasions by foreign aquatic organisms. The dire situation prompted a park ranger in northern Michigan to take matters into her own hands to protect her island paradise from a deadly new invader. The superintendent at Lake Superior's Isle Royale National Park proved that there was a simple way to disinfect ballast water—the most common source of foreign species plaguing the Great Lakes—provided there was a will to do it.

VHS is the most important and dangerous fish virus known worldwide. Its discovery in our fresh water is disturbing and potentially catastrophic.

—Jim Winton, chief of fish health, U.S. Geological Survey

24 SAVING PARADISE

A thin layer of fog hovered above the surface of Lake Superior by the time I set out on a five-mile hike to the top of Greenstone Ridge, the basalt spine of Isle Royale National Park. The archipelago in northwest Lake Superior was one of the Great Lakes' most serene places. Accessible only by boat or seaplane, Isle Royale offered visitors a true nature experience. There were no wheeled vehicles of any kind, no towns, barking dogs, or television. Electricity was only available at a couple of hotels that National Park Service concessionaires operated for those visitors unwilling to go without modern conveniences. Elusive wolves that preyed on moose ruled the island ecosystem. Native lake trout ruled the waters that enveloped the islands. At dawn and dusk, it was more common to hear the haunting call of a loon than the voice of another human being. Solitude was the rule—not the exception.

The trail to Greenstone Ridge snaked through dense stands of spruce, birch, and poplar devoid of man-made structures; across soggy marshes; and over babbling brooks shaded by a thick canopy. I hiked in the shade for a couple of miles before reaching the rocky incline that delivered me to Mount Franklin. The view from the top of the island, 1,074 feet above Superior's cerulean water, was breathtaking. Looking to the north, beyond Pickerel Cove and Amygdaloid Island, my thoughts drifted from the rocky balcony to what lived in Superior's depths. The 475 square miles of clean, cold water within the boundaries of Isle Royale National Park supported 12 species of lake trout, including the rare coaster brook trout. Superior's lake trout fishery was the only one to fully recover from the carnage sea lamprey inflicted on Great Lakes native fish in the mid-1900s. The largest of the Great Lakes was widely considered the healthiest of North America's five freshwater seas at the dawn of the twenty-first century.

The question was whether Superior could hold onto that distinction in the face of increasing threats from new invasive species. As I soaked up the lake's resplendent beauty during my 2005 visit to Isle Royale, scientists 400 miles away were unraveling an ecological nightmare that would put the park's treasured lake trout fishery at great risk. Efforts to protect the fishery would thrust Isle Royale's park superintendent into the raging debate over how best to prevent a deadly fish virus from infesting Lake Superior.

SPRING 2005 BEGAN LIKE SO MANY BEFORE IT IN THE VILLAGES THAT LINED Lake Ontario's sprawling Bay of Quinte. Anglers stored ice-fishing huts and put boats back in the water as the lake released its icy grip. For many in the region known as Quinte Country, the warm months were spent fishing, sailing, and canoeing the huge, sheltered bay on Lake Ontario's north shore. But that year would be different. Overnight, portions of the tranquil bay were transformed into mass graves for dead fish. A massive fish kill blanketed parts of the lake with freshwater drum. Tons of dead fish were pulled from the lake—countless more sank to the bottom. A large fish die-off was always a ghastly site. But the fish kill that year was unlike anything residents of the area, or scientists, had ever observed.

The dead drum, which normally ranged in color from light gray to pale bronze, had red marks along the edges of their scales. Blood was pooled in their bulging eyes and around the gills. Some had large red spots on their sides. The fish appeared to have bled to death. In a way, they had. Scientists at the Ontario Ministry of Natural Resources determined that the fish had died from viral hemorrhagic septicemia, or VHS, one of the world's deadliest and most feared fish viruses.[1] The Bay of Quinte fish kill was the first confirmed case of VHS in the Great Lakes. Many more would follow.

The following year, VHS killed hundreds of thousands of fish in Lake Ontario, Lake Erie, Lake St. Clair, and the Detroit River. Countless tons of dead drums, yellow perch, walleyes, muskellunge, and smallmouth bass piled up on beaches. Windrows of dead fish washed up on Lake Erie beaches in Ohio. All had the telltale signs of VHS: blood in the eyes and around the gills, red spots on their sides. Some scientists called the virus, which multiplied so quickly it caused blood vessels to burst, the fish version of Ebola virus. Ebola virus was an often fatal ailment that killed humans by causing them to hemorrhage and, in many cases, bleed out. VHS didn't affect humans, but its effects on fish were devastating. The virus was known to kill up to 80 percent of fish in contained environments, such as hatcheries. Scientists feared that its rapid spread across the Great Lakes could decimate the lakes' $7 billion sport and commercial fisheries.

The discovery and spread of VHS prompted biologists at the Michigan Department of Natural Resources to reexamine a dead muskellunge that was pulled from Lake St. Clair in 2003. The fish has been stored in a freezer. Biologists who reexamined it found signs of VHS. No one knew, at the time the fish died, that the disease was in the Great Lakes. The Michigan fish that tested positive for VHS meant the virus had become active in the lakes in 2002 or 2003. By the end of 2006, VHS spread to Lake Huron. It killed fish from 40 different species in three of the five Great Lakes that year, including several of the most valuable species: whitefish, walleye, muskellunge, lake trout, salmon, and perch. The deadly plague surprised and terrified anglers, biologists, and policymakers. "I don't want to cry wolf or say the sky is falling, but we have to wonder if it is," said Bill Horns, Great Lakes fisheries coordinator for the Wisconsin Department of Natural Resources, in a January 2007 newspaper interview.[2]

Six months after Horns made that comment, VHS was discovered in fish from the Lake Winnebago chain of lakes, about 25 miles south of Green Bay,

Wisconsin. The virus was discovered later that year in fish in an inland Michigan lake. And in 2008, VHS killed round gobies in western Lake Michigan. Superior was the only Great Lake, as of 2008, to avoid the wrath of VHS.[3]

Worried that the virus might spread beyond the Great Lakes, the U.S. Department of Agriculture in 2006 issued an emergency order restricting the movement of baitfish and other live fish across state borders. The USDA hoped to keep VHS from getting into the Mississippi River and spreading to fisheries from coast to coast. The virus posed a potentially devastating threat to the nation's aquaculture industry, which raised baitfish and larger species for sale in stores and restaurants. There was no way of keeping VHS from spreading to all parts of the Great Lakes. The virus spread from fish to fish, through the water, in bodily fluids, and when healthy fish ate sick fish. Fish could absorb the virus simply by swimming into VHS-infected water. There were only three possible ways to slow the spread of VHS in the Great Lakes: restrict shipments of fish from one lake to another; force freighters to disinfect ballast water to prevent the vessels from spreading the virus; or prohibit ships from taking on ballast in VHS-contaminated waters.

Biologists feared that the virus could become the most serious threat to Great Lakes fish populations since sea lampreys had invaded a half century earlier. Lampreys were eventually controlled through the use of chemicals. There was no known way to control or eradicate VHS once it infected a lake. "It's pretty obvious this is an epidemic, even if it isn't official," said Paul Bowser, a Cornell University professor of aquatic animal medicine, in a 2007 interview.[4]

The source of VHS in the Great Lakes was never confirmed. Some scientists believed a transoceanic freighter imported the virus. That theory was based on evidence that the strain of VHS found in Great Lakes fish was identical to one found in the Gulf of St. Lawrence, near the Canadian maritime province of New Brunswick. That region, where the St. Lawrence River merged with the North Atlantic, was the point of entry for all ocean ships headed for the Great Lakes via the St. Lawrence Seaway. Though ocean freighters were obvious suspects, and a convenient target for those looking to assign blame, scientists conceded that birds or anglers could have carried VHS into the lakes. The question of whether ocean freighters imported VHS was less important in 2007 than figuring out how to halt the avalanche of disease bearing down on Lake Superior.

THAT LAKE SUPERIOR'S FISHERY HAD DODGED THE VHS BULLET FOR THE first few years the virus was in the Great Lakes was little consolation to scientists who understood invasive species and, in particular, how VHS thrived in cold water. Superior was, by far, the largest, deepest, and coldest of the Great Lakes. Spanning 31,700 square miles of surface area, the lake known to early French explorers as Le Lac Superieur—Upper Lake—was inconceivably huge for a body of fresh water. So massive was Superior, the water from Lakes Michigan, Huron, Erie, and Ontario could fit into its basin with room to spare. Not only was Superior huge on its surface, it had an average depth of 483 feet. Its deepest point, off Pictured Rocks National Lakeshore in northern Michigan, plunged to a depth of 1,332 feet. The Sears Tower in Chicago could be placed upright in the lake's deepest hole and only the top of the building and its large antennae would break the surface. The lake's massive size, depth, and location above the 45th parallel—halfway between the equator and the North Pole—contributed to Superior's frigid water temperatures, which rarely exceeded 60 degrees in the summer.

Scientists theorized that the lake's cold water served as a natural defense against some invasive species from warm-water ports in Europe, Asia, and Africa. That hypothesis was based on the fact that Superior had fewer invasive species than the other lakes, even though its largest port, Duluth-Superior Harbor, was the ballast-water-dumping capital of the Great Lakes. Whether it was due to water temperature or chemistry, Lake Superior fared much better in the modern era of invasive species than the other lakes. But VHS played by different rules—it flourished in cold water that ranged from between 40 and 59 degrees Fahrenheit. That put the world's largest lake in the cross hairs of one of the world's deadliest fish viruses.

Something had to be done to keep VHS out of Lake Superior's waters. But by whom, or which government agency? Who would step forward and take up the battle to keep VHS from infecting the greatest of the Great Lakes? The U.S. Coast Guard and Environmental Protection Agency were the most likely candidates to call in the cavalry to keep VHS out of Lake Superior. But neither agency issued a call to arms. No, the battle to keep VHS out of Lake Superior would be led by an unlikely trio of crusaders: a national park ranger, an engineering professor, and a retired government chemist. Collectively, they had the brains and bravado to challenge the status quo on invasive species. They developed a simple, inexpensive method for treating ballast water.

Theirs was a process that shipping interests and federal officials had long ago dismissed as ineffective, harmful to ships, and too dangerous to use.

GARY GLASS SPENT THREE DECADES WORKING AS A CHEMIST FOR THE EPA before semi-retiring to academia. Much of his work focused on ways to reduce the amount of chemical pollutants that cities and industries dumped in the Great Lakes. Based at the EPA's Great Lakes lab in Duluth, Minnesota, Glass conducted science that contributed to some of the lake's greatest environmental improvements since 1970. His research helped force an iron-ore mine to remove tons of asbestos-laced mine tailings from Lake Superior, contributed to a ban on polychlorinated biphenyls, and forced coal-fired power plants to reduce their mercury emissions. The EPA honored Glass with a science achievement award for research that forced power plants to reduce emissions of compounds that cause acid rain. For Glass, working to protect the Great Lakes was more than a career. It was a mission. He grew up on the shores of Lake Superior, a short walk from the channel where freighters entered Duluth Harbor. Glass would not allow Superior's magnificent ecosystem to succumb to VHS without a fight.

Glass became the face of the Izaak Walton League of America's campaign to force all freighters entering Lake Superior to disinfect their ballast water tanks with chlorine. It was ironic that a scientist who had spent a career working to reduce chemical discharges found himself advocating the use of more chlorine in the Great Lakes region. But Glass believed chlorine was the cheapest, quickest way to prevent VHS from sneaking into Lake Superior in the bowels of freighters that inhaled ballast water from other lakes where the virus was prevalent. At a 2007 press conference, during which the Izaak Walton League called for new ballast rules for ships entering Lake Superior, Glass said scientists had long ago identified a magic bullet for the war on shipborne invasive species. The problem was that the weapon wasn't being utilized. EPA scientists had demonstrated three decades earlier that chlorine could be used safely to kill a wide range of organisms, regardless of the setting. "The EPA lab did some research in the early '70s to deal with disinfection of diseases and how you would remediate any negative impact from that," Glass said. "The chlorination process, which documents over 100 years of history in dealing with disinfection of diseases, is one of the approaches that can be used to treat for this disease in any kind of ballast water that might carry it."[5]

Glass maintained that for a few thousand dollars per ship, every freighter entering Lake Superior could be equipped with a chlorine injection system and an accompanying device to neutralize the chlorine before the treated ballast water was discharged into the lake. Chlorine was a tried and true killer of a wide array of aquatic organisms and pathogens. Thousands of cities in the United States and other nations used chlorine to purify drinking water and kill pathogens in sewage treatment facilities. Removing chlorine from treated water also was a relatively simple chemical process, Glass said. Other nations, including Argentina and New Zealand, required transoceanic freighters to disinfect ballast water with chlorine before entering their waters. Glass also noted that some 3,500 freighters and cruise ships worldwide used chlorine to disinfect drinking water and engine-cooling water, and to disinfect sewage on board the vessels. Still, his demand that freighters be required to disinfect ballast water with chlorine before entering Lake Superior fell on deaf ears at the U.S. Coast Guard.

Coast Guard officials who had been studying potential ballast water treatment systems for nearly a decade dismissed the chlorine option years before the VHS scare. Bivan Patnaik, an environmental-affairs manager at Coast Guard headquarters in Washington, D.C., told a radio news program in 2007 that using chlorine to disinfect ballast water could cause other problems. "I think it's been found that chlorine is very effective in killing a variety of critters, but there is some concern with the use of chlorine in ballast water tanks," Patnaik said. "That is, the chlorine could possibly corrode those tanks."[6] The Coast Guard had no intention of requiring freighters to disinfect ballast tanks with chlorine.

Glass and others from the Izaak Walton League appealed to state and federal lawmakers. They sought legislation requiring ships to chlorinate ballast tanks before entering Lake Superior. Their efforts ran into brick walls at every turn. Meanwhile, freighters continued to stream into Lake Superior, dumping about one billion gallons of untreated ballast water annually into Duluth Harbor. Any one of those freighters that took on ballast water at ports on Lakes Ontario, Erie, Huron, or Michigan had the potential to transport VHS to Lake Superior. Those ships, Glass said, were playing a form of roulette with Lake Superior's fisheries. He could not let that stand. "This is my lake," he said.[7] Glass repeatedly appealed to EPA and Coast Guard officials to use their emergency powers to force freighters to treat ballast water with chlorine as they passed through the Soo Locks en route to Lake Superior. In one of his many e-mails to federal officials, Glass said the following:

> Federal law states that "contaminated water may not be used as ballast water," and if that is strictly enforced, then lakers [lake freighters] will have to avoid the contaminated water areas of the lower lakes including Detroit, Cleveland, Toledo, etc. where VHS virus has been shown to contaminate the fishery and water. A "water contamination emergency" could be declared for all waters where VHS virus have been found in fish, and all vessels ordered to be barred from using VHS virus contaminated waters. Existing federal laws need to be strictly enforced with regard to VHS virus contamination, and preventing its further spread into non-contaminated waters.[8]

His comments were not the rantings of an ill-informed environmental zealot. Glass was familiar with federal water-pollution control laws—for three decades he had provided the science that enabled the EPA to enforce those laws. Still, his pleas for immediate action were repeatedly ignored by officials in the EPA, Coast Guard, and the U.S. Department of Agriculture. Karen Phillips, commander of the Coast Guard's Great Lakes environmental division, suggested that Glass read the many international research projects on ballast water treatment. In a February 2007 e-mail to Glass, Phillips said: "Many studies have been conducted based on your hypothesis of using chlorine to treat ballast water, and unfortunately, it does not work because of the high sediment amount in ballast water. Current technology development includes development of efficient filters before use of biocides such as chlorine." Phillips suggested that Glass work with other groups already involved in the ballast treatment issue, such as the Northeast-Midwest Institute. She also invited him to attend the Coast Guard's Marine Community Day. That was the annual meeting where shipping-industry executives discussed regulatory issues with Coast Guard officials by day and socialized with those same officials at night. Ballast water issues, Phillips said, were always discussed at the conference. "Be forewarned, entering the ballast water debate is not for the faint of heart!" Phillips said in her e-mail to Glass. "The engineering/

chemistry/biology complexities, legal dynamics and political arena will make your head swim! Welcome to the arena!"[9]

Glass's chlorine crusade ran into impenetrable bricks walls at every turn in the offices of the Coast Guard, EPA, and USDA. None of those agencies took steps to prevent the spread of VHS via ballast water. His campaign did catch the attention of Phyllis Green, the superintendent at Isle Royale National Park. If VHS could kill millions of fish in the other Great Lakes, Green surmised, it could have similar effects in Lake Superior. "The lake's cold water is perfect habitat for VHS—it thrives in cold water," Green said. Her overriding concern was that VHS could decimate a thriving, diverse lake-trout fishery in the waters surrounding Isle Royale. Freighters headed to Thunder Bay, Ontario, often passed through the 475 square miles of Lake Superior waters within the boundaries of Isle Royale National Park. Green feared that any one of those ships might accidentally deposit VHS into the park's waters, where the virus could inflict huge losses on the lake trout fishery.

Determined to prevent such a disaster, Green tried to find someone, anyone, in a state or federal government agency who could help her develop an emergency plan to prevent VHS from infecting fish near Isle Royale. She found no such person. "There is no one agency in the Great Lakes that has control over all the vectors of aquatic invasive species in the lakes," she said. "There isn't one agency that can convene a team and say, 'We're going to stop this or show me how to stop it.'"[10] Green said fighting forest fires in national parks was easier than combating invasive species in the Great Lakes. "If I have a fire in a national forest I know exactly who to call. If I put in a call for help on invasive species, the answer I get is: 'We're studying it.'"

Frustrated by what she perceived as apathy and regulatory gridlock in the federal bureaucracy, Green chose to exercise the authority granted her as a national park superintendent. She issued an emergency order in September 2007 prohibiting freighters and all other boats from discharging ballast water in Isle Royale's portion of Lake Superior unless the water was disinfected first. It was her way of fighting the spread of VHS. "The entire lake fishery is threatened by this virus," Green said in a newspaper article. "We need to do everything we possibly can to stop it."[11]

Industry officials dismissed Green's order as political grandstanding. Critics said nothing could prevent infected fish from swimming into the

National Park Service

Scientists in 2007 successfully used household bleach to disinfect ballast water tanks on the National Park Service's Isle Royale ferry, the *Ranger III*. Vitamin C was used to neutralize the chlorine before the treated ballast water was discharged into Lake Superior. The treated water met drinking water standards.

waters surrounding Isle Royale if VHS showed up elsewhere in Lake Superior. Shippers also noted that freighters usually discharged ballast while in port, not in the open waters near Isle Royale. "Everybody wants a solution to this [ballast water] problem but you have to look at realities," said Glen Nekvasil, a spokesman for the Cleveland-based Lake Carriers Association. His group represented freighters that operated exclusively within the Great Lakes.[12]

Green further upped the ante by unveiling a chlorine-based system to disinfect ballast water tanks in Isle Royale's passenger ferry, the 165-foot *Ranger III*. For several months prior to issuing the emergency order, Green worked quietly with Michigan Technological University engineering professor John Hand on a chlorine-based ballast water treatment system. Hand, an expert on water and wastewater treatment systems, devised a simple system for treating ballast water on the *Ranger III*. His solution: household bleach and Vitamin C. Hand successfully demonstrated that pouring household bleach into the *Ranger*'s ballast tanks, and then neutralizing it with Vitamin C after

National Park Service

Michigan Technological University civil engineering professor John Hand (*left*) and Mike Hickey, chief engineer of the *Ranger III*, pour chlorine into the ship's ballast tanks.

the chlorine had time to work, killed organisms and left the treated ballast water clean enough to discharge into Lake Superior. The entire system cost $600.

"This may not be directly applicable to Great Lakes freighters, but the willingness to solve the problem should be," Green said during a September 2007 press conference.[13] Green believed that her team had increased pressure on the shipping industry and government agencies to solve the ballast water problem. "We literally painted industry into a box. The other agencies couldn't do what I did because they don't have the power I have as a federal park manager."[14]

Shipping-industry officials remained skeptical of Green's project. Even if the treatment system worked in the 165-foot *Ranger III*, critics said it would never work in ships that were up to five times larger than the *Ranger* and carried millions of gallons of ballast water. Adolph Ojard, director of the

Duluth Seaway Port Authority, summed up the industry's view of chlorine as a tool for treating ballast water. "Chlorine was dismissed as a treatment ten years ago because it was found to be ineffective, caused corrosion in ships, created carcinogenic wastes and posed safety risks to the crew," Ojard said. "Even if that system works on the ferry that goes to Isle Royale, it couldn't be scaled up to the size needed on a freighter."[15]

Professor Hand said the shipping industry dismissed the chlorine-based treatment system because of its utter simplicity. The solution was so simple, Hand said, that shipping companies might be forced to act more quickly to install similar devices on freighters plying the Great Lakes. He said freighters that transited the lakes could be equipped with a chlorine-treatment system for about $10,000 each. His calculations suggested that the chemicals for each ballast water treatment cycle would cost a few hundred dollars—the cost of the chlorine and Vitamin C used to disinfect millions of gallons of water. "A lot of people don't like chlorine. Most of us with city water drink water with chlorine in it all the time. But people are scared of it. I don't understand it—it's ignorance," Hand said. "Chlorine has saved mankind from a lot of disasters, it's a staple in the water treatment industry and it could be for ballast water, too."[16]

Seven years after Michigan officials first proposed using chlorine to disinfect freighter ballast water, the National Park Service—the most unlikely of government agencies—took up the cause and proved that chlorine worked. The Isle Royale ferry project, coupled with Michigan's 2007 ballast water law, prompted other states to consider enacting their own laws regulating ballast water. That prompted industry officials, wary of states enacting different ballast water regulations, to pressure the U.S. Congress to adopt national ballast water standards.

Congress was poised in 2008 to pass the nation's first ballast water regulations—35 years after the EPA made ballast water discharges exempt from the Clean Water Act. But the legislation was sidetracked by presidential-year politics and a crisis in America's financial markets. The congressional action was needed because EPA and Coast Guard officials had failed to produce comprehensive ballast water regulations, despite years of promising a solution.

That it took a deadly fish epidemic—followed by a controversial state law and the relentless efforts of a retired EPA chemist, park ranger, and college professor—to force national ballast water regulations was a sad commentary on the state of federal regulatory agencies in the United States and Canada. The silver lining in the dark cloud of bureaucratic bungling was that the problem of freighters importing foreign species in ballast water seemed to be nearing a resolution, 50 years after the St. Lawrence Seaway invited ocean ships and all their biological wastes into the Great Lakes. That was the hope, anyway.

"In some ways, VHS may be the best thing that's ever happened to the Great Lakes," said Cameron Davis, executive director of the Chicago-based Alliance for the Great Lakes. "It prompted the states and Congress to move like they never had before. VHS was the straw that broke the camel's back and got the states to act. That forced Congress to act."[17]

Congressional movement toward establishing ballast treatment standards was a welcome development in the government's pathetic effort to close the door on shipborne invaders infesting the lakes. Ballast water treatment systems could greatly reduce the number of new invaders entering the lakes in the future. That said, regulating ballast water discharges two decades after zebra mussels opened the world's eyes to the problem of ships spreading biological pollution was a bit like closing the door in a house already infested with disease-carrying mosquitoes. The genie of invasive species was out of the bottle—the damage was done. Precisely how much chaos shipborne invaders would inflict on the incomparable Great Lakes was a question that only the passage of time could answer.

The fact of the matter is we're not driving this bus anymore. Hang on for the ride—we're going to see where this process goes.

—James Baker, Michigan Department of Natural Resources fisheries manager, 2007

EPILOGUE

HOPE AMID THE RUINS

Rarely do professional sporting events provide lessons in ecology. Sports are about competition, teamwork, endurance, and athletes overcoming obstacles to achieve greatness. Once in a while, though, athletic competitions provide a stage for lessons that have little to do with strength and determination, wins or losses. Such was the case on an unusually warm October night in Cleveland, Ohio, in 2007. It was October 5. The Cleveland Indians were battling the mighty New York Yankees in an American League playoff game. The stakes were high—the winner of the series would advance to play in the World Series, the apex of America's national pastime. With the Yankees leading 1–0 in the seventh inning, the Bronx Bombers sent in rookie pitching sensation Joba Chamberlain to preserve the lead. Chamberlain was firing 100-mph fastballs at batters when a cloud of mayflies descended upon Jacobs Field. The insects swarmed around the young pitcher's head and clung to his neck, arms, and jersey. Yankees catcher Jorge Pojada tried to repel the mayflies by spraying insect repellent on Chamberlain. That backfired—more of the bugs clung to the pitcher, rattling his nerves. Chamberlain lost his focus and control of his pitches. He walked two batters, threw two wild pitches, and hit one batter, which allowed Cleveland to tie the game. The Indians went on to win, 2–1, in eleven innings. Cleveland's baseball fans celebrated. So did scientists and anglers who understood the ecological significance of the swarming mayflies that provided a bizarre home-field advantage for Cleveland's beloved Indians.[1]

The return of mayflies to Lake Erie was something of an environmental miracle. Gross pollution in the 1940s and '50s created dead zones in the lake, areas lacking enough oxygen to support aquatic life. Among the victims were mayflies, which were nearly eliminated from the lake by 1960. The insects were scarce over the next three decades. Then, in the early 1990s, mayflies made a surprising, mysterious, and dramatic return to Erie's western basin. Like a bolt from a blue sky, swarms of mayflies took flight over the lake, delighting ecologists and anglers. The annual mayfly hatch was so thick in 1996, some communities along western Lake Erie posted signs warning motorists of slippery conditions created by the oily remains of mayfly road kill. The massive hatch caused a brownout over northwestern Ohio when the insects coated a power plant's substation. Attracted by the substation lights, the mayflies formed a blanket that was so dense, their tiny bodies conducted electricity across the facility's insulators. The power surge knocked the substation out of service, causing lights to dim across the region.

Most researchers believed that water-quality improvements achieved after passage of the federal Clean Water Act, which increased oxygen concentrations in the lake, spurred the resurrection of mayflies. But there was more to the story. Mayflies returned so suddenly, and in such huge numbers, that some scientists believed there were other factors at work. It turned out that zebra mussels, which had inflicted so much ecological pain after invading Lake Erie in 1988, improved conditions for the immature mayflies—called nymphs—that lived in the lake's muddy bottom. The mussels sucked huge quantities of plankton down to the lake bottom, providing a feast for nymphs living in their tiny underwater caves. No one ever proved that zebra mussels were directly responsible for the dramatic resurgence of mayflies. But their recovery four years after zebra mussels were discovered in Lake Erie seemed, to some scientists, like more than a coincidence. Mayfly experts Don Schlosser and Kenneth Krieger explored the factors that contributed to the insects' recovery in a 2000 research report. Their findings were inconclusive but suggestive:

> We do not know the cause for the sudden recolonization of nymphs in large portions of western Lake Erie. Undoubtedly, pollution abatement programs contributed to improved conditions that would have ultimately led to mayfly recovery in the future. However, the explosive growth of the exotic zebra mussel, *Dreissena polymorpha*, undoubtedly diverted plankton foods to bottom substrates which could have increased the speed at which *Hexagenia* nymphs recolonized sediments in western Lake Erie in the 1990s.[2]

That despised zebra mussels contributed to the recovery of mayflies was strangely ironic. It suggested that some native species benefited from the plague of invaders that stormed the Great Lakes in the 50 years after the St. Lawrence Seaway opened. The situation was an inconvenient truth for scientists and environmental advocates who were loath to attribute any ecological benefits to invasive species that colonized the lakes. Would the lakes be better off without zebra and quagga mussels, round gobies and Eurasian ruffe? Without question. But the fact that some native species capitalized on the invaders was a glimmer of hope through the storm of invaders that had conquered the lakes. The lake ecosystems were demonstrating tremendous resilience and adaptability.

LAKE ERIE'S PUT-IN-BAY HAS LONG BEEN ONE OF THE MOST POPULAR vacation destinations in all the Great Lakes. The small bay tucked between South Bass and Gibraltar islands is a place where young children and soccer moms, buttoned-down professionals and college students, jocks and scholars revel in the joys of Midwest summers. South Bass Island is ground zero for Lake Erie's summer party and vacation scene. The island and its surrounding waters are a playground for people whose idea of paradise is a place where they can fish, sail, camp, play golf, track birds or snakes, kayak along the jagged shoreline, or kick back on a boat with a cold drink. For all its human pleasures, and there are many, the island is equally significant for its role in a high-stakes biological drama scripted by invasive species. That was the sole reason I traveled to Put-in-Bay on a June day in 2007. My indulgence in the area's notorious party scene would have to wait—I was headed to a research laboratory that documented the collapse, the resurrection, and great invasion of Lake Erie's ecosystem.

The Jet Express ferry that whisked me and a few other tourists from Port Clinton, on Ohio's mainland, to South Bass Island completed the 12-mile journey in less than a half hour. Once on the island, I walked around the edge of a cove choked with aquatic plants and boarded a water taxi that transported me across the small bay to Ohio State University's Stone Laboratory, on nearby Gibraltar Island. There, I met with Jeffrey Reutter, the lab director and head of Ohio's Sea Grant program. Reutter was a middle-aged scientist and professor who looked like he had just stepped off a hiking trail. Dressed in khaki shorts, a golf shirt, baseball hat, and running shoes, his appearance and engaging personality belied his encyclopedic knowledge of how invasive species affected the Great Lakes—particularly Lake Erie. It was well documented that zebra mussels, quagga mussels, and round gobies transformed Lake Erie's ecosystem, mostly for the worse, and cut short one of the world's greatest examples of ecosystem recovery. To my surprise, the plague of invasives that infected the lakes did cause a few positive changes. I encountered two of those rare success stories while at Gibraltar Island.

I was walking up the boat ramp toward the Stone Laboratory's main building when I noticed something unusual about the building's brick veneer—it seemed to be in motion. Thousands of small tan objects on the building were moving in unison with the wind. The scene resembled a scaled-down version of a field of grain swaying to invisible currents of wind. As I neared the lab, there was a crunching sound underfoot. A noise reminiscent of someone popping bubble wrap was the sound of my shoes crushing mayflies. Hoping to spare as many of the insects as possible, I stepped off the sidewalk and onto the adjacent lawn. That was a mistake. A cloud of mayflies erupted from the grass and enveloped me. The inch-long bugs with oversized transparent wings swarmed my clothes, hair, and seemingly every area of exposed skin. Some tried to sneak into my nose and mouth; others went down my shirt. I covered my nose and mouth with one hand and waved my other arm wildly in a desperate attempt to keep the insects at bay. It was hopeless. The only way to escape the insects—known formally as *Hexagenia*, but commonly called

Jeff Alexander

Mayflies cling to a window on a building at Ohio State University's Stone Laboratory, located on Lake Erie's Gibraltar Island. The return of mayflies in the 1990s was a sign that Lake Erie's ecological health had improved.

June bugs, Canadian sailors, or shad flies—was to go indoors. Annoying as it was, the cloud of mayflies was welcome news.

Mayflies emerge from Lake Erie as nymphs in early June after spending two years burrowing in the lake's muddy bottom. The nymphs shed their exoskeletons and take flight for what amounts to a three-day insect orgy.

Ohio Sea Grant

The Lake Erie water snake was one of few native Great Lakes species to benefit from foreign species that invaded the lakes. The once-endangered snake made a dramatic recovery after round gobies infested Lake Erie around 1990. The snakes feast on the foreign fish.

Clouds of mayflies large enough to be detected on weather radar mate while flying over the lake. After mating, females release up to 8,000 fertilized eggs on the lake's surface before falling to their deaths. Those eggs not eaten by fish sink to the bottom. There, a new generation of nymphs burrows into the lake bottom before emerging two years later to take flight, shed their skins, and engage in the mating ritual. The process is spectacular, despite the hassles it causes for shoreline communities forced to endure mayfly mayhem for a few weeks each summer. Now, the swarming insects were a sign Lake Erie was healthy again. Mayflies were a keystone species, a barometer of water quality and ecosystem health. What did it mean when millions of mayflies took flight and turned shoreline communities into giant insect breeding grounds for weeks at a time? "They're telling us the water's clean out there," said Thomas Edsall, a U.S. Geological Survey biologist.[3]

Mayflies weren't the only native species to flourish in the presence of invasive species. Zebra and quagga mussels that contributed to the collapse of Lake Huron's food chain in 2004 contributed to the decimation of another invader: alewives. That, in turn, allowed two native fish species—walleye and lake trout—to mount long-awaited recoveries. In the absence of alewives, one of Lake Huron's most valuable baitfish species—emerald shiners—returned in huge numbers. And in Lake Erie, the round-goby invasion helped fuel the dramatic recovery of a scorned, but native, snake species.

THE LAKE ERIE WATER SNAKE IS, FOR MY MONEY, ONE OF THE VILEST creatures native to the Great Lakes. It is also one of the most valuable and symbolic. The snake, *Nerodia sipedon insularum*, resides exclusively along the shores of islands that stretch across Erie's south side. It isn't found anywhere else on the planet. Just because the snake has a penchant for swarming beaches and boat docks, urinating and defecating on anyone who dares to move them, does not diminish its inherent ecological value. Is the Lake Erie water snake less valuable than, say, a bald eagle? Not in the eyes of government biologists, who declared the water snake a threatened and endangered species in 1999. Decades of abuse by humans, who refused to share valuable coastal property with the snakes, had driven the creatures to the brink of extinction.

"People killed the snakes because they wanted to build cottages on the islands and didn't want the snakes around," said Kristin Stanford, a Northern Illinois University graduate student who helped foster the water snake's recovery. "I've had people tell me they would find balls of snakes, pour gas on them and set them on fire. Other people shot the snakes with 22-caliber rifles as they swam in the water."[4]

The light brown snakes, which grow to three-and-a-half-feet long, have been persecuted for years by island residents who don't want them nesting on beaches and in rocky structures used to support boat docks. Part of the hatred is due to the fact that water snakes gather by the hundreds in rocky coastal areas. The snake's ill-tempered behavior has solidified its bad reputation. The creatures aren't poisonous. Hostile, yes. Aggressive? Most definitely. The snakes are notorious for warding off aggressors by emitting a noxious musk that can drive a grown human to the brink of vomiting. If that doesn't work, the animals urinate and defecate on their assailants. Their final line of defense is to bite, which is harmless but still hurts like heck.

Despite the snake's nasty disposition and disgusting defense mechanisms, Stanford fell for the scorned water animal. She was a budding, idealistic biologist when she first traveled to the Stone Laboratory in the summer of

2000. Hers was a singular mission: to save a species reviled by residents of Lake Erie's islands. As other college students partied at Put-in-Bay, Stanford was spending long days collecting and studying water snakes. Seven years later, she was credited with saving the species and given the title of Snake Lady. Stanford had a powerful ally in her snake-resurrection campaign. Water snakes, it turned out, feasted on round gobies, the reviled fish from Europe. With an estimated 10 billion gobies in Lake Erie, many of which lived in the rocky areas around the Bass Islands, the water snake was suddenly living large. The snakes began feasting on gobies and eventually stopped hunting their usual prey: native log perch, stone cats, sculpin, and mud puppies. To water snakes, the gobies were a delicacy. "The snakes started growing faster and larger and the females produced 20 percent more offspring," Stanford said. "The gobies helped bring these snakes back from the brink of extinction."[5]

The water-snake population began to rebound in 2003. By 2007, there were about 7,000 water snakes living around the Bass Islands—well above the goal of 5,000 snakes. As their numbers soared, so did tension among some island residents who were forced to share beaches and boat docks with hostile snakes. "We're getting to the point where there could be another public backlash against the snakes," Stanford said in 2007. "We can find hundreds of snakes at one dock." Barring an unforeseen change, the water snake could be taken off the endangered species list by 2010. "It would be a fabulous success story," Stanford said.[6]

Stanford's work turned her into something of a celebrity. She was featured on television shows and in newspaper and magazine articles.[7] An information kiosk placed on South Bass Island, with a photo of Stanford holding a bunch of snakes in her bare hands, prompted hundreds of e-mails. A West Virginia man sent an e-mail asking for Stanford's snake-bitten hand in marriage. She politely declined (she was already engaged to be married). Stanford was bewildered by all the interest in her unconventional line of work. "It's just weird," she said. Still, she was happy to share the story of a beleaguered creature returning from the brink of extinction. The snake's recovery was a symbol of hope.

The water snake's unexpected population explosion also was a conundrum for scientists. It was a success story, without question. But the snake's resurgence was largely due to the arrival of round gobies, an invader that terrorized some native fish species and contributed to the deaths of thousands of fish-eating birds. The water snake's recovery, like that of mayflies and walleyes in Lake Huron, was aided by foreign species that disrupted and destabilized Great Lakes ecosystems. Was it possible that gobies were good for Lake Erie? How about zebra and quagga mussels, which dramatically increased water clarity and created conditions that allowed some endangered species of aquatic plants to flourish in the waters surrounding the Bass Islands? Could the foreign species, which were universally scorned by scientists, improve some aspects of Great Lakes ecosystems? I posed the question to Reutter, a distinguished and widely respected Great Lakes scientist with 30 years of research under his belt. His response came quickly: "When you put all of the effects of these invasive species together," he said, "the negative impacts clearly outweigh the positives."[8]

Invasive species plunged the lakes into a state of biological chaos and created great uncertainty about their ecological future. No one knew how the lakes would react, in the long run, to invaders from distant waters. And the few benefits realized by native species that were able to exploit the invaders came with huge caveats. The fate of the Lake Erie water snake, for instance, could be determined largely by two factors: the availability of food and the willingness of humans to share space with the creatures. If the goby population crashed, water snakes might suffer as well.

Similar uncertainty surrounded the mayfly resurrection in Lake Erie. The foreign mussels that bolstered the mayfly's recovery also clouded its future. Zebra and quagga mussels were suspected of contributing to rising concentrations of soluble reactive phosphorus in Lake Erie. The phosphorus was fertilizer for aquatic vegetation and algae, which contributed to the return of the lake's infamous dead zone. If the mussels were indeed responsible for Erie's latest dead zone, mayflies could suffer in the oxygen-starved waters. Round gobies also were a threat to mayflies—the invaders gorged on mayfly nymphs. The complex dynamics left scientists wondering whether gobies, zebra mussels, and quagga mussels would conspire to diminish the mayfly population. It was a sobering question with no answer in 2008, two decades after the mussels turned up in the Great Lakes.

HAVING WRITTEN ABOUT INVASIVE SPECIES IN THE GREAT LAKES FOR TWO decades, and the failure of government agencies to slow the great invasion, it was hard for me to be optimistic about the future of these imperiled waters. Still, there were major developments in the first decade of the twenty-first century that gave me a genuine sense of hope. Government agencies that had reacted meekly to the problem were finally taking the steps necessary to close the door on new shipborne invaders. The International Maritime Organization had proposed the world's first ballast water treatment standards, and the U.S. Congress was trying to doing the same as I completed this narrative in late 2008. Biologists and economists had produced solid data that forced government officials to discuss the possibility of banning ocean freighters from the lakes if shippers could not clean up their vessels' ballast tanks. On a larger scale, the great invasion of the lakes generated a global discussion on the conveyance of species around the planet by ships and other modes of transportation.

Invasive species weren't part of the public discourse in the United States and Canada before zebra mussels turned up in Lake Erie and demonstrated how destructive invasive species could be to aquatic ecosystems and water-dependent economies. Some researchers, in a desperate effort to find good news in the storm of invasives that reconfigured the Great Lakes ecosystem, noted that zebra mussels put the issue of invasive species on the global map. In my humble estimation, the overriding positive that emerged from the plague of foreign species was the public's improved understanding of how fragile the Great Lakes were, despite their enormity and seeming invincibility. Unfortunately, it took three decades of government negligence—and the resulting biological siege—to make that point forcefully enough to bring about the changes needed to protect the lakes from shipborne invaders.

The United States and Canada in 2008 were finally requiring all ocean freighters to flush ballast tanks with seawater before entering the St. Lawrence Seaway and the Great Lakes. That was a positive step, but it was far from a panacea. Ballast water flushing did not eradicate all organisms lurking in the bowels of ocean ships. Potentially deadly pathogens could survive saltwater flushing. That alone was reason enough to require every ship using the St. Lawrence Seaway to install ballast water treatment systems. Sure, treatment systems were expensive, perhaps as much as $1.5 million per ship. But it would be money well spent. Shipping companies that couldn't afford to disinfect their vessels' ballast tanks didn't deserve the privilege of conducting business on the lakes.

The global shipping industry was not entitled to transport cargo on the Great Lakes. Use of the world's largest freshwater reservoir was a privilege afforded the industry when the United States and Canada built the St. Lawrence Seaway. Clearly, the industry never should have been allowed to dump untreated ballast water into the lakes after U.S. lawmakers passed the federal Clean Water Act in 1972. Because the lakes are a closed ecological system, any chemical or biological pollutants dumped in them will likely remain for decades, if not centuries. That reality made it all the more important to treat these freshwater seas with the utmost care.

A 2008 study by the U.S. National Research Council concluded that the only sure way to keep ships from importing more foreign species into the lakes was to close the St. Lawrence Seaway to all freighters. But the council, an advisory panel to the U.S. Congress, recommended against it. Such a move, the council concluded, would eliminate the potential for future increased global trade in the Great Lakes region. Essentially, the council told Congress that if the United States and Canada wanted international trade to continue on the lakes, the two nations had to accept the risk that transoceanic ships might import more foreign species. Rather than close the Seaway, the council urged the U.S. Congress to immediately adopt ballast water treatment standards proposed by the International Maritime Organization. That was all fine and good, but the fact remained that no ballast water treatment system ready for use in 2009 was 100 percent effective. Moreover, it would take several years to equip all ocean freighters that visited the Great Lakes with ballast treatment systems. Until that day arrived, the lakes remained vulnerable to more shipborne invasions. In the interim, residents of the region could only hope that foreign ships wouldn't deposit more destructive invaders in the lakes.

If we've learned nothing else from the ecological ruin inflicted by shipborne invaders, we should now understand that in the war against invasive species, prevention is the best—perhaps the only—guaranteed protection. No public health official would condone unprotected sex between a person with herpes and an uninfected individual. Why, then, would we allow ships carrying cholera and other pathogens in ballast tanks to engage in unprotected trade in the Great Lakes? It defies logic. What if a transoceanic ship imports another foreign organism that causes billions of dollars of additional ecological damage? What if a ship's ballast water discharge infects a public drinking water system with cholera or another deadly pathogen that kills humans? Would that trigger the kind of dramatic, rapid response needed to end the inoculation of the lakes by shipborne invasive species? Those were the questions that rang in my mind as I headed back to mainland Ohio.

THE SUN WAS SINKING TOWARD THE LAKE ERIE HORIZON WHEN I BOARDED the Jet Express ferry in Put-in-Bay and headed back to the more tranquil community of Port Clinton, Ohio. The steady hum of the ferry's engines and the melodic rhythm of the ship carving through Erie's mild chop set my mind adrift. Passing anglers in small fishing boats that bobbed in the expansive lake, I wondered if future generations would come to know these emerald waters as the Walleye Capital of the World.

After disembarking from the ferry, I hopped in my car and drove a few miles east before settling on a hotel along Lake Erie's southern coast. A cloud of mayflies greeted me as I stepped out of my car. Thousands swarmed around the hotel's exterior lights, while others snapped underfoot as I scurried into the building. I was moving quickly in a desperate attempt to get checked in and deposit my luggage in the hotel room before the sun set. There were few better places than Great Lakes beaches to watch the sun illuminate the evening sky as it sank below the horizon. Lake Erie did not disappoint that night. Moisture that evaporated from the lake formed a thin layer of cirrus clouds that resembled giant horse manes. The clouds provided the canvas upon which the sun cast a glorious pale-red hue as it neared the horizon. Hard as I tried to enjoy the synergistic beauty of the lake, sky, and sun, my thoughts were haunted by an ominous sense of uncertainty.

Would children not yet born in 2009 fall in love with the lakes by playing in clean waters that gently lapped at unspoiled shorelines? Would they come to know the lakes as an intricately woven ecosystem that supported tiny freshwater shrimp, majestic bald eagles, giant lake sturgeon, and delectable whitefish? Would they be able to rely on the lakes as a source of clean drinking water? Would they view the lakes as a priceless resource that warranted protection at any cost? Or would they come to know the lakes as victims of a biological cleansing carried out by an army of invasive species from distant waters? Would they be repulsed by algae blooms that fouled the water, harbored potentially deadly organisms, and turned beaches into avian killing fields? Would they smell the musty odor of blue-green algae when they drew a glass of drinking water from the tap? Would they come to know the simple, profound joy of hearing a common loon cast its tremulous call across the surface of a Great Lakes bay on a tranquil summer morning? In this, the 50th anniversary of the St. Lawrence Seaway's opening, there was no way of knowing what the future held for the lakes. There were many vexing questions but few answers.

Foreign species that transoceanic freighters accidentally deposited in the lakes had caused the most profound changes to the lakes in recorded history. Scientists could not envision how the lakes' ecosystems would function in coming decades—they couldn't even keep pace with profound changes that unfolded before their eyes at the dawn of the twenty-first century. Granted, the lakes had proven remarkably resilient in the past. These freshwater seas—Superior, Michigan, Huron, Erie, and Ontario—staged magnificent ecological recoveries in the 1970s after humans eased up on the gas peddle of exploitation and abuse. But the conquest of the lakes by foreign species was different. There was no chemical, insect, or animal capable of controlling the range or densities of zebra and quagga mussels, round gobies or Eurasian ruffe, bloody red shrimp or spiny water fleas in the lakes. Humans knew how to split atoms and build spacecraft that allowed astronauts to live beyond the reach of Earth's gravity. Yet, once foreign species stormed the lakes, we were largely powerless against the army of invaders that crippled native species, spawned toxic algae blooms, and formed bastardized food chains that delivered toxic chemicals and deadly pathogens to fish, wildlife, and humans.

Looking across Lake Erie on that sultry summer evening, I was struck by the contrast of what my eyes could see and what my mind knew to be true. The shimmering surface of the lake was alluring, exquisite. But its beauty was skin deep. Beneath its winsome veneer, an ecological war raged and the bad guys were winning big. As the sun dipped below the horizon and the sky grew dark, I wondered if Lake Erie—the smallest, shallowest, most productive, and most fragile of all the Great Lakes—would weather the storm of invasive species. Erie's fate was enormously significant. The lake was an ecological guinea pig, a bellwether of future changes in Lakes Ontario, Huron, Michigan, and Superior. That Erie was losing the battle against invasive species was cause for alarm. But its remarkable history provided a glimmer of hope for the beleaguered lakes. Lake Erie, after all, recovered magnificently after it was prematurely declared dead in the 1960s.

Perhaps Erie, along with the other Great Lakes, would survive the reign of ecological terror carried out by invasives and reclaim some of their natural splendor. We could hope for such a turn of events, much like cancer patients who dream of restored health. The alternative emotion, despair, was unworthy of these inimitable waters. There was no changing the fact that the Seaway, and all it allowed into the heart of North America, stripped the lakes of their natural defenses and sickened these freshwater seas. All that remained in 2009 for the Great Lakes, and all whose lives were intricately tied to these priceless waters, were the possibilities that hope afforded. We could hope for a brighter future for the lakes and ourselves—just like the mythical Pandora.

NOTES

PREFACE

1. "European Mussel Seen as Threat to Great Lakes," *Muskegon Chronicle*, Sept. 26, 1989, Muskegon, Mich.
2. David Pimentel, "Aquatic Nuisance Species in the New York State Canal and Hudson River Systems and Great Lakes Basin: An Economic and Environmental Assessment," *Environmental Management* 35, no. 5 (2005): 692–701.
3. Eric Reeves, Cdr., U.S.C.G. (Ret.), "Exotic Policy: An IJC White Paper on Policies for the Prevention of the Invasion of the Great Lakes by Exotic Organisms," July 15, 1999, 10, in author's possession.
4. "The Seaway Opens," *Toronto Globe and Mail*, June 26, 1959.
5. "100,000 Hail Duluth Port," *St. Paul Pioneer Press*, July 12, 1959.

6. David Lodge, "The Impact of Invasive Species on the Great Lakes," testimony before the U.S. House of Representatives Subcommittee on Water Resources and Environment, March 7, 2007, Washington, D.C., 7.
7. "Invasive Species: Clearer Focus and Greater Commitment Needed to Effectively Manage the Problem," U.S. General Accounting Office, Washington, D.C., Oct. 22, 2002, 14.
8. Sarah A. Bailey, David F. Reid, Robert I. Colautti, Thomas Therriault, and Hugh MacIsaac, "Management Options for Control of Nonindigenous Species in the Great Lakes," *Toledo Journal of Great Lakes Law, Science and Policy* 5 (Spring 2004): 102.
9. Reeves, "Exotic Policy," 122.

[PROLOGUE] DEATH OF AN ICON

1. Judith W. McIntyre, *The Common Loon: Spirit of the Northern Lakes* (Minneapolis: University of Minnesota Press, 1989), 6.
2. Irene Mazzochi interview with author, Nov. 15, 2007, Bailey's Harbor, N.Y.
3. Rachel Carson, *Silent Spring* (New York: Houghton Mifflin, 2002), 127.

[01] CONQUERING NATURE

1. Jerry Dennis, *The Living Great Lakes: Searching for the Heart of the Inland Seas* (New York: Thomas Dunne Books, 2003), 182–83.
2. Bob Foley, *Niagara Falls* (Markham, Ont.: Irving Weisdorf, 2007), 4.
3. Paul Gromosiak, *Water over the Falls: 101 of the Most Memorable Events at Niagara Falls* (Toronto: Royal Specialty Sales, 2006), 18.
4. Ibid., 12.
5. Peter L. Bernstein, *Wedding of the Waters: The Erie Canal and the Making of a Great Nation* (New York: W. W. Norton, 2005), 23.
6. Ibid., 23.
7. Stephen R. Powell, ed., "Waterways and Canal Construction, 1700–1825, Buffalo, N.Y., Part III," in *The Buffalonian* (Buffalo, N.Y.: The Peoples History Coalition, n.d.), 7, accessed via www.buffalonian.com.
8. Paul Volpe, "Digging Clinton's Ditch: The Impact of the Erie Canal on America, 1807–1860," University of Virginia, n.d., accessed via http://xroads.virginia.edu, April 7, 2007.

9. William H. Becker, *From the Atlantic to the Great Lakes: A History of the U.S. Army Corps of Engineers and the St. Lawrence Seaway* (Washington, D.C.: U.S. Army Corps of Engineers, 1986), 2.
10. "William Hamilton Merritt, 1861–1870," *Dictionary of Canadian Biography Online*, vol. 9, accessed via www.biographi.ca, July 29, 2007.
11. "Account of the Welland Canal, Upper Canada," *American Journal of Science and Arts*, July 1828, in *Voices from the Great Ditch: The Historic Welland Canals in Pictures, Prose, Poems and Songs*, edited by Robert R. Taylor (St. Catherines, Ont.: Blarney Stone Books, 2004), 10.
12. Kim Todd, *Tinkering with Eden: A Natural History of Exotic Species in America* (New York: W. W. Norton, 2002), 66.
13. William Ashworth, *The Late, Great Lakes: An Environmental History* (Detroit, Mich.: Wayne State University Press, 1986), 112.

[02] VAMPIRES OF THE DEEP

1. Patrick Tivy, *Marilyn Bell: The Heart-Stopping Tale of Marilyn's Record-Breaking Swim* (Canmore, Alb.: Altitude Publishing, 2003), 38.
2. "A Surfeit of Lampreys," *Time*, May 9, 1955, accessed via www.time.com.
3. Tivy, *Marilyn Bell*, 59.
4. "Sea Lamprey: A Great Lakes Invader," Great Lakes Fishery Commission, Ann Arbor, Mich., in author's possession.
5. Vernon C. Applegate, "The Menace of the Sea Lamprey," *Michigan Conservation* 16 (1947): 6–10.
6. Vernon Applegate, "The Sea Lamprey in the Great Lakes," *Scientific Monthly* 72, no. 5 (1951): 276.
7. Rebecca Williams, "Multimillion Dollar Parasite Fight Continues," Great Lakes Radio Consortium, Ann Arbor, Mich., aired Nov. 27, 2006.
8. Great Lakes Fishery Commission, fact sheet on sea lamprey, accessed at www.glfc.org.
9. James Seelye interview with author, Oct. 11, 2007, Millersburg, Mich.
10. "Each Lamprey Killed Costs Nation $7.50," *Muskegon Chronicle*, May 31, 1958.
11. "Extermination of Lamprey Is Certain, but Much Remains to be Accomplished," *Muskegon Chronicle*, Dec. 15, 1956.
12. Great Lakes Fishery Commission, "Sea Lamprey: A Great Lakes Invader," accessed at www.glfc.org.

13. "Sea Lampreys Causing Havoc on Fishery," *Green Bay Press-Gazette*, July 14, 2005.
14. Marc Gaden, Great Lakes Fishery Commission, e-mail correspondence with author, Aug. 9, 2007.

[03] SALT IN THE WOUND

1. Carleton Mabee, *The Seaway Story* (New York: MacMillan, 1961), 131–32.
2. "Remarks of Senator John F. Kennedy on the Saint Lawrence Seaway before the Senate, Washington, D.C., Jan. 14, 1954," from the John F. Kennedy Presidential Library and Museum, Boston, Mass., obtained via www.jfklibrary.
3. "St. Lawrence River Project Under Way," *Massena (N.Y.) Observer*, Aug. 9, 1954.
4. "New St. Lawrence Seaway Opens the Great Lakes to the World," *National Geographic* 115, no. 3 (1959): 299.
5. Mabee, *The Seaway Story*, 253.
6. Dan Egan, "Sinking Treasure: Too Small for Big Ships, Seaway Hasn't Lived Up to Fanfare of the '50s," *Milwaukee Journal-Sentinel*, Oct. 29, 2005.
7. Ibid.
8. Davis Helberg, "Needed: A Bigger Seaway," *Minnesota's World Port* Spring 1993.
9. Daniel J. McConville, "Seaway to Nowhere," *American Heritage* 11, no. 2 (1995).
10. Egan, "Sinking Treasure."
11. "Great Lakes/St. Lawrence Seaway System: An Overview of North America's Most Dynamic Waterway," undated brochure, St. Lawrence Seaway Development Corp., Washington, D.C., accessed via www.greatlakes-seaway.com.
12. "The St. Lawrence Seaway Traffic Report, Historical Tables, 1959–1992" and "Tonnage Information Reports," St. Lawrence Seaway Development Corp., Washington, D.C., accessed via www.greatlakes-seaway.com.
13. "Uncharted Waters: Decisions on the Horizon for the Aged St. Lawrence Seaway," *Milwaukee Journal-Sentinel*, Oct. 31, 2005.

[04] ALEWIFE INVASION

1. "2001–2006 Gaspereau (Alewife) Fishery Eastern New Brunswick Area," Canada Department of Fisheries and Oceans, accessed via www.glf.dfo-mpo.gc.ca.
2. Howard Tanner, Mercer Patriarche, and William Mullendore, "Upper Great Lakes Sportfishery Establishment Has Spectacular Impact," in *Using Our Natural Resources: 1983 Yearbook of Agriculture* (Washington, D.C.: U.S. Department of Agriculture, 1983), 318.

3. "Alewife Explosion," *Time*, July 7, 1967.
4. "Grand Haven Plans an All Out Campaign against Alewives Problem," *Muskegon Chronicle*, Mar. 26, 1968.
5. "Alewife Explosion."
6. "Pee . . ee . . U! Battle with the Alewives Goes On along Lakefront," *Muskegon Chronicle*, June 27, 1967.
7. "They're Coming: Area May Have to Fight Alewife Horde Alone as Federal Project Skips State," *Muskegon Chronicle*, Mar. 29, 1968.
8. Ibid.

[05] A KING IS BORN

1. Bill Taylor interview with author, Jan. 15, 2008.
2. Michael J. Chiarappa and Kristin M. Szylvian, *Fish for All: An Oral History of Multiple Claims and Divided Sentiment on Lake Michigan* (East Lansing: Michigan State University Press, 2003), 566.
3. Daniel J. Miller and Neil H. Ringler, "Atlantic Salmon in New York," State University of New York College of Environmental Science and Forestry, Syracuse, N.Y., 1996, accessed via www.esf.edu.
4. Howard A. Tanner and Wayne H. Tody, "History of the Great Lakes Salmon Fishery: A Michigan Perspective," in *Sustaining North American Salmon: Perspectives across Regions and Disciplines*, ed. Kristine D. Lynch, Michael L. Jones, and William W. Taylor (Bethesda, Md.: American Fisheries Society, 2002), 144.
5. Wayne H. Tody and Howard A. Tanner, "Coho Salmon for the Great Lakes," Michigan Department of Conservation, Fish Management Report No. 1, February 1966, in possession of author.
6. Chiarappa and Szylvian, *Fish for All*, 567.
7. Frank Lupi and Douglas B. Jester, "Uses of Resource Economics in Managing Great Lakes Fisheries: Michigan Examples," in *Sustaining North American Salmon: Perspectives across Regions and Disciplines*, edited by Kristine D. Lynch, Michael L. Jones, and William W. Taylor (Bethesda, Md.: American Fisheries Society, 2002), 200.
8. Ibid., 146.
9. Ibid., 147.
10. Jerry Dennis, *The Living Great Lakes: Searching for the Heart of the Inland Seas* (New York: Thomas Dunne Books, 2003), 191–92.
11. Dave Borgeson interview with author, Aug. 15, 2006.

12. "The Fish That Has Them Hooked," *Business Week*, Oct. 1968, cited in Kristin Szylvian, "Transforming Lake Michigan into the World's Greatest Fishing Hole: The Environmental Politics of Michigan's Great Lakes Sport Fishing, 1965–1985," *Environmental History* (January 2004).
13. Bill Taylor interview with author, Jan. 15, 2008, East Lansing, Mich.
14. Howard Tanner, quoted in Chiarappa and Szylvian, *Fish for All*, 567.
15. Ibid., 565.
16. Gary Fahnenstiel interview with author, Aug. 2, 2006.
17. Michael J. Chiarappa interview with author, Aug. 2, 2006, Kalamazoo, Mich.
18. David N. Cassuto, *Cold Running River* (Ann Arbor: University of Michigan Press, 1994), 104–5.
19. Mark E. Holey, Robert F. Elliott, Susan V. Marcquenski, John G. Hnath, and Kelley D. Smith, "Chinook Salmon Epizootics in Lake Michigan: Possible Contributing Factors and Management Implications," *Journal of Aquatic Animal Health* 10 (1998): 202–10.
20. S. S. Crawford, "Ecological and Genetic Effects of Salmonid Introductions in North America," *Canadian Journal of Fisheries and Aquatic Science* 48, supp. no. 1 (1991): 66–77.
21. S. S. Crawford, "Ecological Effects of Stocking Exotic Salmon in the Great Lakes," in *Salmonine Introductions to the Laurentian Great Lakes: An Historical Review and Evaluation of Ecological Effects*, a special publication of the *Canadian Journal of Fisheries and Aquatic Sciences*, no. 132 (2001).
22. Ibid.

[06] FATAL ERROR

1. Claudia Copeland, "Clean Water Act: A Summary of the Law," Congressional Research Service Report to Congress, Congressional Research Service, Washington, D.C., Jan. 20, 1999, 1.
2. Order Granting Plaintiffs' Motion for Summary Judgment Denying Defendants' Motion for Summary Judgment, in *Northwest Environmental Advocates, et al. v. United States Environmental Protection Agency*, U.S. District Court for the Northern District of California, U.S. District Judge Susan Illston, March 30, 2005, 2.
3. "Aquatic Nuisance Species in Ballast Water Discharges: Issues and Options, Draft Report for Public Comment," Sept. 10, 2001, U.S. Environmental Protection Agency, Office of Water, Office of Wetlands, Oceans and Watersheds, Office of Wastewater Management, Washington, D.C., 33, in author's possession.

4. "Great Lakes Aquatic Nonindigenous Species List," National Oceanic and Atmospheric Administration, National Center for Research on Aquatic Invasive Species, Ann Arbor, Mich., in author's possession.

[07] DANGEROUS CARGO

1. Martin Associates, Lancaster, Pa., "Economic Impact Study of the Great Lakes St. Lawrence Seaway System," prepared for the U.S. St. Lawrence Seaway Development Corp., Aug. 1, 2001, E-2.
2. John C. Taylor and James L. Roach, *Ocean Shipping in the Great Lakes: Transportation Cost Increases That Would Result from a Cessation of Ocean Vessel Shipping* (Grand Rapids, Mich.: Grand Valley State University, 2005), 19.
3. Port of Duluth-Superior, Summary of 2001 Season Impacts, Duluth, Minn., in author's possession.
4. Adolph Ojard interview with author, Sept. 24, 2007, Duluth, Minn.
5. "Marine Investigation Report Breakup and Sinking: The Bulk Carrier 'FLARE,' Cabot Strait, 16 January 1998," Transportation Safety Board of Canada, Gatineau, Quebec, 39–40.
6. "Carrier Court Date Set," *Canadian Press*, March 2, 2004, in author's possession.
7. Kristen T. Holeck, Edward L. Mills, Hugh J. MacIsaac, Margaret R. Dochoda, Robert I. Colautti, and Anthony Ricciardi, "Bridging Troubled Waters: Biological Invasions, Transoceanic Shipping and the Laurentian Great Lakes," *BioScience* 54, no. 10 (2004): 919–29.
8. Allegra Cangelosi and Nicole Mays, "Great Ships for the Great Lakes? Commercial Vessels Free of Invasive Species in the Great Lakes–St. Lawrence Seaway System: A Scoping Report for the Great Ships Initiative," Northeast Midwest Institute, Washington, D.C., May 2006, 136–39.
9. Bio-Environmental Services, "The Presence and Implication of Foreign Organisms in Ship Ballast Waters Discharged into the Great Lakes," for *Environment Canada*, March 1981, iv–ix, in author's possession.
10. Ibid, 51.
11. "Ottawa Ignored Zebra Mussel Warning, Scientist Says," *Toronto Star*, Oct. 2, 1989.
12. Holeck et al., "Bridging Troubled Waters."
13. Gregory M. Ruiz and David F. Reid, "Current State of Understanding about the Effectiveness of Ballast Water Exchange in Reducing Aquatic Nonindigenous

Species Introductions to the Great Lakes Basin and Chesapeake Bay, USA: Synthesis and Analysis of Existing Information," National Oceanic and Atmospheric Administration Technical Memorandum GLERL-142, September 2007, Ann Arbor, Mich., 8.

14. Cangelosi and Mays, "Great Ships for the Great Lakes?" 136.

[08] THE RECKONING

1. David F. Reid, "Dreissenids in North America: 20 Years of Consequences," presentation at International Association of Great Lakes Research annual conference, Trent University, Peterborough, Ontario, May 22, 2008.
2. International Joint Commission letter to Joe Clark, Canadian Secretary of State for External Affairs, Aug. 9, 1988, in author's possession.
3. "Wish You Weren't Here," *(Paonia, Colo.) High Country News,* March 5, 2007.
4. "Ottawa Ignored Zebra Mussel Warning, Scientist Says," *Toronto Star,* Oct. 2, 1989.
5. Wilfred Laurier LePage, "The Impact of *Dreissena polymorpha* on Waterworks Operations at Monroe, Michigan: A Case History," in *Zebra Mussels: Biology, Impact and Control,* ed. Thomas F. Nalepa and Donald W. Schloesser (Boca Raton, Fla.: Lewis Publishers, 1993), 333–58.
6. "Zebra Mussels Invade Eastern U.S. Waterways," in *On Tap,* newsletter of the National Drinking Water Clearinghouse, vol. 9, no. 2 (2000): 14.
7. "Water Crisis Cripples Area," *Monroe Evening News,* Dec. 15, 1989.
8. "Coping—Boon for Grocers, Bust for Restaurants," *Monroe Evening News,* Dec. 16, 1989.
9. "Mussels Worsen Water Crisis," *Monroe Evening News,* Dec. 16, 1989.
10. Ibid.
11. LePage, "The Impact of Dreissena Polymorpha," 352.
12. "From Tough Ruffe to Quagga: Intimidating Invaders Alter Earth's Largest Freshwater Ecosystem," *Science News,* July 25, 1992.
13. Jeffery Reutter interview with author, Put-in-Bay, Ohio, June 7, 2007.
14. LePage, "The Impact of *Dreissena polymorpha,*" 359–77.
15. "Attack of the Zebra Mussels? Controversies about *Dreissena polymorpha* in North America," in *Science Lives!,* accessed via www.sciencelives.com.
16. "Exotic Species and the Shipping Industry: The Great Lakes–St. Lawrence Ecosystem at Risk; A Special Report to the Governments of the United States

and Canada," International Joint Commission, Washington, D.C., and Great Lakes Fishery Commission, Ann Arbor, Mich., September 1990, 3–4.

17. "Nonindigenous Aquatic Nuisance Prevention and Control Act of 1990," Sec. 1002, U.S. Congress, Washington, D.C., 2.

[09] RUFFE SEAS

1. "Bone Cold Cafe Suits Ruffe," *Minnesota Sea Grant Newsletter*, June 2000, Duluth, Minn., accessed via www.seagrant.umn.edu.
2. "Nonindigenous Aquatic Nuisance Prevention and Control Act of 1990," Sec. 1002, U.S. Congress, Washington, D.C., 2.
3. "From Tough Ruffe to Quagga: Intimidating Invaders Alter Earth's Largest Freshwater Ecosystem," *Science News*, July 25, 2002.
4. Thomas R. Busiahn, "Ruffe Control: A Case Study of an Aquatic Nuisance Species Control Program," in *Zebra Mussels and Aquatic Nuisance Species*, ed. Frank D'Itri (Chelsea, Mich.: Ann Arbor Press, 1997), 69–85.
5. Ibid.
6. Ibid.
7. R. M. Newman, "Ruffe—A Problem or Just a Pest?" *Aquatic Nuisance Species Digest* 3, no. 4 (1999): 44–46.
8. "Ruffe in the Great Lakes: A Threat to North American Fisheries," Great Lakes Fishery Commission Ruffe Task Force, Ann Arbor, Mich., Sept. 1992, 7.
9. Busiahn, "Ruffe Control," 75.
10. "Chemicals May Fight Lake's Pests," *Muskegon Chronicle*, May 21, 1993.
11. "Fishery Management Officials Re-Examine Ruffe Control Strategy after the Recent Appearance of Ruffe in Lake Huron," Council of Lake Committees press release issued by the Great Lakes Fishery Commission, Nov. 15, 1995, accessed via www.glfc.org.
12. Ibid.
13. "Ruffe Outlook for Lake," *Cleveland Plain Dealer*, May 18, 1997.
14. Ibid.
15. Ibid.
16. Roger Bergstedt interview with author, Oct. 11, 2007, Millersburg, Mich.
17. "LCA Applauds U.S. Coast Guard's Marine Protection Efforts," testimony of George J. Ryan to House Subcommittee on Coast Guard and Marine Transportation, July 1998, Washington, D.C., in author's possession.

18. "Ruffe," U.S. Geological Survey Nonindigenous Aquatic Species fact sheet for ruffe, accessed via http://nas.er.usgs.gov.
19. Busiahn, "Ruffe Control," 84.
20. "Overview of the International Symposium on Eurasian Ruffe Biology, Impacts and Control," *Journal of Great Lakes Research* 24, no. 2 (1998): 165–69.
21. Ibid.
22. Dave Zentner interview with author, Sept. 24, 2007, Duluth, Minn.
23. Janet Raloff, "From Tough Ruffe to Quagga: Intimidating Invaders Alter Earth's Largest Freshwater Ecosystem," *Science News*, July 25, 1992.

[10] SMOKE AND MIRRORS

1. Eric Reeves interview with author, Aug. 22, 2007, Erie, Penn.
2. Nonindigenous Aquatic Nuisance Prevention and Control Act of 1990, Sec. 1101, 2A.
3. Eric Reeves, "Exotic Politics: An Analysis of the Law and Politics of Exotic Invasions of the Great Lakes," *Toledo Journal of Great Lakes Law, Science & Policy* 2, no. 2 (2000): 125–206, 145.
4. Gregory M. Ruiz and David F. Reid, "Current State of Understanding about the Effectiveness of Ballast Water Exchange in Reducing Aquatic Nonindigenous Species Introductions to the Great Lakes Basin and Chesapeake Bay, USA: Synthesis and Analysis of Existing Information," National Oceanic and Atmospheric Administration Technical Memorandum GLERL-142, September 2007, Ann Arbor, Mich., 48.
5. "Effectiveness of Mid-Ocean Exchange in Controlling Freshwater and Coastal Zooplankton in Ballast Water," Canadian Fisheries and Aquatic Sciences Report, Great Lakes Laboratory, Burlington, Ontario, 9, cited in Eric Reeves, "Analysis of Laws & Policies Concerning Exotic Invasions of the Great Lakes," March 15, 1999, Michigan Department of Environmental Quality, 17.
6. Eric Reeves, "Exotic Policy: An IJC White Paper on Policies for the Prevention of the Invasion of the Great Lakes by Exotic Organisms," Toledo, Ohio, July 15, 1999, 26.
7. Nonindigenous Aquatic Nuisance Prevention and Control Act of 1990.
8. "Effectiveness of Mid-Ocean Exchange in Controlling Freshwater and Coastal Zooplankton in Ballast Water," cited in Reeves, "Analysis of Laws."

9. Katherine Weathers and Eric Reeves, "The Defense of the Great Lakes against the Invasion of Nonindigenous Species in Ballast Water," *Marine Technology* 33, no. 2 (1996): 92–100.
10. Eric Reeves, "Exotic Politics: An Analysis of the Law and Politics of Exotic Invasions in the Great Lakes," *Toledo Journal of Great Lakes' Law, Science & Policy*, 2000: 145.
11. Chris Wiley, "Ballast Water Management in Canada: National Direction, Regional Realities," *Toledo Journal of Great Lakes Law, Science & Policy* 2, no. 2 (2000): 253–54.
12. Ibid.
13. "Ballast Water Management for Vessels Entering the Great Lakes That Declare No Ballast On Board," U.S. Coast Guard notice of proposed rulemaking, *Federal Register* 70, no. 5, Jan. 7, 2005.
14. "U-M Led Study: Rules to Protect Great Lakes from Ship-borne Organisms Are Inadequate; Stronger Measures Advocated," University of Michigan News Service press release, Ann Arbor, Mich., July 10, 2007.
15. "Great Lakes Aquatic Nonindigenous Species List," National Oceanic and Atmospheric Administration, National Center for Research on Aquatic Invasive Species, Ann Arbor, Mich., in author's possession.
16. Claudia Copeland, "Clean Water Act: A Summary of the Law," Congressional Research Service Report to Congress, Congressional Research Service, Washington, D.C., Jan. 20, 1999.
17. "Research and Management Priorities for Aquatic Invasive Species in the Great Lakes," unpublished paper prepared by the International Association for Great Lakes Research, Ann Arbor, Mich., in author's possession.
18. "Great Lakes Aquatic Nonindigenous Species List."
19. Reeves, "Exotic Policy," 41.

[11] MELTDOWN

1. David Jude interview with author, Sept. 12, 2007, South Haven, Mich.
2. Stephen R. Hensler and David J. Jude, "Diel Vertical Migration of Round Goby Larvae in the Great Lakes," *Journal of Great Lakes Research* 33 (2007): 295–302.
3. Jeffrey S. Schaeffer, Anjanette Bowen, Michael Thomas, John R. P. French III, and Gary L. Curtis, "Invasion, History, Proliferation and Offshore Diet of Round Goby in Western Lake Huron, USA," *Journal of Great Lakes Research* 31 (2005): 414–25.

4. David Jude, "Round Gobies: Cyberfish of the Third Millennium," *Great Lakes Research Review* 3, no. 1 (1997): 32–33.
5. Ibid.
6. David Jude, "The Galloping Goby Blues," unpublished and undated lyrics, in author's possession.
7. "Invaders Create Meltdown," *Muskegon Chronicle*, March 7, 2000.
8. Henry A. Vanderploeg, Thomas F. Nalepa, David J. Jude, Edward L. Mills, Kristen T. Holeck, James R. Leibig, Igor A. Grigorovich, and Henn Ojaveer, "Dispersal and Emerging Ecological Impacts of Ponto-Caspian Species in the Laurentian Great Lakes," *Canadian Journal of Fisheries and Aquatic Sciences* 59 (2002): 1224.
9. Jeffery Reutter interview with author, Put-in-Bay, Ohio, June 7, 2007.
10. Testimony of David Lodge before the U.S. House of Representatives Committee on Transportation and Infrastructure Subcommittee on Water Resources and Environment hearing on "The Impact of Aquatic Invasive Species on the Great Lakes," March 7, 2007, Washington, D.C., 1–2.
11. Transcript of U.S. Coast Guard's public hearing "Public Scoping Meeting on the Environmental Impact Statement for the Ballast Water Discharge Standards Rulemaking," Oct. 31, 2003, Cleveland, Ohio, 62–63.
12. "The Fish We Love to Hate: Foreign Invaders Prompt Hundreds to Commit 'Gobycide,'" *Muskegon Chronicle*, June 11, 2005.
13. "Goby Population Explodes in Lakes," *Muskegon Chronicle*, Nov. 13, 2004.

[12] SOMETHING AMUCK

1. "Lake Ontario Algae Cause and Solution Workshop Proceedings," May 30, 2002, Rochester, New York, organized by New York Sea Grant. The $20 billion figure was cited by Murray Charlton of Environment Canada, a presenter at the workshop.
2. Scott N. Higgins, E. Todd Howell, Robert E. Hecky, Stephanie J. Guildford, and Ralph E. Smith, "The Wall of Green: The Status of *Cladophora glomerata* on the Northern Shores of Lake Erie's Eastern Basin, 1995–2002," *Journal of Great Lakes Research* 31 (2005): 547–63.
3. "A Greener Lake Erie," International Association for Great Lakes Research press release based on the study by Joseph D. Conroy, Douglas D. Kane, David M. Dolan, William J. Edwards, Murray N. Charlton, and David A. Culver, "Temporal Trends in Lake Erie Plankton Biomass: Roles of External Phosphorus Loading and Dreissenid Mussels," *Journal of Great Lakes Research* 31, no. 2 (2005): 89–110.

4. Victoria A. Harris, "*Cladophora* Confounds Coastal Communities—Public Perceptions and Management Dilemmas," paper contained in "*Cladophora* Research and Management in the Great Lakes: Proceedings of a Workshop Held at the Great Lakes WATER Institute, University of Wisconsin-Milwaukee," Dec. 8, 2004, accessed via www.glwi.uwm.edu.
5. Joseph Martin interview with author, Oct. 30, 2007.
6. "*Cladophora* Research and Management in the Great Lakes: Proceedings of a Workshop Held at the Great Lakes WATER Institute, University of Wisconsin-Milwaukee."
7. Greg Pien interview with author, Oct. 30, 2007.
8. Wendy S. Stankovich, "The Interaction of Two Nuisance Species in Lake Michigan: *Cladophora glomerata* and *Dreissena polymorpha*," paper contained in "*Cladophora* Research and Management in the Great Lakes: Proceedings of a Workshop Held at the Great Lakes WATER Institute, University of Wisconsin-Milwaukee," Dec. 8, 2004.
9. Henry Singer interview with author, Oct. 12, 2007, Cross Village, Mich.
10. Ibid.
11. "Floating Algae Mess Takes the Bloom off Nice Beach Weekend," *Muskegon Chronicle*, June 18, 2007.

[13] BLUE, GREEN, AND DEADLY

1. "A Lake Full of Mysteries," *Burlington Free Press*, Sept. 21, 2006.
2. Greg Boyer presentation on cyanobacteria at the annual meeting of the International Association of Great Lakes Research, Penn State University, State College, Pa., May 29, 2007.
3. Gregory Boyer, "Toxic Cyanobacteria in the Great Lakes: More Than Just the Western Basin of Lake Erie," *Great Lakes Research Review* 7 (2006): 2.
4. Ibid.
5. Gregory Boyer and Julie Dyble, "Harmful Algal Blooms: A Newly Emerging Pathogen in Water," Great Lakes Environmental Research Laboratory, Ann Arbor, Mich., unpublished manuscript accessed via http://cws.msu.edu/documents/HarmfulAlgalBloomsWhitePaper_Boyer_Dyble.pdf).
6. Boyer, "Toxic Cyanobacteria in the Great Lakes," 2.
7. Henry A. Vanderploeg, James R. Leibig, Wayne W. Carmichael, Megan A. Agy, Thomas H. Johengen, Gary L. Fahnenstiel, and Thomas F. Nalepa, "Zebra Mussel Selective Filtration Promoted Toxic Microcystis Blooms in Saginaw Bay (Lake

Huron) and Lake Erie," *Canadian Journal of Fisheries and Aquatic Sciences* 58 (2001): 1208–21.

8. "Lakes with Zebra Mussels Have Higher Levels of Toxins, MSU Research Finds," *Science Daily*, March 11, 2004, accessed via www.sciencedaily.com.
9. "Scientists Warn of Toxic Algae Spread," *Muskegon Chronicle*, March 4, 2005.
10. Tim Keilty and M. Megan Woller, "A Community Partnership Approach to Zebra Mussel Control," Leelanau Watershed Council and Leelanau Conservancy, 2005, accessed via www.theconservancy.com.
11. "Danger That Floats—Toxic Algae Blooming in Area Lakes," *Muskegon Chronicle*, Oct. 10, 2004.
12. "Bear Lake Algae Draws First Ever Health Warning," *Muskegon Chronicle*, July 12, 2005.
13. Harmful Algal Bloom Workshop, Bay City, Mich., May 8, 2007, comments by Rick Rediske, water resources professor, Grand Valley State University, Allendale, Mich.
14. Boyer and Dyble, "Harmful Algal Blooms," 9–10.

[14] A CRUEL HOAX

1. "Outbreaks of Rare Botulism Strain Stymie Scientists," *New York Times*, Oct. 22, 2002.
2. Bob Wellington interview with author, Sept. 16, 2007, Presque Isle State Park, Erie, Pa.
3. "Botulism in Lake Erie Workshop Proceedings," New York Sea Grant, Ohio Sea Grant, and Pennsylvania Sea Grant, March 25, 2004, Erie, Pa. Proceedings by Helen M. Domske, September 2004, Buffalo, N.Y.
4. Ibid.
5. Ibid., 10.
6. "Birds Feeding at Lake Erie Die in Botulism Outbreak," *New York Times*, Nov. 19, 2002.
7. Ward Stone telephone interview with author, Sept. 27, 2007.
8. David Adams, Kenneth Roblee, and Ward Stone, "Type E Botulism Impact on Waterbirds, Lake Erie and Lake Ontario, New York State," New York Department of Environmental Conservation, Division of Fish, Wildlife and Marine Resources, unpublished data in author's possession.
9. "Loons' Water Turns Deadly," *Rochester Times Union*, Oct. 27, 2006.

10. Irene Mazzochi interview with author, Nov. 15, 2007, Bailey's Harbor, N.Y.
11. Tom Cooley interview with author, Nov. 8, 2007, East Lansing, Mich.
12. Ibid.
13. Henry Singer, personal communication with author, Nov. 6, 2007.

[15] CASPIAN SEA DIET

1. Chuck Madenjian interview with author, Sept. 20, 2007, Grand Haven, Mich.
2. Ibid.
3. "Prey Fish Dwindling in Lake Michigan," *Milwaukee Journal Sentinel*, Jan. 12, 2008.
4. Jim Johnson presentation at Lake Huron Fishery Workshop, Apr. 19, 2008, Alpena, Mich.
5. "Scientists See Trouble Ahead for Big Lakes," *Muskegon Chronicle*, May 30, 2007.
6. "Great Lakes Fish Community Impacted by *Diporeia* Disappearance," fact sheet prepared by National Oceanic and Atmospheric Administration's Great Lakes Environmental Research Laboratory, Ann Arbor, Mich., accessed via http://www.glerl.noaa.gov.
7. "Aquatic Invasive Species and Recent Food Web Disruptions in the Great Lakes," presentation by Thomas F. Nalepa, Great Lakes Environmental Research Laboratory, National Oceanic and Atmospheric Administration, Ann Arbor, Mich., in author's possession.
8. "Important Food Source for Fish Disappearing from Great Lakes," Associated Press, Feb. 1, 2005, in author's possession.
9. Ibid.

[16] WHITEFISH AND GREEN SLIME

1. "Great Lakes Landings by Lake," National Marine Fisheries Service, National Oceanic and Atmospheric Administration, Washington, D.C., in author's possession.
2. "Great Lakes Whitefish," Michigan Sea Grant, accessed via www.greatlakeswhitefish.com.
3. Brady Baker interview with author, June 13, 2007, Naubinway, Mich.
4. Ibid.
5. Bob King interview with author, June 13, 2007, Naubinway, Mich.

[17] PARADOX

1. Tammy Newcomb, "The Changing Nature of Lake Huron's Fishery—Pandemonium or Promise," presentation at Lake Huron Fishery Workshop, April 19, 2008, Alpena, Mich.
2. Jim Johnson presentation at Lake Huron Fishery Workshop, April 19, 2008, Alpena, Mich.
3. Jim Seelye interview with author, Oct. 11, 2007, Millersburg, Mich.
4. Jim Johnson interview with author, July 29, 2007, Alpena, Mich.
5. Tammy Newcomb, "The Changing Nature of Lake Huron's Fishery—Pandemonium or Promise," presentation at Michigan State University, March 27, 2008.
6. Ed Retherford interview with author, Oct. 10, 2007, Rogers City, Mich.
7. "Lake Huron's New Ecosystem and Food Web," unpublished report by Michigan Department of Natural Resources, Lansing, Mich., in author's possession.
8. Jim Johnson interview with author, July 19, 2007, Alpena, Mich.
9. "Lake Huron's New Ecosystem and Food Web, Spring 2007," Michigan Department of Natural Resources, accessed via www.michigan.gov/documents/LakeHuronNewEcosystem-foodweb_122463_7.pdf.
10. Ed Retherford interview with author, Oct. 10, 2007, Rogers City, Mich.
11. Ibid.
12. David Jude, personal communication with author, Oct. 18, 2007.
13. Ed Retherford interview with author, Oct. 10, 2007, Rogers City, Mich.
14. James Baker speech at Lake Huron Fishery Workshop, April 19, 2008, Alpena, Mich.
15. Jim Johnson interview with author, July 29, 2007, Alpena, Mich.
16. Jeff Schaefer, U.S. Geological Survey, presentation at Lake Huron Fishery Workshop, April 19, 2008.
17. "Fishing Industry Suffers as Lakes Shift: Commercial Fishermen Watch Prices Drop for Their Catch As Invasive Species Grow in Number," *Detroit News*, Aug. 14, 2005.
18. "Big Lake's Fish Population Plummeting," *Muskegon Chronicle*, Jan. 5, 2008.
19. Johnson presentation at Lake Huron Fishery Workshop.

[18] FEAR THIS

1. Stephen J. Dubner and Steven D. Levitt, "The Jane Fonda Effect," *New York Times*, Sept. 16, 2007.

2. Ibid.
3. "Ontario Power Generation's Pickering Station Deals with Algae," Ontario Power Generation press release, Aug. 19, 2005, in author's possession.
4. "Ontario Power Generation's Darlington Station Deals with Algae and Silt," Ontario Power Generation press release, Sept. 29, 2005, in author's possession.
5. Ibid.
6. "Seaweed Again Shuts Down Nuclear Plant," *Syracuse Post-Standard*, Oct. 30, 2007.
7. "Algae Mesh Barrier: Presentation to Pickering Community Advisory Council, Dec. 12, 2006," PowerPoint presentation in author's possession; and "Algae Prompt Reactor Shutdown," *Toronto Star*, Aug. 10, 2007.
8. "James A. FitzPatrick Nuclear Power Plant License Renewal Application," U.S. Nuclear Regulatory Commission, accessed via www.nrc.
9. "How a Mussel Shut Down a Nuke," *Syracuse Post-Standard*, Nov. 12, 2007.
10. "Nuke Plant Struggles against Seaweed," *Syracuse Post-Standard*, Oct. 16, 2007.
11. U.S. Nuclear Regulatory Commission Review of Emergency Shutdown at James A. FitzPatrick Nuclear Power Plant, January 2006, in author's possession.

[19] DIRTY SECRETS

1. Bruce Babbitt, "Launching a Counterattack against the Pathogens of Global Commerce," keynote address to First National Conference on Marine Bioinvasions, Massachusetts Institute of Technology, Sea Grant College, Cambridge, Mass., Jan. 26, 1999, speech in author's possession.
2. Ibid.
3. "Executive Order 13112 of Feb. 3, 1999, President William Jefferson Clinton, in author's possession.
4. "Petition for Repeal of 40 C.F.R. 122.3(a)," Northwest Environmental Advocates et al., Jan. 13, 1999, in author's possession.
5. Gregory M. Reid and David F. Reid. "Current State of Understanding about the Effectiveness of Ballast Water Exchange in Reducing Aquatic Nonindigenous Species Introductions to the Great Lakes Basin and Chesapeake Bay, USA: Synthesis and Analysis of Existing Information." National Oceanic and Atmospheric Administration Technical Memorandum GLERL-142, September 2007, Ann Arbor, Mich.

6. Dave Knight, Great Lakes Commission, "As Ballast Water Discharge Standards Loom Focus on Treatment Technology Intensifies," personal communication with author, Aug. 15, 2008.
7. Northwest Environmental Advocates' petition for repeal of 40 C.F.R. 122.3(a), Jan. 15, 1999, Portland. Ore., in author's possession.
8. J. Charles Fox letter to Pacific Environmental Advocacy Center, April 6, 1999, cited in Complaint of Pacific Environmental Advocacy Center and Earthjustice Legal Defense Fund against the U.S. Environmental Protection Agency, April 2, 2001, U.S. District Court, San Francisco, Calif., in author's possession.
9. Sandra Zellmer, "Vanquishing Exotic Species in the Great Lakes," in state of Michigan's 1999 State of the Great Lakes Report, Michigan Department of Environmental Quality, Lansing, Mich., 30–32.
10. "EPA Urged to Act against Lake Invaders," *Muskegon Chronicle*, Nov. 14, 2002.
11. "EPA Won't Regulate Ship Ballast Discharge," *San Jose Mercury News*, Sept. 3, 2003.
12. Ibid.
13. Ibid.
14. "Order Granting Plaintiffs' Motion for Permanent Injunctive Relief," *Northwest Environmental Advocates et al. v. U.S. Environmental Protection Agency*, U.S. District Judge Susan Illston, Sept. 18, 2006, U.S. District Court, Northern District of California, San Francisco, Calif.
15. Ibid.
16. G. Tracy Mehan interview with author, March 5, 2008.
17. Ibid.
18. "Order Granting Plaintiffs' Motion for Permanent Injunctive Relief," *Northwest Environmental Advocates et al. v. U.S. Environmental Protection Agency*, U.S. District Judge Susan Illston, Sept. 18, 2006, U.S. District Court, Northern District of California, San Francisco, Calif., 9–14.
19. Ibid., 15.
20. "Great Lakes Aquatic Nonindigenous Species List," National Oceanic and Atmospheric Administration's Great Lakes Environmental Research Laboratory, Ann Arbor, Mich., accessed via www.noaa.glerl.gov on June 15, 2008.
21. "Invasive Species: Their Impact and Our Response," Jeffrey Reutter testimony at the Great Lakes Regional Public Meeting of the U.S. Commission on Ocean Policy, Sept. 25, 2002, Chicago, Ill.

22. Frank Lichtkoppler, "Twenty Years Post-Invasion: Overview of Impacts from Dreissenids on Recreational Users in Lake Erie," presented by Ohio Sea Grant at International Association of Great Lakes Research annual conference, Trent University, Peterborough, Ontario, May 22, 2008.
23. Jeffery Reutter interview with author, June 7, 2007, Put-in-Bay, Ohio.

[20] WHO'S IN CHARGE?

1. Matt Edwards interview with author, Sept. 17, 2007, Massena, N.Y.
2. "Coast Guard Commandant Visits Cleveland for Marine Community Day," U.S. Coast Guard news release, Feb. 23, 2008, in author's possession.
3. Ibid.
4. Eric Reeves interview with author, Oct. 15, 2007, Whitehall, Mich.
5. Ibid.
6. Congressman Vernon Ehlers interview with author, Aug. 9, 2007, Muskegon, Mich.
7. Arlington National Cemetery website, http://www.arlingtoncemetery.org/
8. Eric Reeves, personal communication with author, Sept. 21, 2007.
9. "Shipping Industry Attracts Ex-Coast Guard Officers: Some See Conflict When Enforcers Join the Regulated," *Seattle Post-Intelligencer*, Nov. 21, 2002.
10. Duluth Seaway Port Authority press release, Duluth, Minn., March 10, 2003.
11. "Dry Cargo Residue Discharges in the Great Lakes: Notice of Proposed Rulemaking and Availability of Draft Environmental Impact Statement," *Federal Register* 73, no. 101, May 23, 2008, 30014–24.
12. "Coast Guard Study Will Look at Effects of Ship 'Sweepings,'" *Muskegon Chronicle*, March 14, 2006.
13. David F. Reid and Guy A. Meadows, "Proceedings of the Environmental Implications of Cargo Sweeping in the Great Lakes," NOAA Technical Memorandum ERL GLERL-114, Ann Arbor, Mich., 1999, 48.
14. "Protecting Our Great Lakes: Ballast Water and the Impact of Invasive Species," a hearing before the Subcommittee on Regulatory Affairs of the Committee on Government Reform, U.S. House of Representatives, Sept. 9, 2005, Fair Haven, Mich., Serial no. 109–98, p. 90.
15. "Seaway Acts to Flush Great Lakes Invaders," *Muskegon Chronicle*, May 6, 2008.
16. Ibid.

[21] MISSION IMPOSSIBLE

1. "Freshwater Ballast Testing Facility Opens," Minnesota Sea Grant newsletter, Duluth, Minn., August 2007.
2. Allegra Cangelosi interview with author, Sept. 25, 2007, Duluth, Minn.
3. "Introduction to the Clean Water Act: History, Objective, Goals and Scope of the Act," National Science and Technology Center Bureau of Land Management, 3, in author's possession.
4. "Report to Congress on the Great Lakes Ecosystem," U.S. Environmental Protection Agency, Washington, D.C., February, 1994, 6, in author's possession.
5. Nicole Mays, "Ballast Water Research and Funding: A History and Analysis," Northeast Midwest Institute, Washington, D.C., June 2001, 15.
6. Helen Brohl and Ivan Lantz, "Perspective of Commercial Maritime Regarding Exotic Species," paper presented at the "Exotics and Public Policy in the Great Lakes" workshop, Sept. 21, 1999, Milwaukee, Wis., and reprinted in the *Toledo Journal of Great Lakes Law, Science and Policy* 2, no. 2 (2000): 245–47.
7. Chris Wiley, "Ballast Water Management in Canada: National Direction, Regional Realities," *Toledo Journal of Great Lakes Law, Science and Policy* 2, no. 2 (2000): 249–60.
8. Eric Reeves, "Exotic Politics: An Analysis of the Law and Politics of Exotic Invasions in the Great Lakes," *Toledo Journal of Great Lakes' Law, Science & Policy* (2000): 133–34.
9. Eric Reeves interview with author, Oct. 15, 2007, Whitehall, Mich.
10. "Sixth Biennial Report under the Great Lakes Water Quality Agreement of 1978," International Joint Commission, Washington, D.C., 15–16.
11. Frank D'Itri, ed., *Zebra Mussels and Aquatic Nuisance Species* (Chelsea, Mich.: Ann Arbor Press, 1997), 288.
12. "Aquatic Nuisance Species in Ballast Water Discharges: Issues and Options," draft report for public comment, Sept. 10, 2001, U.S. EPA Office of Water, Office of Wetlands, Oceans and Watersheds, Office of Wastewater Management, Washington, D.C., 11.
13. Eric Reeves, "Exotic Politics: An Analysis of the Law and Politics of Exotic Invasions in the Great Lakes," *Toledo Journal of Great Lakes' Law, Science & Policy* 2, no. 2 (2000): 197.
14. Eric Reeves interview with author, Oct. 15, 2007, Whitehall, Mich.
15. Mark Coscarelli interview with author, Sept. 21, 2007, Lansing, Mich.
16. Ibid.

17. "Studies to Address the Issues Raised by the Michigan Environmental Science Board (2002): Critical Review of a Ballast Water Biocide Treatment Demonstration Project Using Sodium Hypochlorite," submitted to the State of Michigan by BMT Fleet Technology Ltd, Victoria, British Columbia, and Stantec Consulting Ltd, Guelph, Ontario, 32.
18. Mark Coscarelli interview with author, Sept. 21, 2007, Lansing, Mich.
19. International Maritime Organization Ballast Water Management Guidelines, Resolution A.868(20), adopted Nov. 27, 1997, International Maritime Organization, London, England, accessed via http://globallast.imo.org.
20. Joseph J. Cox, "Ballast Water Management: New International Standards and National Invasive Species Act Reauthorization," testimony on behalf of the Shipping Industry Ballast Water Coalition, to the Coast Guard Maritime Transportation and Water Resources and Environment Subcommittees of the House Transportation and Infrastructure Committee, March 25, 2004, 2–5, in author's possession.
21. *Fednav Limited, Canadian Forest Navigation Co. Ltd., Nicholson Terminal and Dock Co., et al. v. Steven E. Chester, director, Michigan Department of Environmental Quality*, U.S. District Court Eastern District of Michigan, Case No. 07-cv-1116, Detroit, Michigan, March 14, 2007.
22. Cangelosi, Allegra, and Nicole Mays. "Great Ships for the Great Lakes? Commercial Vessels Free of Invasive Species in the Great Lakes-St. Lawrence Seaway System: A Scoping Report for the Great Ships Initiative." Washington, D.C.: Northeast Midwest Institute, May 2006, 96–97.
23. "Ballast Project Enlists Industry," *Duluth News-Tribune*, July 12, 2006.
24. "Great Lakes Aquatic Nonindigenous Species List," National Oceanic and Atmospheric Administration, National Center for Research on Aquatic Invasive Species, Ann Arbor, Mich., in author's possession.
25. "European Invader Found in Channel," *Muskegon Chronicle*, Dec. 22, 2006.
26. A. Ricciardi and J. B. Rasmussen, "Predicting the Identity and Impact of Future Biological Invaders: A Priority for Aquatic Resource Management," *Canadian Journal of Fisheries and Aquatic Sciences* 55 (1998): 1759–63.
27. "Noxious Cargo: Loophole in Ballast Law Lets Invasive Species In," *Milwaukee Journal-Sentinel*, Oct. 30, 2005.
28. Adolph Ojard, "Federal Solution to Ballast Water Problem Would Serve Industry and Environment," *Duluth News Tribune*, Aug. 12, 2007.
29. Cangelosi, "Great Ships Initiative Scoping Report," 68.

[22] SEAWAY HERETICS

1. Gary Fahnenstiel speech to media at the National Oceanic and Atmospheric Administration's Lake Michigan Field Station, Aug. 9, 2007, Muskegon, Mich.
2. "Intruders at the Gate," *Milwaukee Journal-Sentinel*, Dec. 26, 2004.
3. Ibid.
4. "NOAA Scientist: Close Door on Lake Invaders," *Muskegon Chronicle*, Dec. 26, 2004.
5. Helen Brohl and Michael Broad, letter to Vice Admiral Conrad C. Lautenbacher Jr., National Oceanic and Atmospheric Administration, Washington, D.C., Dec. 24, 2004, in author's possession. At the time, Brohl was executive director of the U.S. Great Lakes Shipping Association; Broad was president of the Shipping Federation of Canada.
6. "Some Ships Could Lose Lake Access," *Detroit Free Press*, Jan. 2, 2005.
7. Gail Krantzberg telephone interview with author, Nov. 12, 2007.
8. Ibid.
9. Ibid.
10. John C. Taylor and James L. Roach, *Ocean Shipping in the Great Lakes: Transportation Cost Increases That Would Result from a Cessation of Ocean Vessel Shipping* (Grand Rapids, Mich.: Grand Valley State University, 2005), 1–5.
11. Ibid., 4.
12. Taylor and Roach, *Ocean Shipping*, 21.
13. Ibid., 19; and personal communication with Collister Johnson Jr., director of the St. Lawrence Seaway Development Corp., July 14, 2008.
14. Albert Jacquez, "Reporting on Reports," in *Seaway Compass*, newsletter of the St. Lawrence Seaway Development Corp., Washington, D.C., April 2006, 1.
15. "Sinking Treasure: Too Small for Big Ships, Seaway Hasn't Lived up to the Fanfare of the '50s," *Milwaukee Journal-Sentinel*, Oct. 29, 2005.
16. Ibid.
17. "St. Lawrence Seaway Security Focus Insufficient, Analyst Says," *Watertown (N.Y.) Daily Times*, March 15, 2005.
18. "Scientists: Great Lakes Ecosystem in Danger of Collapse," *Water & Wastewater News*, Dec. 1, 2005.
19. "Testimony of Frank Ettawageshik, tribal chairman of the Little Traverse Bay Bands of Ottawa Indians, before the House Transportation and Infrastructure Subcommittee on Water Resources and Environment, March 7, 2007, Washington, D.C.

20. "Testimony of Adolph Ojard, president of the American Great Lakes Ports Association and executive director of the Duluth Seaway Port Authority before the House Transportation and Infrastructure Subcommittee on Water Resources and Environment, March 7, 2007, Washington, D.C.
21. "Lock the Lakes, Groups say," *Milwaukee Journal-Sentinel,* May 23, 2007.
22. John C. Taylor and James L. Roach, "Ocean Shipping in the Great Lakes: An Analysis of Issues," Grand Rapids, Mich., October, 2007, 6–7.
23. "Study: Great Lakes 'Salties' Ban May Create Jobs," *Muskegon Chronicle,* March 25, 2008.
24. Ibid.
25. "Great Lakes Shipping, Trade and Aquatic Invasive Species," National Research Council of the National Academies, Transportation Research Board, Special Report 291, Washington, D.C., July 2008, 104.
26. Ibid., 81.
27. "Treatment of Ballast Water Deemed Crucial," *Muskegon Chronicle,* July 18, 2008.
28. Gail Krantzberg interview with author, Nov. 12, 2007.

[23] WESTWARD HO!

1. "Fatal Mistakes Lead to Divers' Deaths," KGO-TV, San Francisco, Calif., Feb. 27, 2008.
2. "California Department of Water Resources Review Report of Serious Accident, San Luis Division Dive Fatalities," Dec. 10, 2007, in author's possession.
3. "Review Report of Serious Accident," California Department of Water Resources, Dec. 10, 2007, appendix S, p. 16.
4. "Wish You Weren't Here," *Paonia (Colo.) High Country News,* 39, no. 4 (March 5): 2007.
5. "Zebra Mussel's Arrival Threatens Lake Mead's Ecosystem," *Las Vegas Sun,* Jan. 11, 2007.
6. "California's Response to the Zebra/Quagga Mussel Invasion in the West: Recommendations of the California Science Advisory Panel," prepared for the California Incident Command, May 2007, 11.
7. "The 100th Meridian Initiative: A Strategic Approach to Prevent the Westward Spread of Zebra Mussels and Other Aquatic Nuisance Species," U.S. Fish and Wildlife Service, Washington, D.C., 1998, 5, in author's possession.
8. Ibid.

9. Ibid.
10. "Mussels on the Move," zebra/quagga mussel informational brochure, 100th Meridian Initiative, accessed via www.100thmeridian.org.
11. "The 100th Meridian Initiative," 8.
12. "Zebra and Quagga Mussel—California Distribution," California Department of Water Resources fact sheet, in author's possession.
13. "Lake Mead Mussels Identified as Quagga, Not Zebra," *Las Vegas Sun*, Jan. 13, 2007.
14. "California's Response to the Zebra/Quagga Mussel Invasion in the West: Recommendations of the California Science Advisory Panel," prepared for the California Incident Command, May 2007, i.
15. "Potential Economic Impacts of Zebra Mussels on the Hydropower Facilities in the Columbia River Basin," prepared for the Bonneville Power Administration by Stephen Phillips, Pacific States Marine Fisheries Commission, February 2005, 2.
16. "California's Response to the Zebra/Quagga Mussel Invasion in the West," 23.
17. Andrew Cohen interview with author, Feb. 4, 2008.
18. "Invasive Quagga Mussels Identified in California," California Department of Fish and Game press release, Jan. 19, 2007, in author's possession.
19. "Meltdown of the Lakes: Great Lakes in Crisis," *St. Catherines Standard*, St. Catherines, Ontario, June 19, 2003.
20. "Quagga Mussel Running History," California Department of Fish and Game fact sheet, in author's possession.
21. Zebra Mussel Found in California Reservoir," *Los Angeles Times*, Jan. 16, 2008.
22. Gary Britton correspondence with author, Nov. 11, 2007, notes in author's possession
23. Ibid.
24. Kevin Len Burch and Robert F. McMahon, "Assessment of Potential for Dispersal of Aquatic Nuisance Species by Recreational Boaters into the Western United States," prepared for the Western Regional Panel of the Aquatic Nuisance Species Task Force by the University of Texas at Arlington, May 29, 2001, iii.
25. "Zebra Mussels Making Their Way Downstream," *Albany Times Union*, Aug. 6, 1991.
26. "Wish You Weren't Here," *Paonia (Colo.) High Country News*.
27. Andrew Cohen interview.

[24] SAVING PARADISE

1. "Viral Hemorrhagic Septicemia: A New Invader in the Great Lakes," *Fish Lines*, U.S. Fish and Wildlife Service, vol. 5, no. 5, 2007.
2. "Fish-Killing Virus Nears," *Milwaukee Journal-Sentinel*, Jan. 31, 2007.
3. "Viral Hemorrhagic Septicemia in the Great Lakes: July 2006 Emerging Disease Notice," U.S. Department of Agriculture press release, Washington, D.C., in author's possession.
4. "Cornell Lab Confirms Deadly Fish Virus Spreading to New Species," *Science Daily*, accessed via www.sciencedaily.com.
5. Bob Kelleher, "Group Warns of Approaching Fish Disease," *Minnesota Public Radio*, Feb. 14, 2007, transcript obtained at http://minnesotapublicradio.org.
6. Ibid.
7. Gary Glass interview with author, Sept. 25, 2007, Duluth, Minn.
8. Emergency Declaration Needed to Keep VHS Virus out of Twin Ports," Gary Glass e-mail to office of U.S. Rep. James Oberstar, Jan. 12, 2007, in author's possession.
9. "Ballast Water Reporting: VHS Virus Contaminated Waters," Karen Phillips e-mail to Gary Glass, Feb. 8, 2007, in author's possession.
10. Phyllis Green speech at Michigan State University, Mar. 27, 2008.
11. "Park Boss Bans Ballast Water Dumping Near Isle Royale," *Muskegon Chronicle*, Sept. 18, 2007.
12. Ibid.
13. "Lawsuits, Legislation Aim to Regulate Ships' Ballast Water," *Duluth News Tribune*, Sept. 18, 2007.
14. Phyllis Green speech at Michigan State University, Mar. 27, 2008.
15. Adolph Ojard interview with author, Sept. 24, 2007, Duluth, Minn.
16. John Hand telephone interview with author, Feb. 12, 2008.
17. Cameron Davis telephone interview with author, Apr. 30, 2008.

[EPILOGUE] HOPE AMID THE RUINS

1. "A Plague on the Yankees," *New York Times*, Oct. 6, 2007; and "Last Night's Action: Big Trouble in Yankeeland," www.gothamist.com.
2. Don W. Schlosser, Kenneth A. Krieger, Jan J. H. Ciborowski, and Lynda D. Corkum, "Recolonization and Possible Recovery of Burrowing Mayflies in Lake Erie of the Laurentian Great Lakes," *Journal of Aquatic Ecosystem Stress and Recovery* 8, no. 2 (2000): 125–41.

3. "Burrowing Mayfly Swarms Signal a Healthier Ecosystem, USGS Scientist Says," U.S. Geological Survey press release, U.S. Department of Interior, Reston, Va., July 13, 1999, in author's possession.
4. Kristin Stanford interview with author, June 7, 2007, Put-in-Bay, Ohio.
5. Ibid.
6. "Snake Charmer: Researcher Has Made the Survival of Endangered Lake Erie Species Her Career, Passion," *Columbus Dispatch*, Aug. 14, 2007.
7. Ibid.
8. Jeffery Reutter interview with author, Put-in-Bay, Ohio, June 7, 2007.

GLOSSARY

algae Aquatic photosynthetic organisms, ranging in size from microscopic single-celled forms to large filamentous and colonial forms.

amphipods Crustaceans with laterally compressed bodies and leg-like appendages. Small (ca. 5–10 mm) shrimp-like organisms associated with the bottom. The most important species is *Diporeia*.

ballast Material such as dirt, rocks, and water, used to adjust the trim, stability, and draft of ships.

ballast tank A specially designed cargo hold used to store ballast water on board a ship. Individual ballast tanks can range in size from one cubic meter to thousands of cubic meters each on freighters.

ballast tank flushing, or "swish and spit" A process by which a ship with nearly empty ballast tanks takes in sea water and allows it to slosh around briefly before discharging it. The process removes residual ballast water and sediment, which often harbor aquatic organisms, from the bottom of ballast tanks.

ballast water exchange A process in which a transoceanic freighter destined for the Great Lakes replaces freshwater ballast with ocean water before entering the St. Lawrence River. The process purges or kills some organisms in ballast tanks.

benthic zone The bottom area of lakes. Creatures that live in and near the bottom sediments are called benthos.

Cladophora A noxious type of large, green, filamentous algae that forms large mats in the nearshore regions of the Great Lakes.

copepods Tiny freshwater crustaceans (0.2–2 mm) that are very abundant in the pelagic region of the Great Lakes. One of the two most common types of zooplankton in the Great Lakes; the other is Cladocerans.

crustaceans Aquatic animals that have a segmented body, a hard exoskeleton, and paired, jointed limbs. The three major groups of crustaceans in the Great Lakes are Copepods, Cladocerans, and Malacostracans (*Mysis* and *Diporeia*).

cubic meter A volume equal to 264 gallons of water.

cyanobacteria (blue-green algae) A species of microorganisms that are related to bacteria but are capable of photosynthesis. Generally, this group of algae is considered undesirable because it produces noxious blooms and is an unwanted source of food for most invertebrates. Many species of cyanobacteria produce a variety of toxins.

diatom Microscopic form of single-celled algae. These organisms are the important base of the food web and are a desirable source of food for all invertebrates.

Diporeia A benthic amphipod that once was the most abundant source of fish food in the Great Lakes.

***Dreissena* (dreissenids)** A genus of filter-feeding mussels that includes zebra mussels (*D. polymorpha*) and quagga mussels (*D. bugenis*).

efficacy Effectiveness. In the case of ballast water exchange, a low efficacy rate means the process replaced less ballast water, or purged fewer aquatic organisms, than desired.

euryhaline Species capable of tolerating a wide range of saltwater concentrations. Euryhaline organisms can live in freshwater or saltwater; they are usually found in estuaries, the interface between rivers and the ocean.

exotic species An early term assigned to foreign species that invaded the Great Lakes from other ecosystems. Invasive species is considered a more precise term.

extirpated Exterminated, eliminated. Usually refers to a species in a single environment.

invader A nonindigenous species that has successfully established a reproducing population.

invasive, invasive species Nonindigenous species likely to cause economic or environmental harm or threaten human health.

invertebrate Mussels, insects, and other animals that lack a backbone or spinal column.

lakers Freighters that haul cargo exclusively in the Great Lakes. Many of the ships are too large to fit in the Seaway's locks in the Welland Canal and the St. Lawrence River.

metric ton 2,204 pounds.

Microcystis A naturally occurring genus of blue-green algae, or cyanobacteria. Many species of *Microcystis* produce the potent toxin microcystin, which is often found in high concentrations in bright green surface scums in the summer months.

NoBOB An acronym for "No Ballast On Board," a term assigned to freighters carrying no pumpable ballast water in their ballast tanks. However, these ships may still contain thousands of gallons of water and sediment from other environments.

nonindigenous species Any species or viable biological material that enters an ecosystem beyond its historic range or is transferred from one region into another.

pelagic zone The open water, or upper layer, of a lake or ocean.

phytoplankton Tiny, free-floating aquatic plants. Most of these organism are microscopic (< 0.1 mm) and they form the base of the food web. The most desirable forms of phytoplankton are diatoms.

plankton Microscopic organisms, including algae, protozoans, and zooplankton, that drift with wind and wave currents and form the base of the Great Lakes food web.

Ponto-Caspian region The area surrounding and including the Black, Azov and Caspian Sea basins in Eastern Europe. *Dreissena* mussels and several other invasive species in the Great Lakes originated in this region.

prey fish Small fish species, such as bloaters, chubs, and alewives, that are food for larger fish species, such as lake trout, whitefish and salmon.

salties Ocean freighters that visit the Great Lakes.

seiche A fluctuation in water levels caused by sudden changes in atmospheric pressure or strong, sustained winds. The phenomenon can cause rapid changes in Great Lakes water levels.

upwelling Cold water that is pushed to the surface to replace warmer water that has moved offshore. Great Lakes upwellings can cause water temperatures at beaches to drop several degrees in one day.

vector The mechanism or activity by which a species is transported from its present native geographic range to a new ecosystem. Most vectors for nonindigenous species are related to human trade activity, such as maritime commerce, canals and waterways that connect previously separated ecosystems, trade in live organisms (aquaria, live bait, live food, ornamental plants/horticulture), recreational activities, and aquaculture.

veligers Microscopic larval form of zebra and quagga mussels.

zooplankton Plankton that consists of animals, most of which are self-propelled but cannot swim large distances. The two most common zooplankton in the Great Lakes are Copepods and Cladocerans.

BIBLIOGRAPHY

Applegate, Vernon C. "The Menace of the Sea Lamprey." *Michigan Conservation* 16 (1947): 6–10.

Applegate, Vernon. "The Sea Lamprey in the Great Lakes." *Scientific Monthly* 72, no. 5 (1951): 276.

Ashworth, William. *The Late, Great Lakes: An Environmental History*. Detroit: Wayne State University Press, 1986.

Babbitt, Bruce. "Launching a Counterattack against the Pathogens of Global Commerce." Keynote address to the First National Conference on Marine Bioinvasions. Massachusetts Institute of Technology, Sea Grant College, Cambridge, Mass., January 26, 1999.

Bailey, Sarah A., David F. Reid, Robert I. Colautti, Thomas Therriault, and Hugh MacIsaac. "Management Options for Control of Nonindigenous

Species in the Great Lakes." *Toledo Journal of Great Lakes Law, Science and Policy* 5 (Spring 2004): 102.

"Ballast Water Management for Vessels Entering the Great Lakes That Declare No Ballast On Board." U.S. Coast Guard notice of proposed rulemaking. *Federal Register* 70, no. 5 (January 7, 2005).

Becker, William H. *From the Atlantic to the Great Lakes: A History of the U.S. Army Corps of Engineers and the St. Lawrence Seaway*. Historical Division, Office of Administrative Services, Office of the Chief of Engineers, 1986.

Bernstein, Peter L. *Wedding of the Waters: The Erie Canal and the Making of a Great Nation*. New York: W. W. Norton, 2005.

Boyer, Gregory. "Toxic Cyanobacteria in the Great Lakes: More Than Just the Western Basin of Lake Erie." *Great Lakes Research Review* 7 (2006): 2–7.

Brohl, Helen, and Ivan Lantz. "Perspective of Commercial Maritime Regarding Exotic Species." *Toledo Journal of Great Lakes Law, Science and Policy* 2, no. 2 (2000): 245–47.

Brown, Andrew H. "New St. Lawrence Seaway Opens the Great Lakes to the World," *National Geographic* 115, no. 3 (1959): 299–339.

Burch, Kevin Len, and Robert F. McMahon. "Assessment of Potential for Dispersal of Aquatic Nuisance Species by Recreational Boaters into the Western United States." Prepared for the Western Regional Panel of the Aquatic Nuisance Species Task Force. Arlington: University of Texas, May 29, 2001.

Busiahn, Thomas R. "Ruffe Control: A Case Study of an Aquatic Nuisance Species Control Program." In *Zebra Mussels and Aquatic Nuisance Species*, edited by Frank D'Itri. Chelsea, Mich.: Ann Arbor Press, 1997.

Cangelosi, Allegra, and Nicole Mays. "Great Ships for the Great Lakes? Commercial Vessels Free of Invasive Species in the Great Lakes–St. Lawrence Seaway System: A Scoping Report for the Great Ships Initiative." Northeast Midwest Institute, Washington, D.C., May 2006.

Carson, Rachel. *Silent Spring*. New York: Houghton Mifflin, 2002.

Cassuto, David N. *Cold Running River*. Ann Arbor: University of Michigan Press, 1994.

Chiarappa, Michael J., and Kristin M. Szylvian. *Fish for All: An Oral History of Multiple Claims and Divided Sentiment on Lake Michigan*. East Lansing: Michigan State University Press, 2003.

Cohen, Andrew N., et al. "California's Response to the Zebra/Quagga Mussel

Invasion in the West: Recommendations of the California Science Advisory Panel." Prepared for the California Incident Command, May 2007.

Committee on Ships' Ballast Operations, Marine Board, Commission on Engineering and Technical Systems, National Research Council. *Stemming the Tide: Controlling Introductions of Nonindigenous Species by Ships' Ballast Water*. Washington, D.C.: National Academy Press, 1996.

Conroy, Joseph D., et al. "A Greener Lake Erie." *Journal of Great Lakes Research* 31, supplement 2 (2005).

Copeland, Claudia. "Clean Water Act: A Summary of the Law." Congressional Research Service Report to Congress, Congressional Research Service, Washington, D.C., January 20, 1999.

Cox, Joseph J.: "Ballast Water Management: New International Standards and National Invasive Species Act Reauthorization." Testimony on behalf of the Shipping Industry Ballast Water Coalition, to the Coast Guard Maritime Transportation and Water Resources and Environment Subcommittees of the House Transportation and Infrastructure Committee, March 25, 2004.

Crawford, S. S. "Ecological Effects of Stocking Exotic Salmon in the Great Lakes." *Canadian Journal of Fisheries and Aquatic Sciences* 132 (2001).

Crawford, S. S. "Ecological and Genetic Effects of Salmonid Introductions in North America." *Canadian Journal of Fisheries and Aquatic Science* 48, supplement 1 (1991).

D'Itri, Frank, ed. *Zebra Mussels and Aquatic Nuisance Species*. Chelsea, Mich.: Ann Arbor Press, 1997.

Dempsey, Dave. *On the Brink: The Great Lakes in the Twenty-First Century*. East Lansing: Michigan State University Press, 2004.

Dennis, Jerry. *The Living Great Lakes: Searching for the Heart of the Inland Seas*. New York: Thomas Dunne Books, 2003.

Domske, Helen. "Botulism in Lake Erie Workshop Proceedings." New York Sea Grant, Ohio Sea Grant, and Pennsylvania Sea Grant, March 25, 2004, Erie, Pa.

"Dry Cargo Residue Discharges in the Great Lakes: Notice of Proposed Rulemaking and Availability of Draft Environmental Impact Statement." *Federal Register* 73, no. 101 (May 23, 2008): 30014–24.

Egan, Dan. "Sinking Treasure: Too Small for Big Ships, Seaway Hasn't Lived Up to Fanfare of the '50s." *Milwaukee Journal-Sentinel,* October 29, 2005.

Egan, Dan. "Troubled Waters: The Great Invasion." *Milwaukee Journal-Sentinel,* December 12, 19, and 26, 2004.

Ettawageshik, Frank. Testimony before the House Transportation and Infrastructure Subcommittee on Water Resources and Environment, Washington, D.C., March 7, 2007.

"Exotic Species and the Shipping Industry: The Great Lakes–St. Lawrence Ecosystem at Risk; A Special Report to the Governments of the United States and Canada." International Joint Commission, Washington, D.C., and Great Lakes Fishery Commission, Ann Arbor, Mich.

Foley, Bob. *Niagara Falls.* Markham, Ont.: Irving Weisdorf, 2007.

"Great Lakes Aquatic Nonindigenous Species List." National Oceanic and Atmospheric Administration, National Center for Research on Aquatic Invasive Species, Ann Arbor, Mich., 2007.

Great Lakes Fishery Commission. Fact sheet on sea lamprey. Accessed at www.glfc.org.

Great Lakes Fishery Commission. "Sea Lamprey: A Great Lakes Invader." Accessed at www.glfc.org.

"Great Lakes Shipping, Trade, and Aquatic Invasive Species." Special Report 291. National Research Council of the National Academies, Transportation Research Board, Washington, D.C., July 2008.

"Great Lakes St. Lawrence Seaway Study." Transport Canada, U.S. Army Corps of Engineers, U.S. Department of Transportation, St. Lawrence Seaway Management Corp., St. Lawrence Seaway Development Corp., Environment Canada, U.S. Fish and Wildlife Service, Fall 2007.

Gromosiak, Paul. *Water over the Falls: 101 of the Most Memorable Events at Niagara Falls.* Toronto: Royal Specialty Sales, 2006.

Gunderson, Jeffrey L., et al. "Overview of the International Symposium on Eurasian Ruffe Biology, Impacts and Control." *Journal of Great Lakes Research* 24, no. 2 (1998): 165–69.

Hecht, Roger W. *The Erie Canal Reader: 1790–1950.* Syracuse, N.Y.: Syracuse University Press, 2003.

Helberg, Davis. "Needed: A Bigger Seaway." *Minnesota's World Port* (magazine of the Seaway Port Authority of Duluth), Spring 1993.

Hensler, Stephen R., and David J. Jude. "Diel Vertical Migration of Round Goby Larvae in the Great Lakes." *Journal of Great Lakes Research* 33 (2007).

Higgins, Scott N., et al. "The Wall of Green: The Status of *Cladophora glomerata* on the Northern Shores of Lake Erie's Eastern Basin, 1995–2002." *Journal of Great Lakes Research* 31, no. 4 (2005).

Holeck, Kristen T., et al. "Bridging Troubled Waters: Biological Invasions, Transoceanic Shipping and the Laurentian Great Lakes." *BioScience* 54, no. 10 (2004).

Holey, Mark E., et al. "Chinook Salmon Epizootics in Lake Michigan: Possible Contributing Factors and Management Implications." *Journal of Aquatic Animal Health, American Fisheries Society* 10 (1998).

"Invasive Species: Clearer Focus and Greater Commitment Needed to Effectively Manage the Problem." United States General Accounting Office, Washington, D.C., October 22, 2002.

Jackson, John N., and Fred A. Addis. *The Welland Canals: A Comprehensive Guide*. St. Catherines, Ont.: The Welland Canals Foundation, 1982.

Jacquez, Albert. "Reporting on Reports." *Seaway Compass* (newsletter of the St. Lawrence Seaway Development Corp.), April 2006.

Jude, David. "Round Gobies: Cyberfish of the Third Millennium." *Great Lakes Research Review* 3, no. 1 (1997).

Kennedy, John F. "Remarks of Senator John F. Kennedy on the Saint Lawrence Seaway before the Senate, Washington, D.C., Jan. 14, 1954." From the John F. Kennedy Presidential Library and Museum, Boston, Mass., accessed via .

LePage, Wilfred. "The Impact of *Dreissena polymorpha* on Waterworks Operations at Monroe, Michigan: A Case History." In *Zebra Mussels: Biology, Impact and Control*, edited by Thomas F. Nalepa and Donald W. Schloesser. Boca Raton, Fla.: Lewis Publishers, CRC Press, 1993.

Lesstrang, Jacques. *Seaway: The Untold Story of North America's Fourth Seacoast*. Vancouver, B.C.: Superior Publishing Co., 1976.

Lodge, David. "The Impact of Invasive Species on the Great Lakes." Testimony before the U.S. House of Representatives Subcommittee on Water Resources and Environment, Washington, D.C., March 7, 2007.

Lupi, Frank, and Douglas B. Jester. "Uses of Resource Economics in Managing Great Lakes Fisheries: Michigan Examples." In *Sustaining North American Salmon: Perspectives across Regions and Disciplines*, edited by Kristine D. Lynch, Michael L. Jones, and William W. Taylor. Bethesda, Md.: American Fisheries Society, 2002.

Mabee, Carleton. *The Seaway Story*. New York: Macmillan, 1961.

"Marine Investigation Report Breakup and Sinking: The Bulk Carrier 'FLARE,' Cabot Strait, 16 January 1998." Transportation Safety Board of Canada, Gatineau, Quebec, 39–40.

Martin Associates. "Economic Impact Study of the Great Lakes St. Lawrence Seaway System." Prepared for the U.S. St. Lawrence Seaway Development Corp., August 2001.

Mays, Nicole. "Ballast Water Research and Funding: A History and Analysis." Prepared for Northeast Midwest Institute, Washington, D.C., June 2001.

McConville, Daniel J. "Seaway to Nowhere." *American Heritage* 11, no. 2 (1995).

McIntyre, Judith W. *The Common Loon: Spirit of the Northern Lakes*. Minneapolis: University of Minnesota Press, 1989.

Meersman, Tom. "Invaded Waters." *Minneapolis Star-Tribune*, June 13–15, 2004.

Miller, Daniel J., and Neil H. Ringler. "Atlantic Salmon in New York." State University of New York College of Environmental Science and Forestry, Syracuse, N.Y., 1996, accessed via www.esf.edu.

Mills, Edward L., Joseph H. Leach, James T. Carlton, and Carol L. Secor. "Exotic Species in the Great Lakes: A History of Biotic Crises and Anthropogenic Introductions." *Journal of Great Lakes Research* 19, no. 1 (1993).

Nalepa, Thomas F., and Donald W. Schloesser, eds. *Zebra Mussels: Biology, Impact, and Control*. Boca Raton, Fla.: Lewis Publishers, CRC Press, 1993.

Newcomb, Tammy. "The Changing Nature of Lake Huron's Fishery: Pandemonium or Promise." Michigan Department of Natural Resources, presentation at Michigan State University, March 27, 2008.

Newman, R. M. "Ruffe: A Problem or Just a Pest?" *Aquatic Nuisance Species Digest* 3, no. 4 (1999).

Nijhuis, Michelle. "Wish You Weren't Here." *Paonia (Colo.) High Country News* 39, no. 4 (2007).

Ojard, Adolph. "The Impact of Aquatic Invasive Species on the Great Lakes." Testimony before the House Transportation and Infrastructure Subcommittee on Water Resources and Environment, Washington, D.C., March 7, 2007.

Pimentel, David. "Aquatic Nuisance Species in the New York State Canal and Hudson River Systems and Great Lakes Basin: An Economic and Environmental Assessment." *Environmental Management* 35, no. 5 (2005): 692–701.

"The Presence and Implication of Foreign Organisms in Ship Ballast Waters Discharged into the Great Lakes." Prepared by Bio-Environmental Services, Georgetown, Ontario, for Environment Canada, March 1981.

"Public Scoping Meeting on the Environmental Impact Statement for the Ballast Water Discharge Standards Rulemaking." Transcript of U.S. Coast Guard's public hearing, October 31, 2003, Cleveland, Ohio.

Raloff, Janet. "From Tough Ruffe to Quagga: Intimidating Invaders Alter Earth's Largest Freshwater Ecosystem." *Science News*, July 25, 1992.

Reeves, Eric. "Analysis of Laws and Policies Concerning Exotic Invasions of the Great Lakes." Prepared for the Michigan Department of Environmental Quality, March 1999.

Reeves, Eric. "Exotic Policy: An IJC White Paper on Policies for the Prevention of the Invasion of the Great Lakes by Exotic Organisms." Prepared for the International Joint Commission, July 15, 1999.

Reeves, Eric. "Exotic Politics: An Analysis of the Law and Politics of Exotic Invasions in the Great Lakes." *Toledo Journal of Great Lakes Law, Science and Policy* 2, no. 2 (2000): 125–206.

Reid, David F., and Guy A. Meadows. "Proceedings of the Environmental Implications of Cargo Sweeping in the Great Lakes." National Oceanic and Atmospheric Administration Technical Memorandum ERL GLERL-114, 1999.

Reutter, Jeffrey: "Invasive Species: Their Impact and Our Response." Testimony at the Great Lakes Regional Public Meeting of the U.S. Commission on Ocean Policy, Chicago, Ill., September 25, 2002.

Ricciardi, A., and J. B. Rasmussen. "Predicting the Identity and Impact of Future Biological Invaders: A Priority for Aquatic Resource Management." *Canadian Journal of Fisheries and Aquatic Sciences* 55 (1998).

Ruiz, Gregory M., and David F. Reid. "Current State of Understanding about the Effectiveness of Ballast Water Exchange in Reducing Aquatic Nonindigenous Species Introductions to the Great Lakes Basin and Chesapeake Bay, USA: Synthesis and Analysis of Existing Information." National Oceanic and Atmospheric Administration Technical Memorandum GLERL-142, September 2007, Ann Arbor, Mich.

Ryan, George J. "LCA Applauds U.S. Coast Guard's Marine Protection Efforts." Testimony to House Subcommittee on Coast Guard and Marine Transportation, Washington, D.C., July 1998.

Schaeffer, Jeffrey S., et al. "Invasion, History, Proliferation and Offshore Diet of Round Goby in Western Lake Huron, USA." *Journal of Great Lakes Research* 31, no. 4 (2005).

Schloesser, Don W., et al. "Recolonization and Possible Recovery of Burrowing Mayflies in Lake Erie of the Laurentian Great Lakes." *Journal of Aquatic Ecosystem Stress and Recovery* 8, no. 2 (2000).

"The St. Lawrence Seaway Traffic Report, Historical Tables, 1959–1992"; and "Tonnage Information Reports." St. Lawrence Seaway Development Corp., Washington, D.C., accessed via www.greatlakes-seaway.com.

Szylvian, Kristin. "Transforming Lake Michigan into the World's Greatest Fishing Hole: The Environmental Politics of Michigan's Great Lakes Sport Fishing, 1965–1985." *Environmental History* (January 2004).

Tanner, Howard, et al. "Upper Great Lakes Sportfishery Establishment Has Spectacular Impact." In "Using Our Natural Resources, 1983 Yearbook of Agriculture." U.S. Department of Agriculture, Washington, D.C., 1983.

Taylor, John C., and James L. Roach. Ocean Shipping in the Great Lakes: An Analysis of Issues." Grand Valley State University, Grand Rapids, Mich., October 2007.

Taylor, John C., and James L. Roach. "Ocean Shipping in the Great Lakes: Transportation Cost Increases That Would Result from a Cessation of Ocean Vessel Shipping." Grand Valley State University, Grand Rapids, Mich., August 2005.

Taylor, Robert R., ed. *Voices from the Great Ditch: The Historic Welland Canals in Pictures, Prose, Poems and Songs.* St. Catherines, Ont.: Blarney Stone Books, 2004.

Tivy, Patrick. *Marilyn Bell: The Heart-Stopping Tale of Marilyn's Record-Breaking Swim.* Canmore, Alb.: Altitude Publishing, 2003.

Todd, Kim. *Tinkering with Eden: A Natural History of Exotic Species in America.* New York: W. W. Norton, 2002.

U.S. Environmental Protection Agency. "Aquatic Nuisance Species in Ballast Water Discharges: Issues and Options; Draft Report for Public Comment." U.S. Environmental Protection Agency, Office of Wastewater Management, Washington, D.C., September 10, 2001.

Vanderploeg, Henry A., et al. "Dispersal and Emerging Ecological Impacts of Ponto-Caspian Species in the Laurentian Great Lakes." *Canadian Journal of Fisheries and Aquatic Sciences* 59 (2002).

Vanderploeg, Henry A., et al. "Zebra Mussel Selective Filtration Promoted Toxic Microcystis Blooms in Saginaw Bay (Lake Huron) and Lake Erie." *Canadian Journal of Fisheries and Aquatic Sciences* 58 (2001).

"Viral Hemorrhagic Septicemia: A New Invader in the Great Lakes." *Fish Lines* 5, no. 5 (U.S. Fish and Wildlife Service, 2007).

Weathers, Katherine, and Eric Reeves. "The Defense of the Great Lakes against the Invasion of Nonindigenous Species in Ballast Water." *Marine Technology* 33, no. 2 (1996).

Wiley, Chris. "Ballast Water Management in Canada: National Direction, Regional Realities." *Toledo Journal of Great Lakes Law, Science and Policy* 2, no. 2 (2000): 249–60.

Zellmer, Sandra. "Vanquishing Exotic Species in the Great Lakes." In State of the Great Lakes Report, Michigan Department of Environmental Quality, Lansing, Mich., 1999.

COURT RULINGS

Fednav Limited, Canadian Forest Navigation Co. Ltd., Nicholson Terminal and Dock Co., et al. v. Steven E. Chester, director, Michigan Department of Environmental Quality, U.S. District Court Eastern District of Michigan, Case no. 07-cv-1116, Detroit, Michigan, March 14, 2007.

Illston, Susan. "Order Granting Plaintiffs Motion for Summary Judgment Denying Defendants Motion to Summary Judgment," in *Northwest Environmental Advocates, et al. v. United States Environmental Protection Agency*. U.S. District Court for the Northern District of California, U.S. District Judge Susan Illston, March 30, 2005.

Illston, Susan. "Order Granting Plaintiffs' Motion for Permanent Injunctive Relief." *Northwest Environmental Advocates, et. al v. U.S. Environmental Protection Agency*. U.S. District Court, Northern District of California, San Francisco, Calif., September 18, 2006.

INTERNET RESOURCES

St. Lawrence Seaway: www.greatlakes-seaway.com

Healing Our Waters–Great Lakes Coalition: www.healthylakes.org

Shipping Federation of Canada: www.shipfed.ca
U.S. Great Lakes Port Association: www.greatlakesports.org
Lake Carriers Association: www.lcaships.com
National Wildlife Federation: www.nwf.org/greatlakes
Great Lakes United: www.glu.org
Aquatic Nuisance Species Task Force: www.anstaskforce.gov
NOAA Great Lakes Environmental Research Laboratory: www.glerl.noaa.gov
U.S. Great Lakes Science Center: www.glsc.usgs.gov
Environment Canada: www.on.ec.gc.ca/greatlakes/
U.S. Environmental Protection Agency: www.epa.gov/greatlakes

INDEX

B

C

D

F

G

M

S